Content Review Panel

Mark D. Clark
Associate Professor of Mathematics,
Palomar College,
San Marcos, CA

John E. Daggett
Math Lab Coordinator,
DeAnza College,
Cupertino, CA

Math Learning Center Coordinator,
Mira Costa College,
Mira Costa, CA

Tim Hempleman
Adjunct Faculty,
Palomar College and California
StateUniversity,
San Marcos, CA

Harris S. Shultz
Professor of Mathematics,
California State University,
Fullerton, CA

Authors

Paula Allison
Michelle Fior
Mark Bredin
Lana Chow,
Gloria Conzon
Brent Corrigan
Jackie Fierbach
Connie Goodwin
Theresa Gross
Barry Gruntman
Christine Henzel
Peter Kerr
Chris Kirkpatrick
J. Martin McGeough
Greg McInulty
Roxane Menssa
W. Keith Molyneux
Loretta Morhart
David Nutbean
Robert Payne
Brent Pfeil
C. David Pilmer
Grant L. Plett
Joan E. Rashid
Caleb Reppenhagen
Rod Rysen
Rick Sept
Joe Shenher
Don Sparkes
Katharine D. Tetlock
Ina Tomiyama
Colleen Tong
Stella Tossell
Roxann Trouth
Dale Weimer
Ketri Wilkes

Product Development

Garry Popowich
Ron Blond
Milt Petruc
Sharon Tappe
Dale Weimer
Lisa Wright
Angus McBeath
Tom Winkelmans
Paula Allison
Duncan McCaig
Katharine D. Tetlock
Barry Mitschke
Larry Markowski
Eleanor Milne
Maxine Stinka
Steven Daniels
Lee Kubica
Barb Morrison
Connie Goodwin
Theresa Gross
Terri Hammond
Carol Besteck Hope
Ted Keating
Chris Kirkpatrick
Susan Woollam

Math Explorers

Design and Programming
Ron Blond

Windows Versions
Grant Arnold
Vladislav Hala

Editorial Development

First Folio Resource Group, Inc.

Project Manager

David Hamilton

Senior Editor

Eileen Jung

Print Materials

Fran Cohen, Project Manager
Robert Templeton, Supervising Editor
Brenda McLoughlin, Editor
Tom Dart, Bruce Krever, Design and Layout

Storyboard Editors

Shirley Barrett *Mark Bryant* *Angel Carter* *Jackie Lacoursiere* *Anna Marsh*
Darren McDonald *Don Rowsell* *Robert Templeton* *Jackie Williams* *Susan Woollam*

Picture Research and Copy Editing

Robyn Craig *Bruce Krever* *Jonathan Lampert* *Catherine Oh* *Mike Waters*

Multimedia Development

Calian eLearning

Vice-President – eLearning

Justin Ferrabee

Director, Delivery

Lenka Jordanov

Project Manager

Kevin Kernohan

Project Administration

Sylvia Panaligan

Technical Manager

Collin Chenier

Testing Manager

Geoffrey Heaton

Programming

Reed Carriere, Main Engine Programmer
Collin Chenier *Kevin Kernohan* *Andrée Descôteaux* *Geoffrey Heaton*

Graphic Design

Mike Martel, Creative Lead

Illustration, Photo Adaptation

Mike Martel *Catherine Fitzpatrick*

Authoring

Bill Currie, Senior Author *Collin Chenier* *Geoffrey Heaton* *Jo-anne Landriault*
Kevin Kernohan *Lianne Zitzelsberger* *Madelyn Hambly* *Mike Cherun*
Catherine Fitzpatrick *Mike Martel*

Programming

Andrée Descoteaux, Main Engine Programmer
Collin Chenier *Kevin Kernohan* *Reed Carriere*

Testing

Diana Guy *Geoffrey Heaton*

Audio Production

Kevin Kernohan

Narration Voices

June Dewetering *Patrick Fry* *Ron Purvis* *Genevieve Spicer*

Contents

PREFACE

The Learning Equation (TLE) is interactive multimedia courseware and student workbooks for developmental mathematics, from basic math/arithmetic through intermediate algebra. Developed in Macromedia Director®, TLE is attractive with a professional look and feel, and runs fast and reliably on all Windows and Macintosh platforms. Delivered on CD-ROMs, it can run on a stand-alone computer, over a LAN, and over the internet. The Learning Equation ships with a sophisticated, browser-based course management system and assessment package.

TLE uses interactive, activity-directed learning to involve the student in their own education. Each TLE lesson is designed to build skills in algebra and problem solving. The entire series of lessons has a sound, curriculum-based foundation. As you progress through the lessons, you will learn the vocabulary of mathematics, practice key concepts, and develop your skills in reasoning, modeling, and analysis.

Each TLE lesson contains a wealth of application problems. Numerous applications are included from disciplines such as business, entertainment, science and technology, and history.

The Learning Equation

STUDENT USER'S GUIDE

Table of Contents

Introduction

There are two general places or installations where *The Learning Equation* (TLE) may be used: **on campus**, where you use TLE on your school's computers, and **at-a-distance** or **off campus**, where you use TLE on your own computer at home or somewhere other than on campus. This guide provides instructions on how to use TLE for both on campus and off campus installations. This guide does not provide installation instructions. Please refer to the <readme.txt> file on your TLE CD #1 for installation instructions.

Furthermore, when using TLE off campus, you can choose to either remain connected to the Internet or work off line to free up your phone line (if you have a dial up modem connection to the Internet).

Make sure to follow any specific instructions given by your instructor to the letter. If you don't follow your instructor's specific instructions, there's a good chance that you will not receive a grade for your work.

Using The Learning Equation

The Learning Equation (TLE) is complete "courseware" designed to help you learn mathematics. TLE is made up of two parts: a course management system that, among other things, tracks your progress and provides you with an up-to-the-minute look at your progress in the course; and the learning content, which is an interactive guide to your learning. The course management system runs in a web browser and can operate on any computer, whether connected to the Internet or not. The learning content must run "locally," meaning either from your computer's hard disk (using a "Full local install") or CD-ROM drive (using a "Minimal local install"), so you must install TLE on your own computer off campus.

Follow the steps below the first time you use TLE. Once you've logged on for the first time and created your user name and password, you'll use that user name and password every time thereafter to use TLE.

Using The Learning Equation for the Very First Time

The first time you log on to TLE, you'll need to be connected to the Internet. This will probably be done during your first day of class on campus or the first time you sit down to use TLE if you are learning at-a-distance (using your home computer, for instance).

NOTE: If you are using TLE off campus and do not have an Internet connection, see "Using TLE Without an Internet Connection".

Registering to Use TLE

Follow the steps below to register and use TLE.

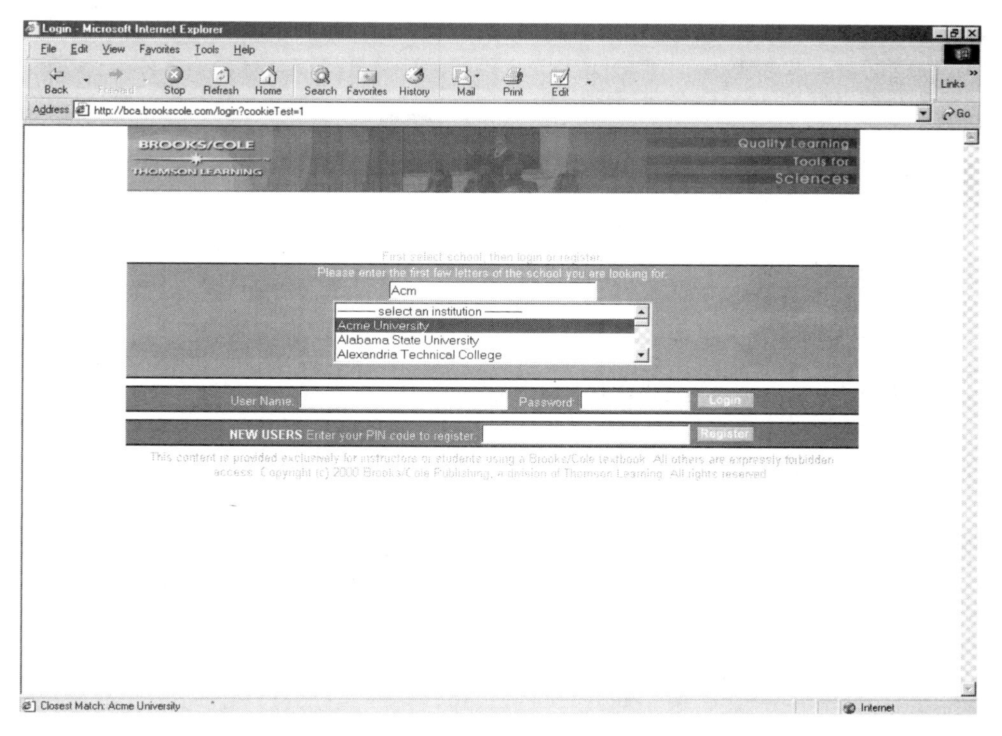

1. Log on to the Internet.
2. Double click the TLE icon on the desktop. This will launch your web browser. Type in the following address: <u>http://bca.brookscole.com</u>. This will bring up the TLE log in screen.
3. In the upper input box, above the pick list that includes all the schools using TLE, type the first few letters of your school. Scroll down if necessary. Locate your school in the pick list and select it by clicking once. Your school name will be inserted into the input box above.

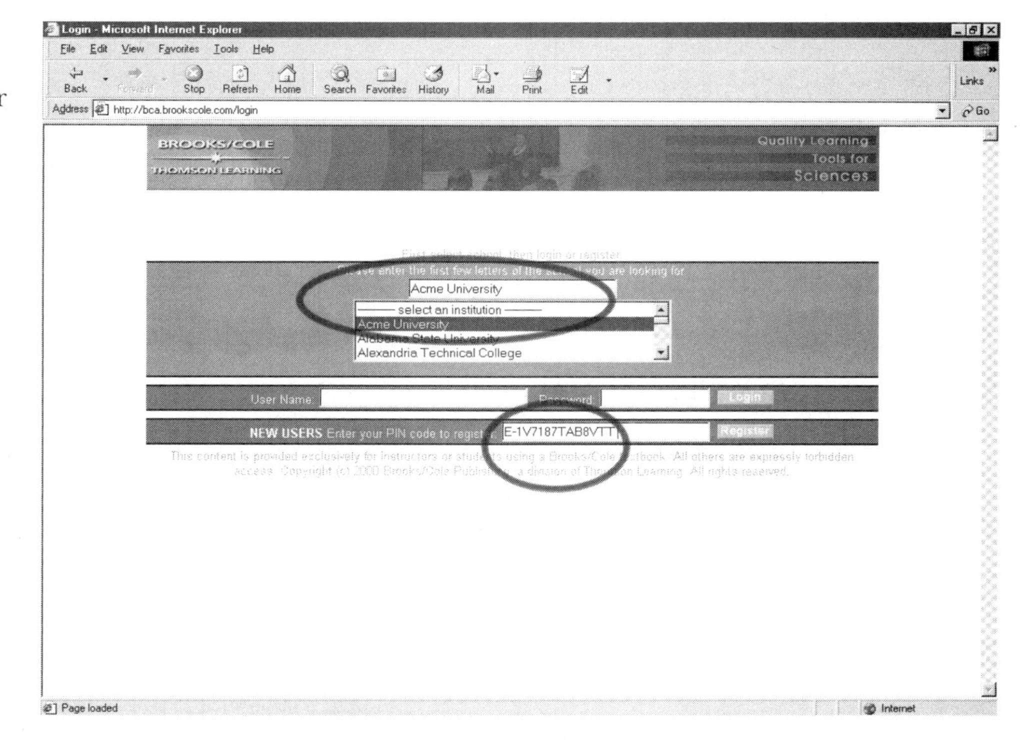

4. In the "New Users" input box, type in the PIN given to you by your instructor and click the "Register" button. Your instructor may have already created a user name and password for you, in which case you do not have to register. See your instructor for more information.

5. Complete the registration form that appears next. You must fill out the fields in the form that are denoted with an asterisk "*"—these fields are required. If you do not fill out the registration form correctly, you will not be registered correctly and will not receive a grade for the course. Create your user name and password. You may have received specific instructions from your instructor such as what to use for your user name and password. Follow your instructor's special instructions to the letter. If you do not follow your instructor's special instructions to the letter, you will not be correctly registered and will not receive a grade for the course.

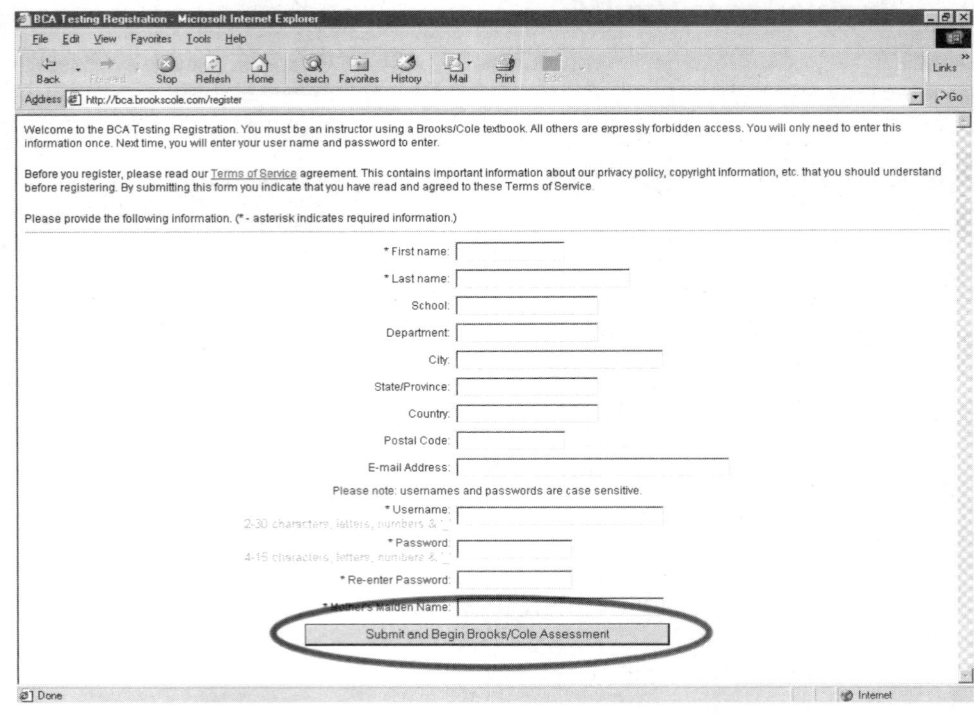

6. Once you've completed the registration form, click the "Submit" button. You'll then be brought to the Activity Page. You are now automatically enrolled into your instructor's class roster and grade book.

7. Click the "Administration" link either on the page to the right or under the Activity Page tab.

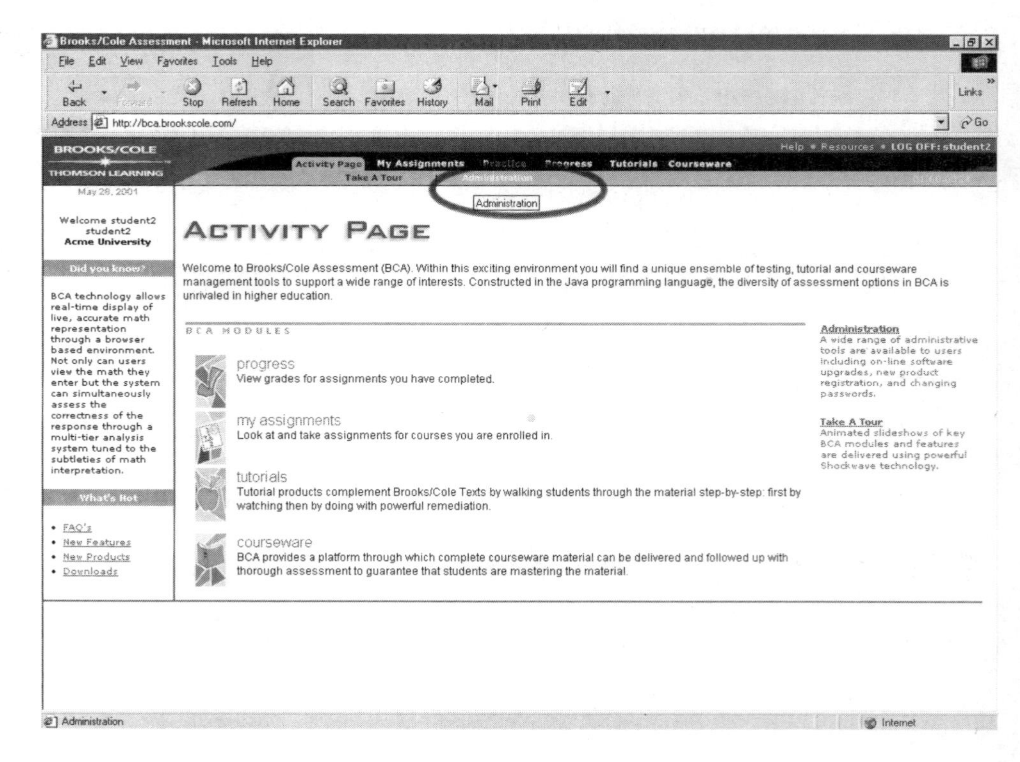

8. Click the "Register to use new books" under "My Account".

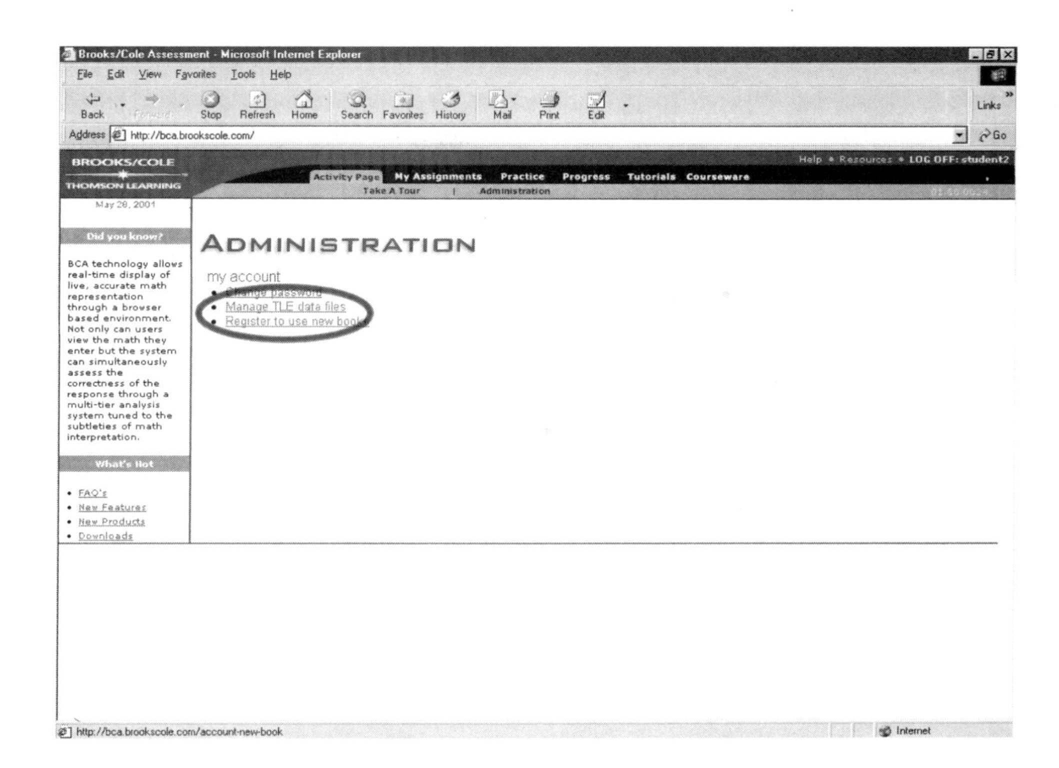

9. Type in the PIN registration code into the input box. This registration code is the ISBN number found on the back cover of your TLE Student Workbook. Click the "Submit" button. You now have completed access to the appropriate TLE course, indicated by the "Congratulations…"

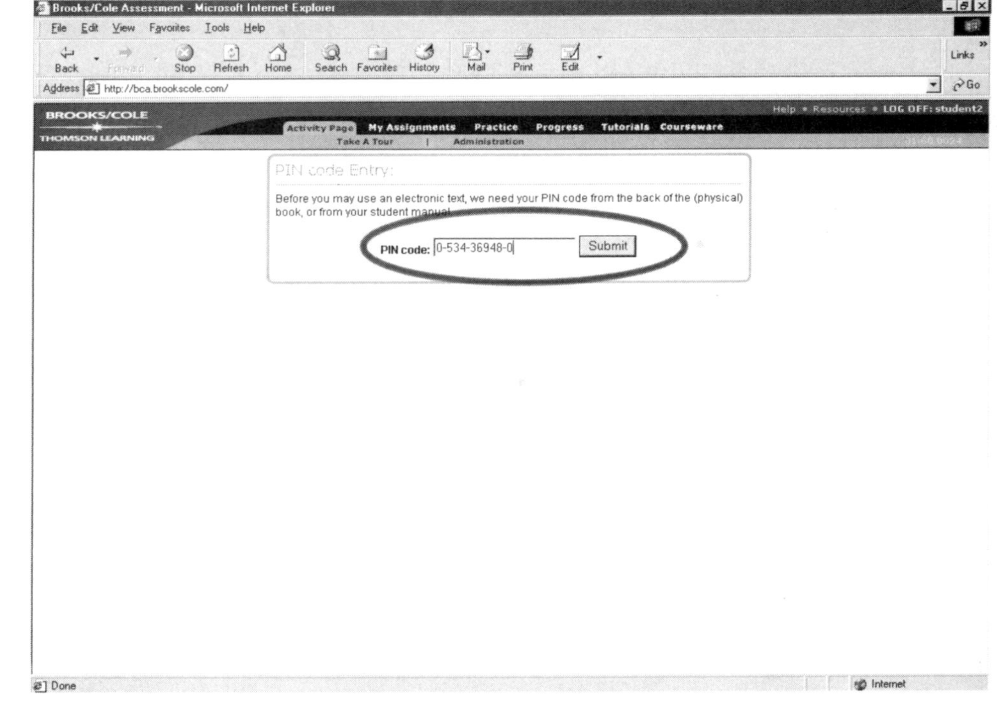

10. Click the "log off" link in the upper right of the page to be brought back to the log in screen. You must always log off to end your TLE session. Failing to log off will record incorrect time-spent information and could result in a poor grade for the work you have done.

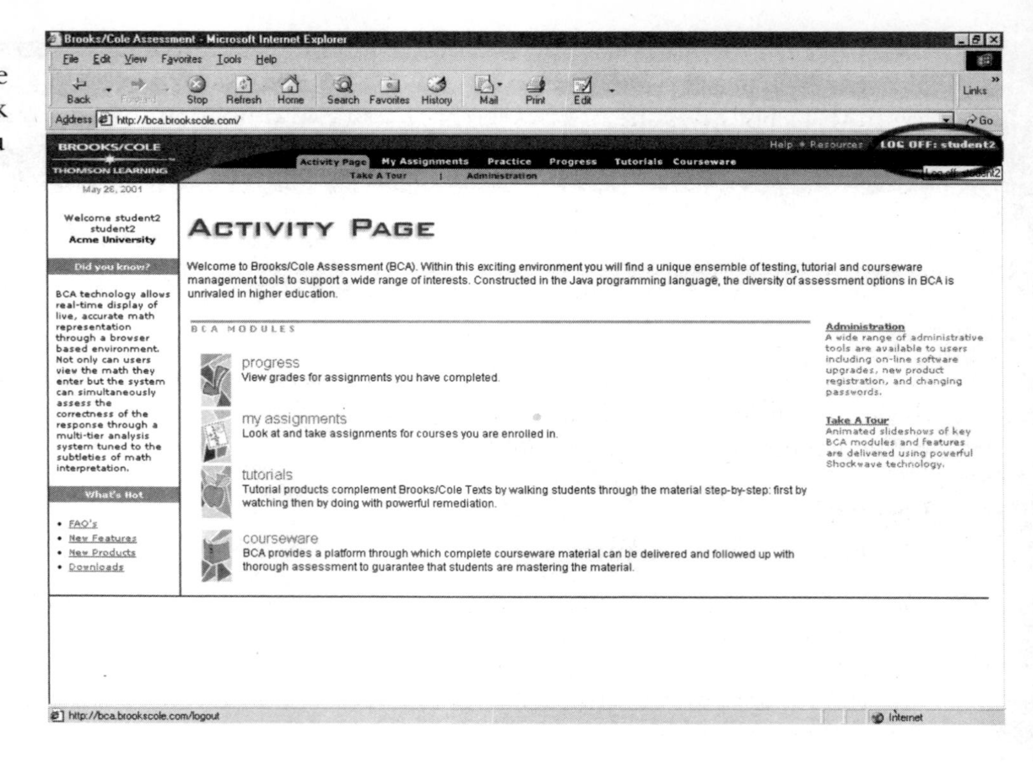

Follow the steps in the next sections for using TLE every time from now on.

Using The Learning Equation Every Time After the Very First Time

If you have not registered to use TLE, go back to the section titled "Using TLE for the Very First Time".

When using TLE after your first time and at a location other than on campus (at home, with your own computer for instance), you'll need to decide:

1. whether you don't mind remaining connected to the internet during your TLE session (ideal for those who have a dedicated phone line for the Internet) or
2. if you want to remain off-line as you work with TLE, connecting to the Internet only at the beginning and end of your TLE session (to free up your phone line during the session).

If you remain connected to the Internet during your session, you will be connecting to the http://bca.brookscole.com site for tracking purposes. If you remain off line, you'll still need to connect to the Internet at the beginning and end of your TLE session to update your computer and the web site your instructor uses to track your progress.

When you launch TLE on your computer at home, the web browser that automatically opens will point to a location on your hard disk (http://localhost). To go out to the Internet and the bca.brookscole.com web site, you'll need to type that address into your browser. If you do not have a connection to the Internet at home, you can save tracking information to a floppy disk. See "Using TLE Without an Internet Connection".

Using TLE While Remaining Connected to the Internet

Now that you are registered, you will use your user name and password to log in to TLE. This is the only way to access TLE and have all your work tracked and graded. If you do not log in using your user name and password, your progress will not be tracked and you will not receive a grade for the course.

Follow the steps below to log in and start using TLE while remaining connected to the Internet.

Note: you must register to use TLE. If you have not registered, go back to the section titled "Using TLE for the Very First Time" and follow the steps for registration and to create your user name and password.

Logging In

1. Log on to the Internet.
2. Double click the TLE icon on your desktop. Your web browser will open. The log on screen that you see is for using TLE off line. You must change the address in the input box of your browser. Type in the following address: http://bca.brookscole.com. You will be brought to the TLE log in screen.
3. Locate your school in the pick list, type in your user name and password into the appropriate input box, and click the "Log In" button. You will be brought to your Activity Page.
4. Click on "My Assignments" to display the list of assignments created by your instructor.

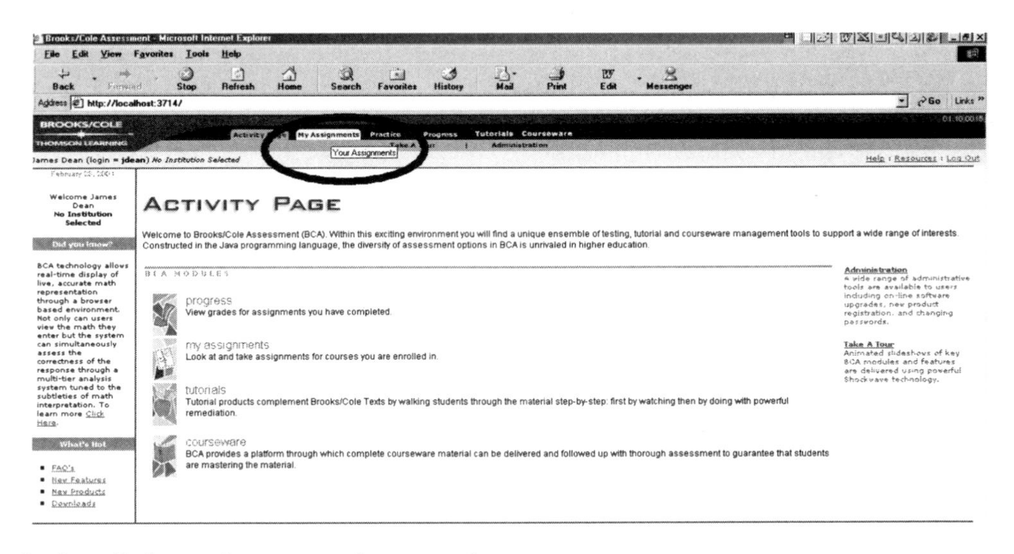

5. Your Assignments page includes all the assignments that your instructor has created for the course. The assignments are listed by assignment category in the order in which they are due. Start with the first assignment and work your way down the list. Make sure to pay particular attention to any specific instructions from your instructor.

6. Go to a TLE lesson assignment by clicking the "Take" button associated with that assignment (ask your instructor if this is unclear).

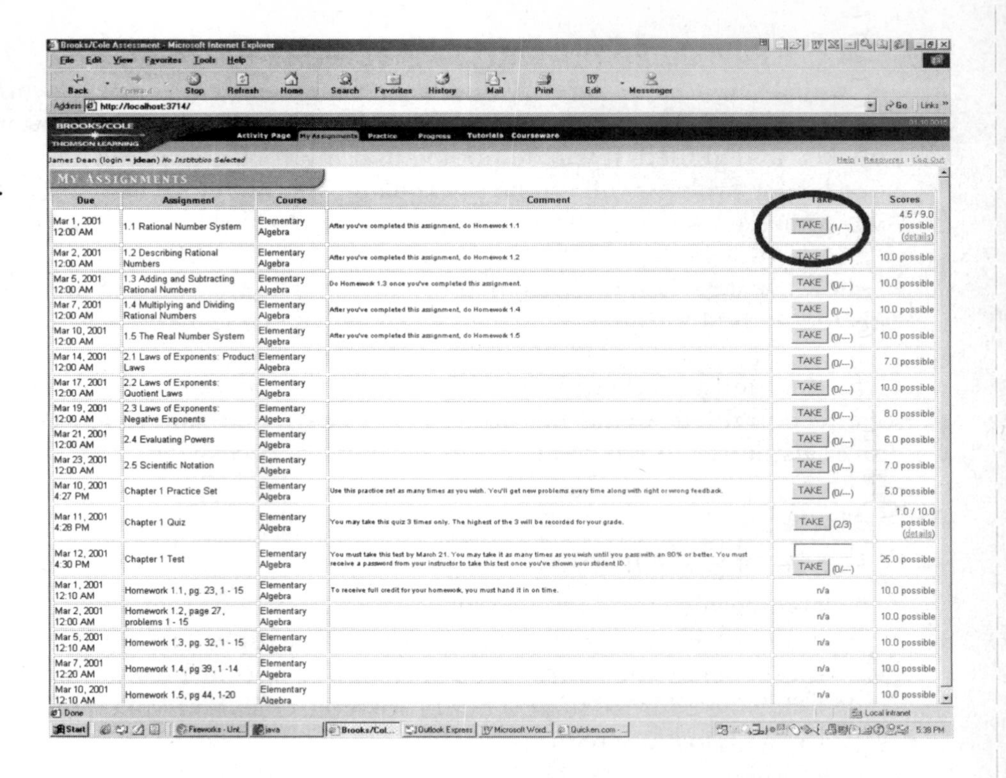

7. A "File Download" dialogue window will appear, asking you "What would you like to do with this file?" Select "Open this file from its current location". You will then be brought to the TLE Tour (the first time you access a TLE lesson assignment) or directly to the TLE lesson assigned (every time after the first time you access a TLE lesson assignment).

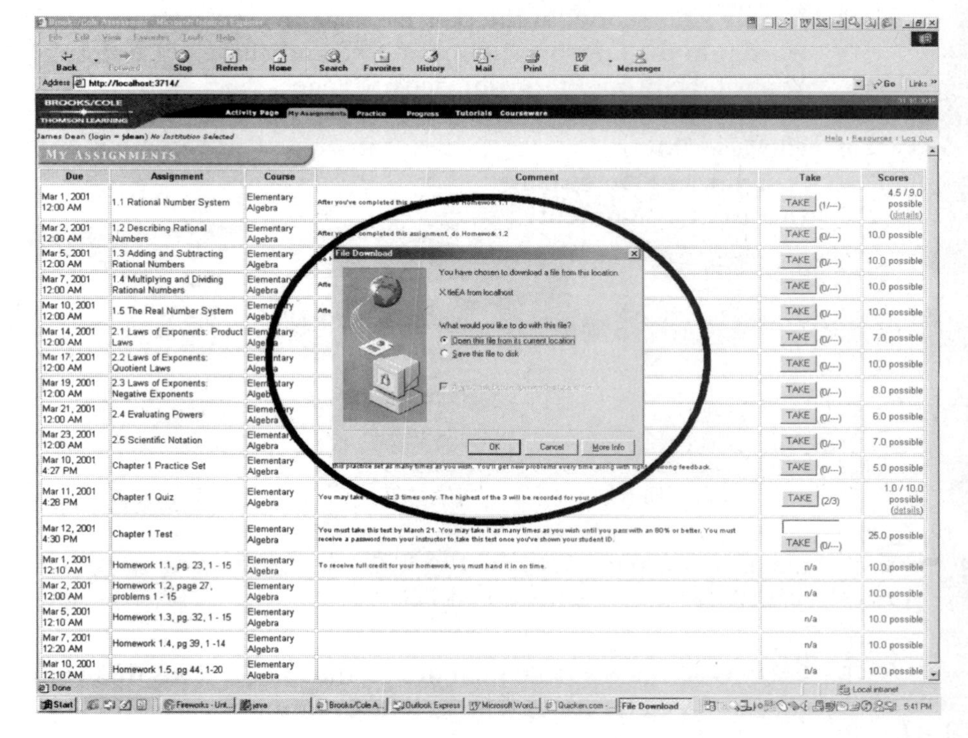

8. If this is your first time accessing a TLE assignment, click through the tour to familiarize yourself with how TLE works. Once you are done with the Tour, click the "close" button in the lower right of the screen. This will bring you to the TLE lesson assignment your instructor has assigned for you to work on.

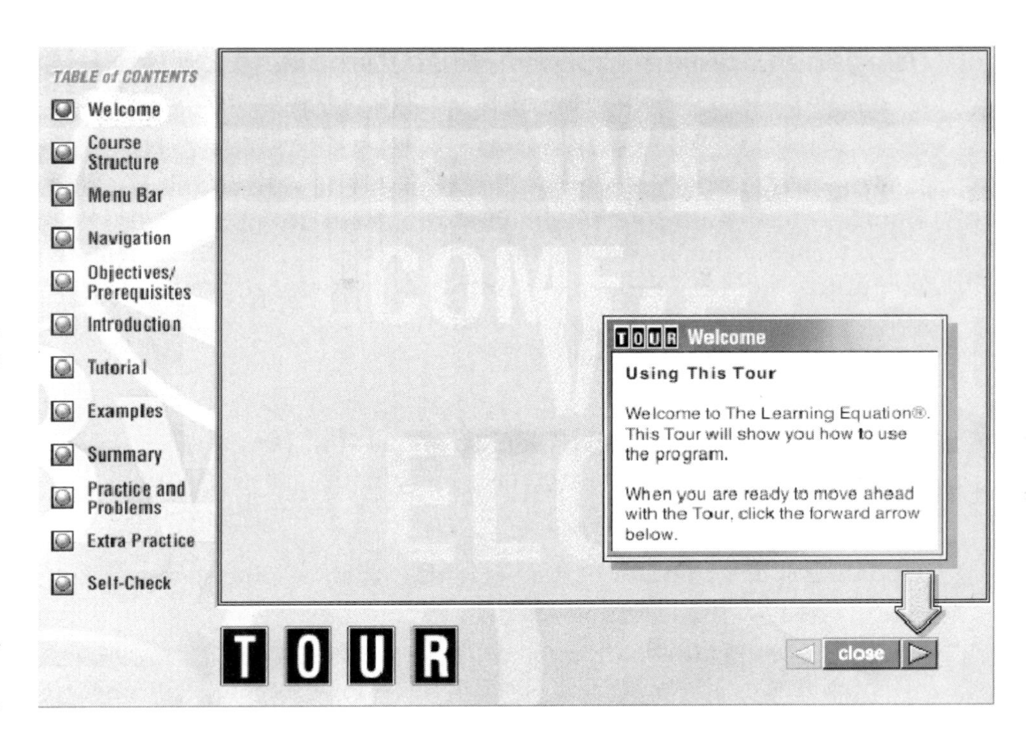

9. When you want to quit your TLE lesson assignment session, click the "Quit" button at the bottom of the TLE screen. Confirm quit by clicking "Quit" in the window that appears, and you'll be brought back to the course management system running in your web browser. Note: Do not click any "Menu" buttons at any time. This will take you to the TLE table of contents, allowing you to go to another TLE lesson. If you go to another TLE lesson from the table of contents, your progress will not be accurately recorded. Always access a TLE lesson from the Assignments page when using TLE. See the "Using the Assignments Page" section for more information on using this component of the TLE software.

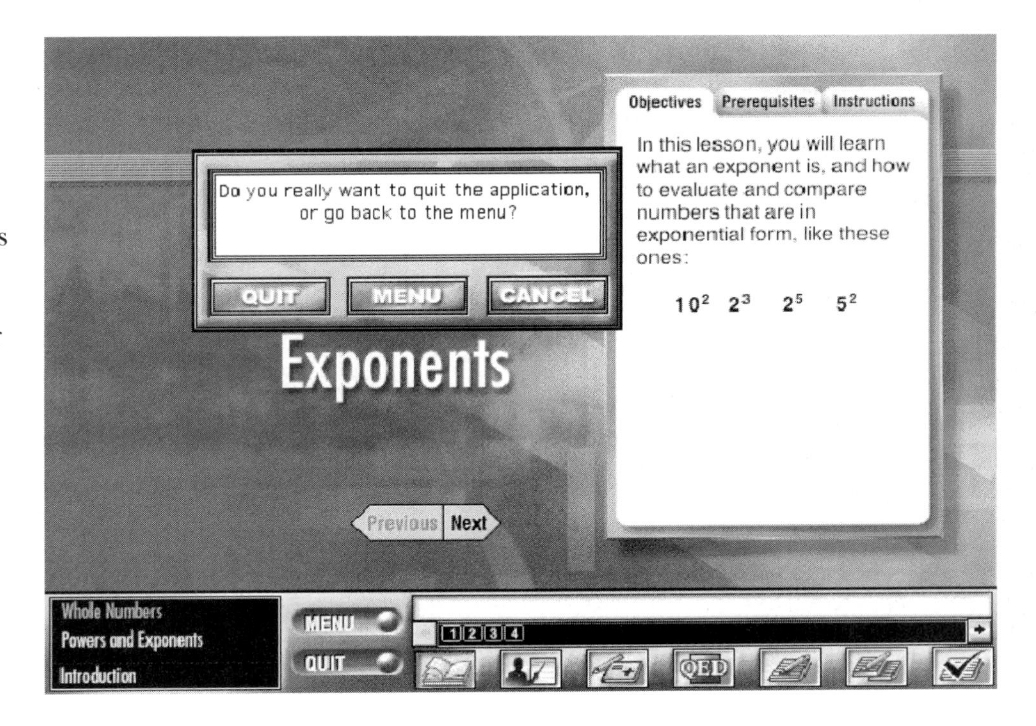

10. Either choose another assignment from "My Assignments" or log off the web browser to end your TLE session.

Using The Learning Equation While Remaining Off Line

If you have not registered to use TLE, go back to the section titled "Using TLE for the Very First Time".

You can use TLE off campus and remain off line (not connected to the Internet) to free up your phone line. You must, however, briefly connect to the Internet at the beginning and end of your session to update both your computer and the web site your instructor uses to track your progress. If you do not have an Internet connection, refer to "Using TLE Without an Internet Connection" below.

As stated above, you will need to connect to the Internet both at the beginning and end of your TLE session to update your computer and have your progress tracked so you may receive a grade for your work done off campus. Follow the steps below to use TLE while remaining off line.

Updating your computer at the beginning of an off line session

1. Log on to the Internet.
2. Double click the TLE icon on your desktop. Your web browser will open. The log in screen that you see is for using TLE off line. Note the address http://localhost is automatically inserted into the address field. This means your browser is now connecting to your hard disk and remaining off line. **You are not connected to the Internet at this time.** You must first update your computer by going on line. To go on line, change the address in the input box of your browser. Type in the following address: http://bca.brookscole.com. You will be brought to the TLE log in screen on the Internet.
3. Locate your school in the pick list, type in your user name and password into the appropriate input box, and click the "Log In" button. You will be brought to your Activity Page.
4. Click on the Administration sub-tab (part of the Activity Page tab).
5. Click the "Manage TLE data files" link. This displays the "Import/Export TLE Data" page.

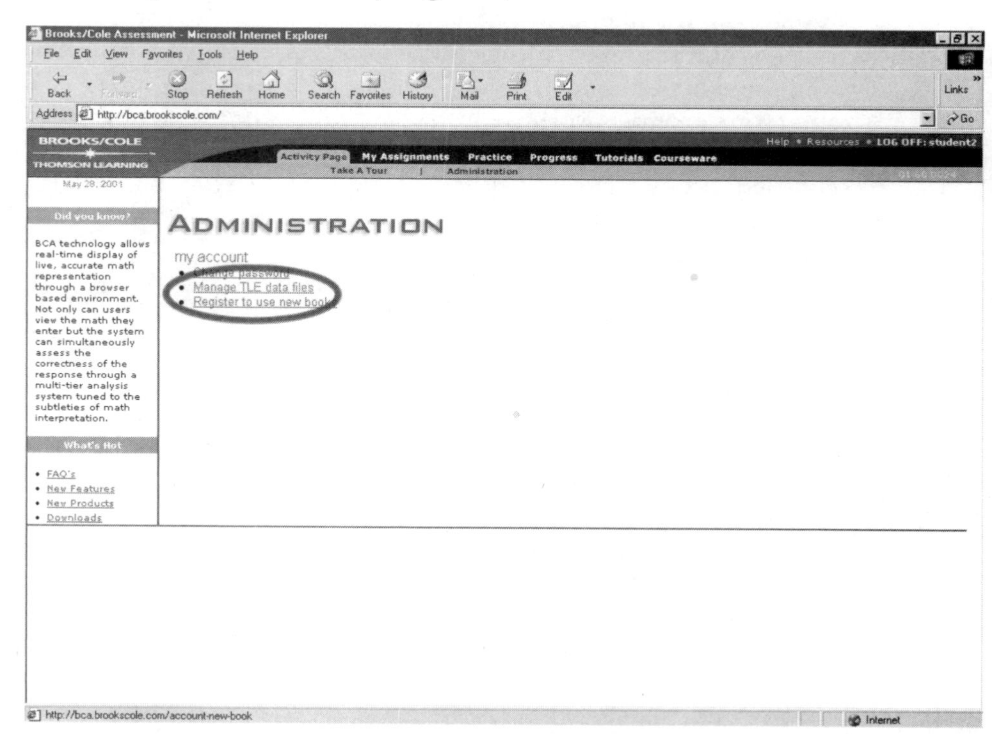

6. Click the “Save TLE data to disk” button.

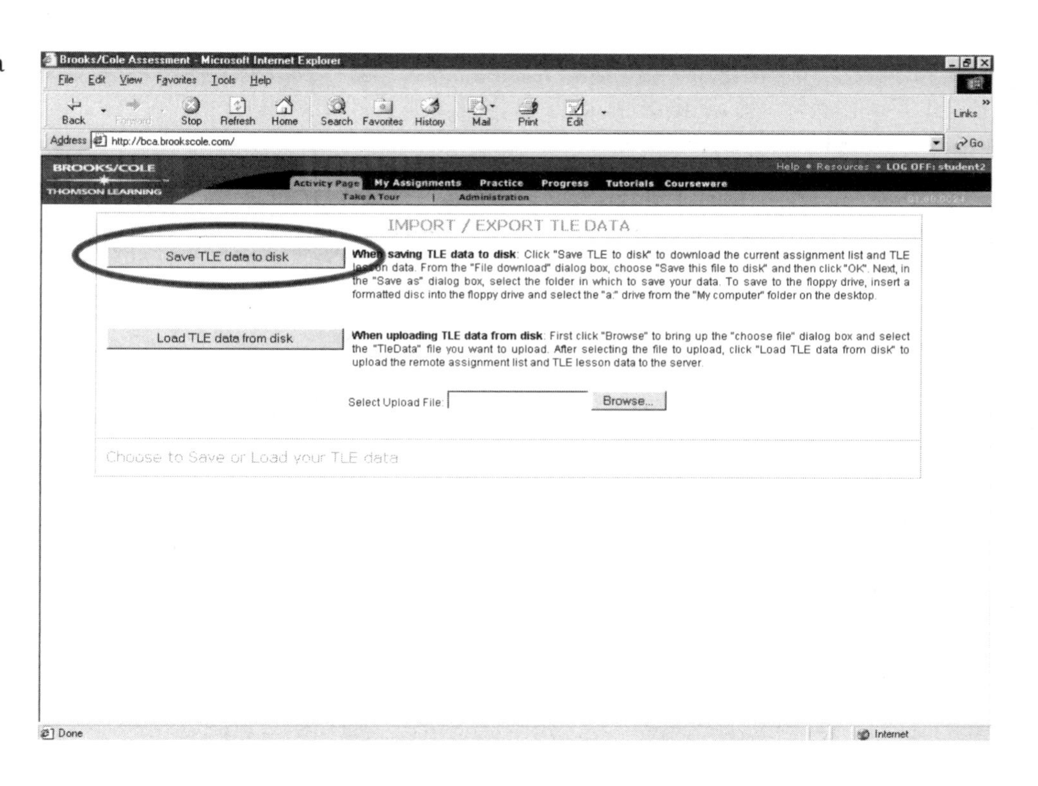

7. This will display a “File download” dialogue box.

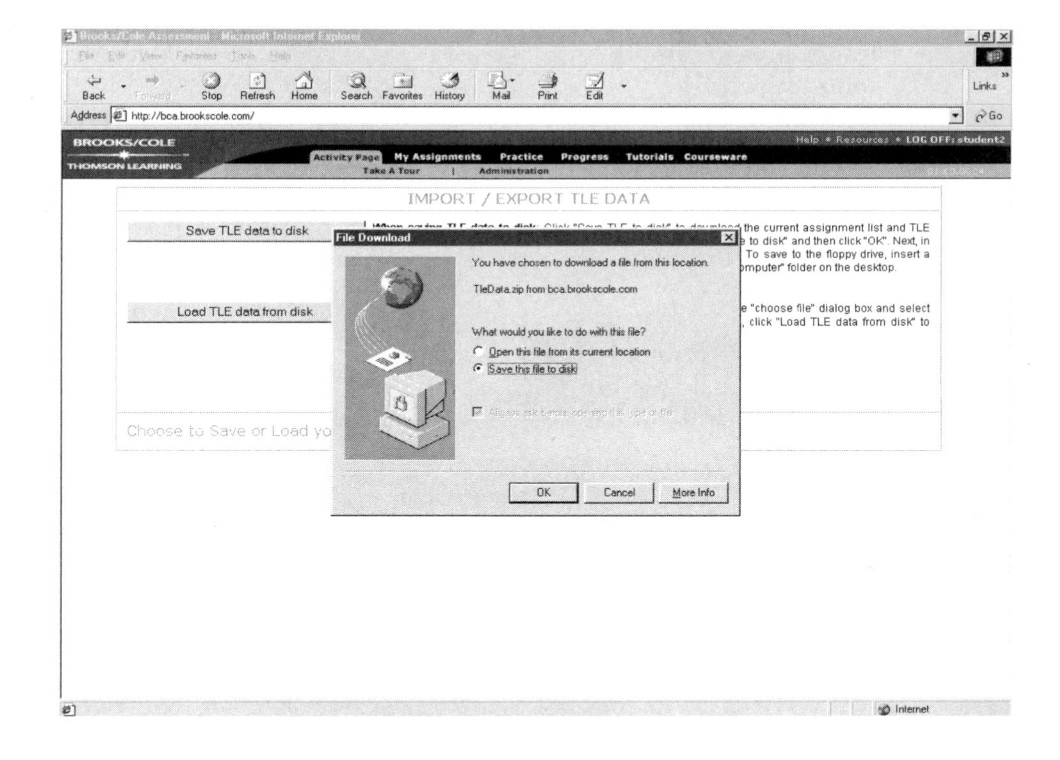

8. Click the “Save file to disk” option and click “OK”. A “Save as” dialogue box will be displayed.

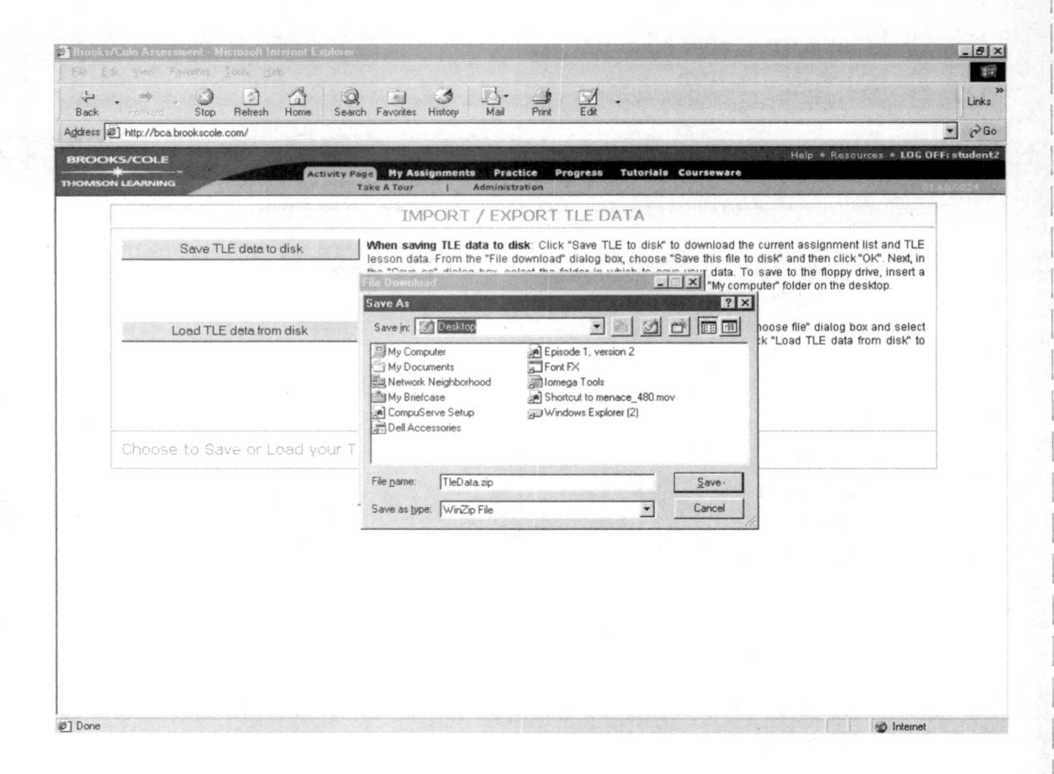

9. Locate and select the folder into which you want to save the <TleData> file. “My Documents” is an easy location to remember and locate later. Click the “Save” button. If the file already exists, choose to overwrite it. Close the “Download complete” dialogue box once the download has finished.

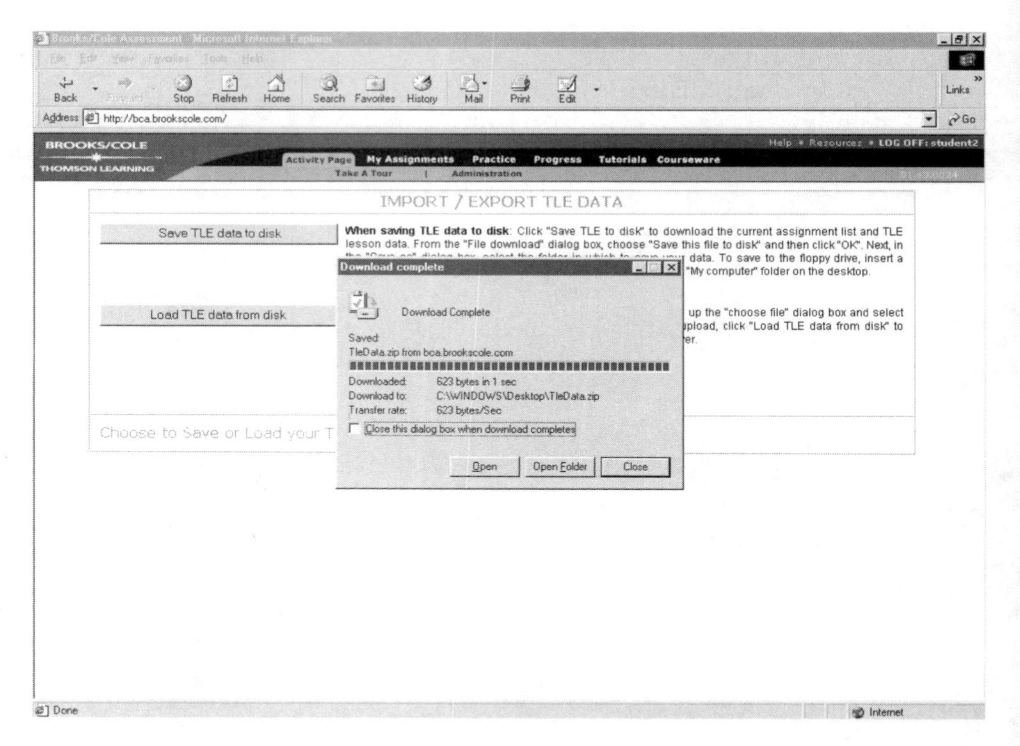

10. Log off by clicking the “Log off” link in the upper right corner of the browser window.

11. Close the web browser session (click the X in the uppermost right hand corner) and close your connection to the Internet.

12. Double click the TLE icon on your desktop to re-launch TLE and your web browser. Note the address http://localhost is automatically inserted into the address field. This means your browser is now connecting to your hard disk and remaining off line. **You are not connected to the Internet at this time.**

13. Log in to TLE by first locating your school, then typing in your user name and password. This will bring you to the Activity page.

14. Click on Administration.

15. Click on the "Manage TLE data files" link. This displays the "Import/ Export TLE Data" page.

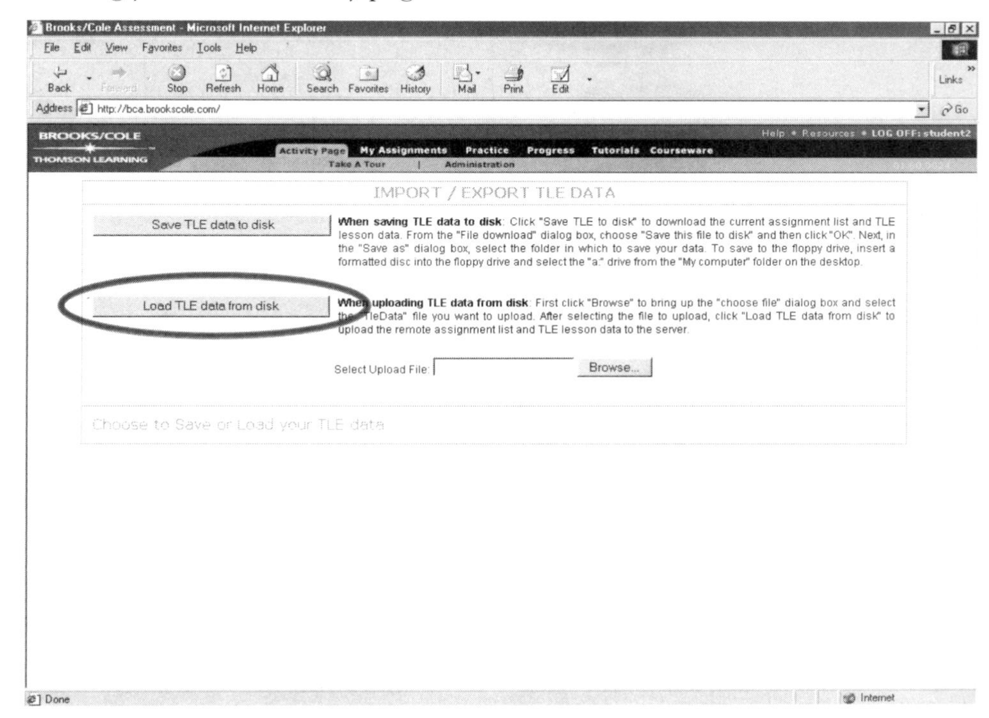

16. Click the "Browse" button next to the "Select Upload File" input box. This will display a "Choose file" dialogue box.

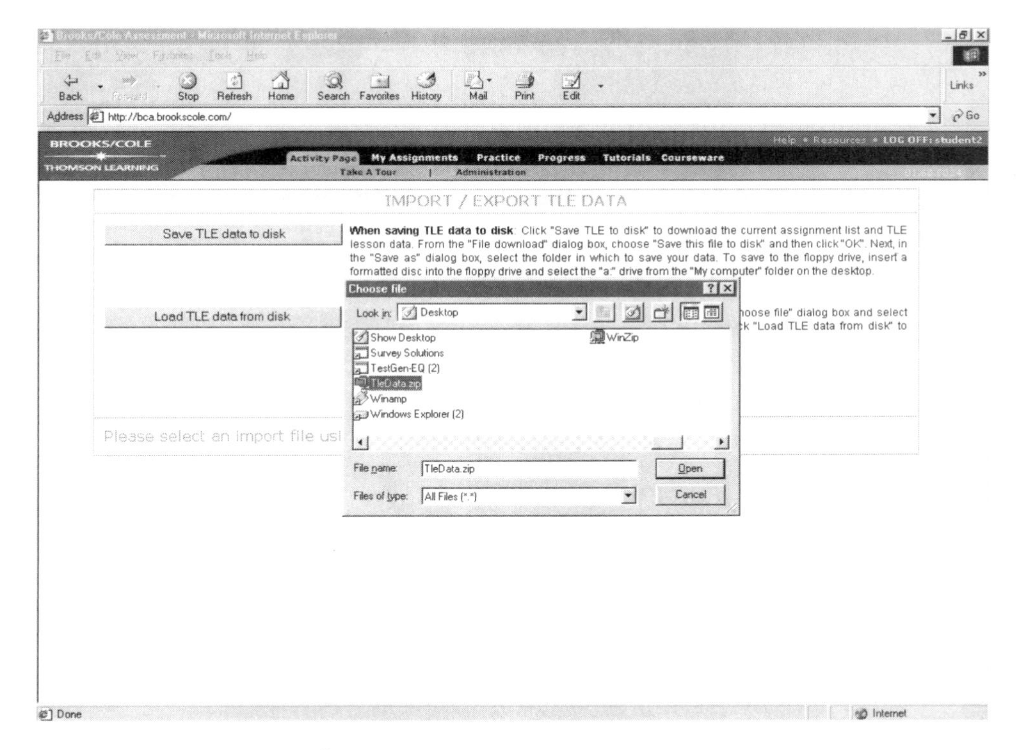

17. Locate and select the <TleData> file you saved in step #9 above and click the "Open" button. This will insert the pathname for the file in the "Select Upload File" input box.

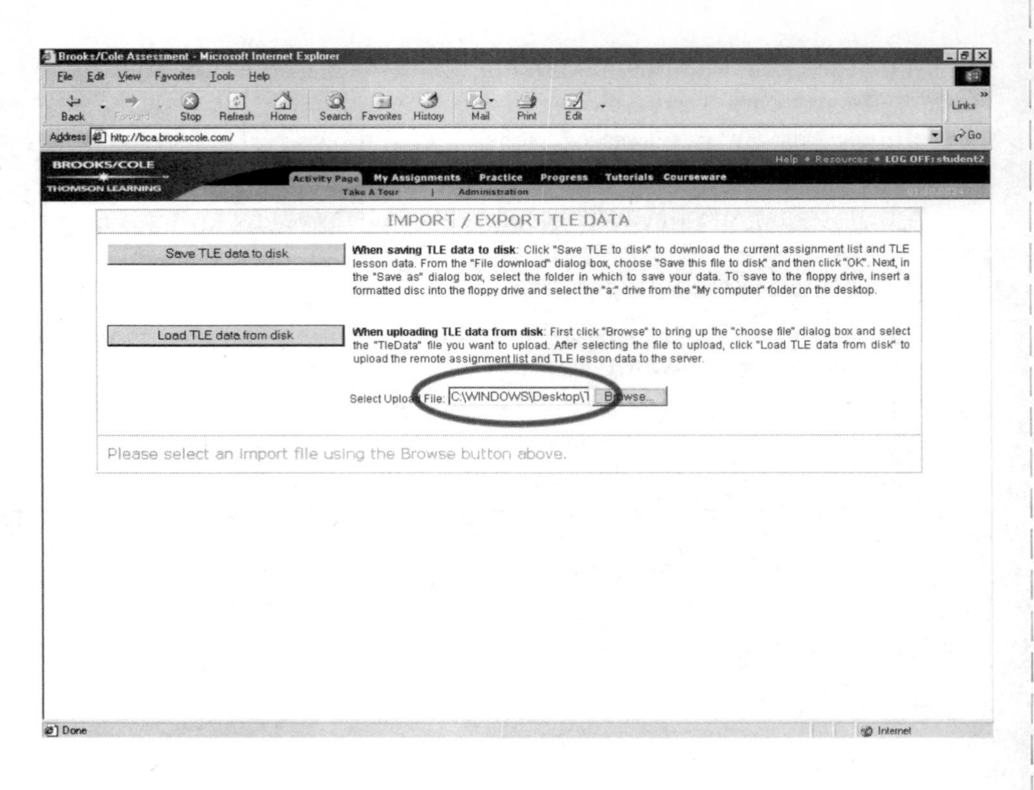

18. Click the "Load TLE data from disk" button. You'll see an "Importing from file" notification. Wait until you see "Done" in the lower left corner of your web browser. This will update your computer with the latest tracking information, grade book progress, and any messages that may have been sent to you by your instructor.

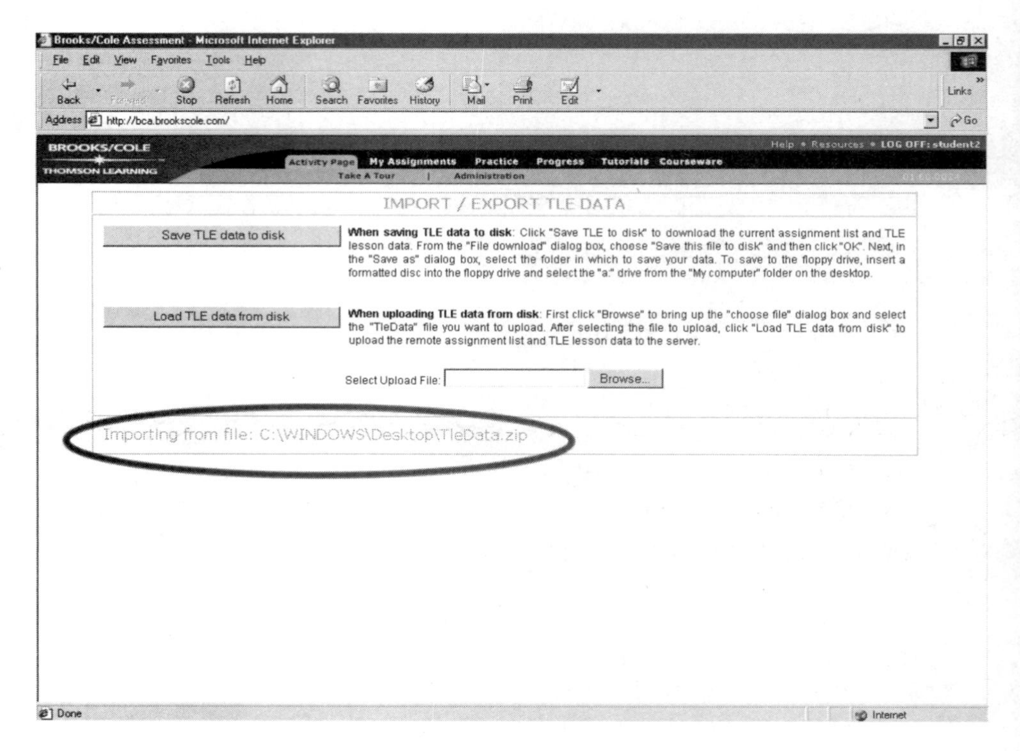

19. Click the "My Assignments" tab to display a list of your assignments. Click the "Take" button to access any TLE content assignment.

Updating your instructor's web site at the end of an offline session

When you decide to end your TLE session, you need to update your computer again and get it ready to update the web site your instructor uses to track your progress. Follow these steps before logging off.

1. Click on the Administration sub-tab (part of the Activity Page tab).
2. Click the "Manage TLE data files" link. This displays the "Import/Export TLE Data" page.
3. Click the "Save TLE data to disk" button. This will display a "File download" dialogue box.
4. Click the "Save file to disk" option and click "OK". A "Save as" dialogue box will be displayed.
5. Locate and select the folder into which you want to save the <TleData> file. "My Documents" is an easy location to remember and locate later. Click the "Save" button. If the file already exists, choose to overwrite it. Close the "Download complete" dialogue box once the download has finished.
6. Log off by clicking the "Log off" link in the upper right corner of the browser window.
7. Close the web browser session (click the X in the uppermost right hand corner).
8. Log on to the Internet.
9. Double click the TLE icon on your desktop to re-launch TLE and your web browser. The log on screen that you see is for using TLE off line. Note the address http://localhost is automatically inserted into the address field. This means your browser is now connecting to your hard disk and remaining off line. **You are not connected to the Internet at this time.** You must now update your instructor's web site by going on line. To go on line, change the address in the input box of your browser. Type in the following address: http://bca.brookscole.com. You will be brought to the TLE log in screen on the Internet.
10. Log in to TLE by first locating your school, then typing in your user name and password. This will bring you to the Activity page.
11. Click on Administration.
12. Click on the "Manage TLE data files" link.
13. Click the "Browse" button next to the "Select Upload File" input box.
14. Locate and select the <TleData> file you saved in step #5 above and click the "Open" button. This will insert the pathname for the file in the "Select Upload File" input box.
15. Click the "Load TLE data from disk" button. Wait until you see "Done" in the lower left corner of your browser. This will update your instructor's web site with the latest tracking information, grade book progress, and any messages that you may have sent to your instructor.
16. Log off your session, close your web browser, and close your connection to the Internet.

Using The Learning Equation Without an Internet Connection

If you do not have an Internet connection, you can still update your home computer and your instructor's web site to accurately track and report your progress. You will need a floppy disk to do so. You must remember to "Save TLE data to disk" before ending each and every of your on campus sessions if you use TLE at your school and then use TLE at home without an Internet connection available. Follow the steps below.

Saving your TLE data file to floppy disk at the end of an on campus session

1. At the end of your TLE session on campus, and before you log off, go to the Administration sub-tab.
2. Insert a $3\frac{1}{2}$ in. floppy disk into the floppy drive.
3. Click the "Manage TLE data files" link. This displays the "Import/Export TLE Data" page.
4. Click the "Save TLE data to disk" button. This will display a "File download" dialogue box.

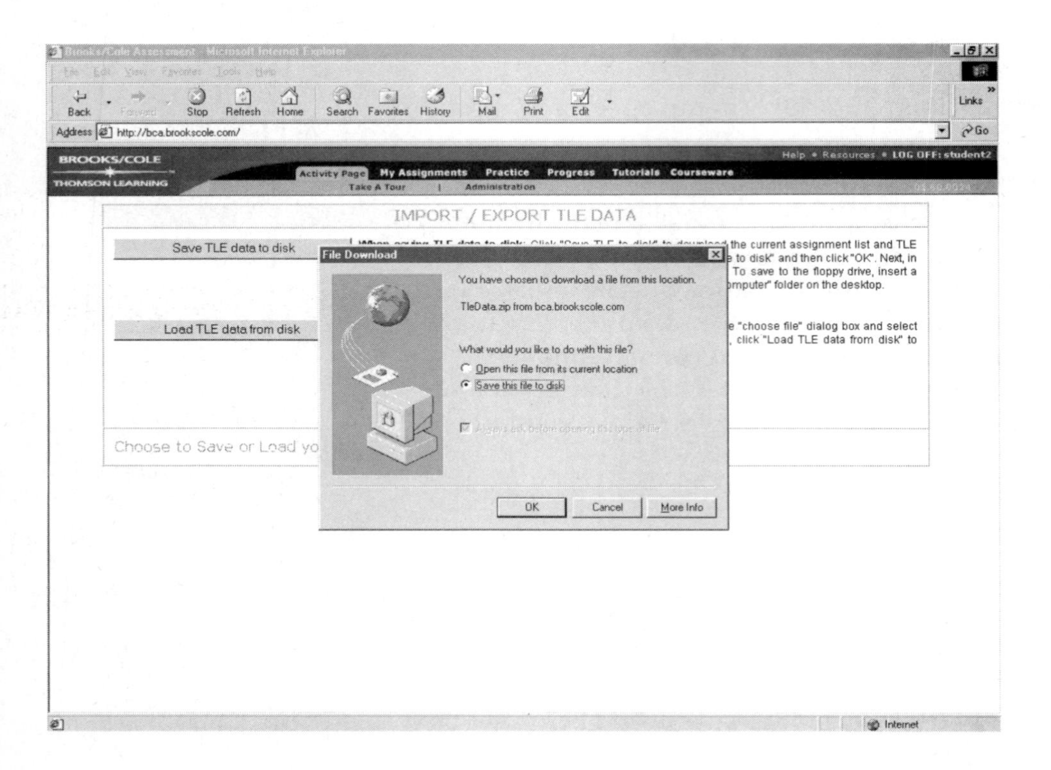

5. Click the "Save file to disk" option and click "OK". A "Save as" dialogue box will be displayed.

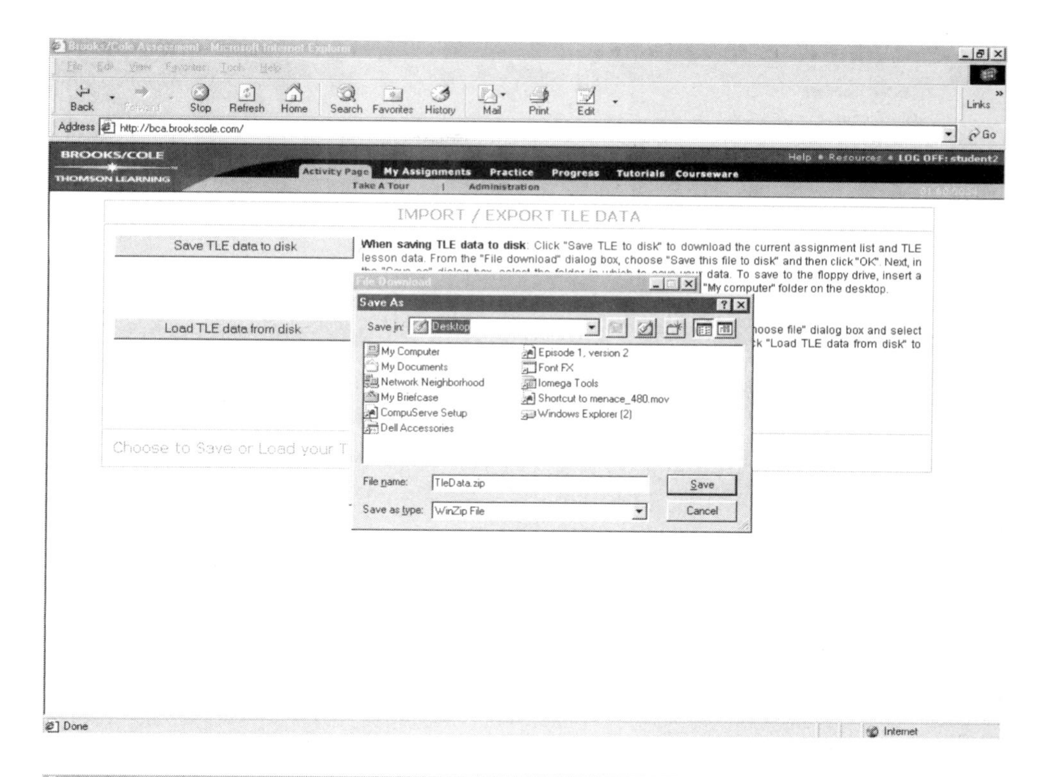

6. Locate and select the "$3\frac{1}{2}$ in. floppy drive" (usually the A: drive) and click "Open". You will find the A: drive under My Computer. Click the "Save" button. If the file already exists, choose to overwrite it. Close the "Download complete" dialogue box.

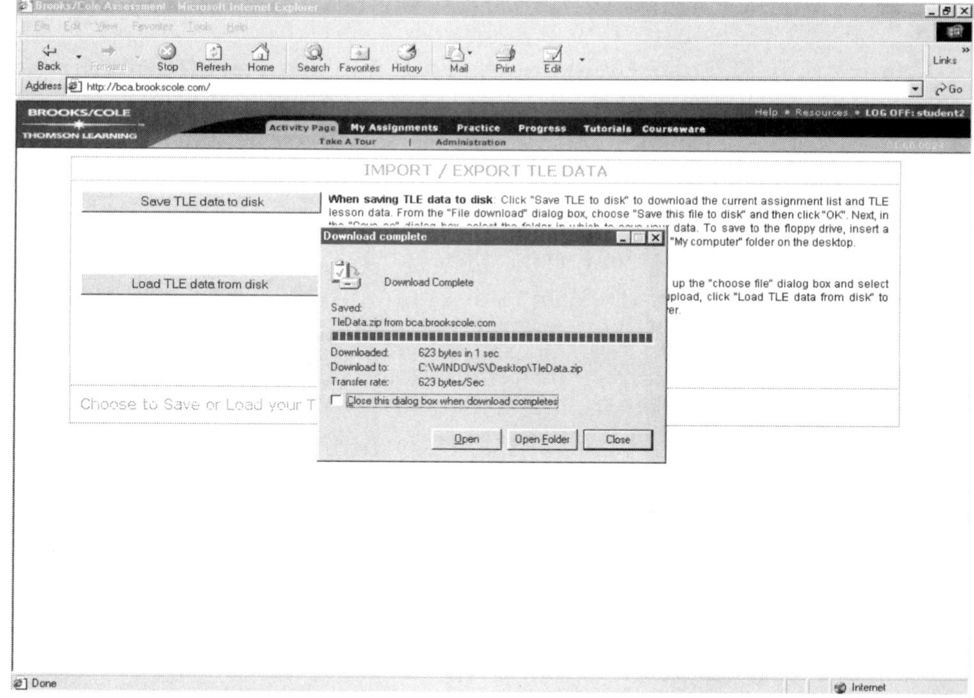

7. Log off by clicking the "Log off" link in the upper right corner of the browser window.
8. Close the web browser session (click the X in the uppermost right hand corner).
9. Take the floppy disk out of the floppy drive and take it with you to update your computer off campus.

Updating your computer from floppy disk at the beginning of an off campus session

1. Double click the TLE icon on your desktop to re-launch TLE and your web browser. Note the address http://localhost is automatically inserted into the address field. This means your browser is now connecting to your hard disk and remaining off line. **You are not connected to the Internet at this time.**
2. Log in to TLE by first locating your school, then typing in your user name and password. This will bring you to the Activity page.
3. Click on Administration.
4. Click on the "Manage TLE data files" link. This displays the "Import/Export TLE Data" page.
5. Click the "Browse" button next to the "Select Upload File" input box.
6. Locate and select the <TleData> file you saved to floppy disk above and click the "Open" button. This will insert the pathname for the file in the "Select Upload File" input box.
7. Click the "Load TLE data from disk" button. This will update your computer with the latest tracking information and grade book progress from earlier in the day on campus.
8. Click the "My Assignments" tab to display a list of your assignments. Click the "Take" button to access any TLE content assignment.

Saving your TLE data file to floppy disk at the end of an off campus session

1. At the end of your TLE session off campus, and before you log off, go to the Administration sub-tab.
2. Insert a $3\frac{1}{2}$ in. floppy disk into the floppy drive.
3. Click the "Manage TLE data files" link. This displays the "Import/Export TLE Data" page.
4. Click the "Save TLE data to disk" button. This will display a "File download" dialogue box.
5. Click the "Save file to disk" option and click "OK". A "Save as" dialogue box will be displayed.
6. Locate and select the "$3\frac{1}{2}$ in. floppy drive" (usually the A: drive) and click "Open". Click the "Save" button. If the file already exists, choose to overwrite it. Close the "Download complete" dialogue box.
7. Log off by clicking the "Log off" link in the upper right corner of the browser window.
8. Close the web browser session (click the X in the uppermost right hand corner).
9. Take the floppy disk out of the floppy drive and take it with you to campus (or find any computer with an Internet connection) to update your instructor's web site for tracking your progress.

Updating your instructor's web site from floppy disk at the beginning of an on campus session

1. Double click the TLE icon on the desktop of the computer at school (or any computer connected to the Internet) to launch TLE and the web browser. This should take you automatically to http://bca.brookscole.com. You are connected to the Internet at this time.

2. Log in to TLE by first locating your school, then typing in your user name and password. This will bring you to the Activity page.
3. Click on Administration.
4. Click on the "Manage TLE data files" link. This displays the "Import/Export TLE Data" page.
5. Click the "Browse" button next to the "Select Upload File" input box.
6. Locate and select the <TleData> file you saved to floppy disk above and click the "Open" button. This will insert the pathname for the file in the "Select Upload File" input box.
7. Click the "Load TLE data from disk" button. This will update your instructor's web site with the latest tracking information, grade book progress, and any messages that you may have sent to your instructor. You can now proceed by accessing you assignments, checking your progress, or logging off.
8. To continue using TLE, click the "My Assignments" tab to display a list of your assignments. Click the "Take" button to access any TLE content assignment.

Using the Assignments Page

1. Launch and log in to TLE using your user name and password if you have not already done so. You will automatically be brought to your Activity page. By clicking on "Administration" (either as a sub menu of the Activity Page tab or as a link on the right side of the page), you can change your password or register to use new books.

2. Click on the "My Assignments" tab at the top of the page. This will display all the assignments that your instructor has created for the course. The assignments are listed by assignment category (assignment categories are created by your instructor) in the order in which they are due. Start with the first assignment and work your way down the list. Make sure to pay particular attention to any specific instructions from your instructor.

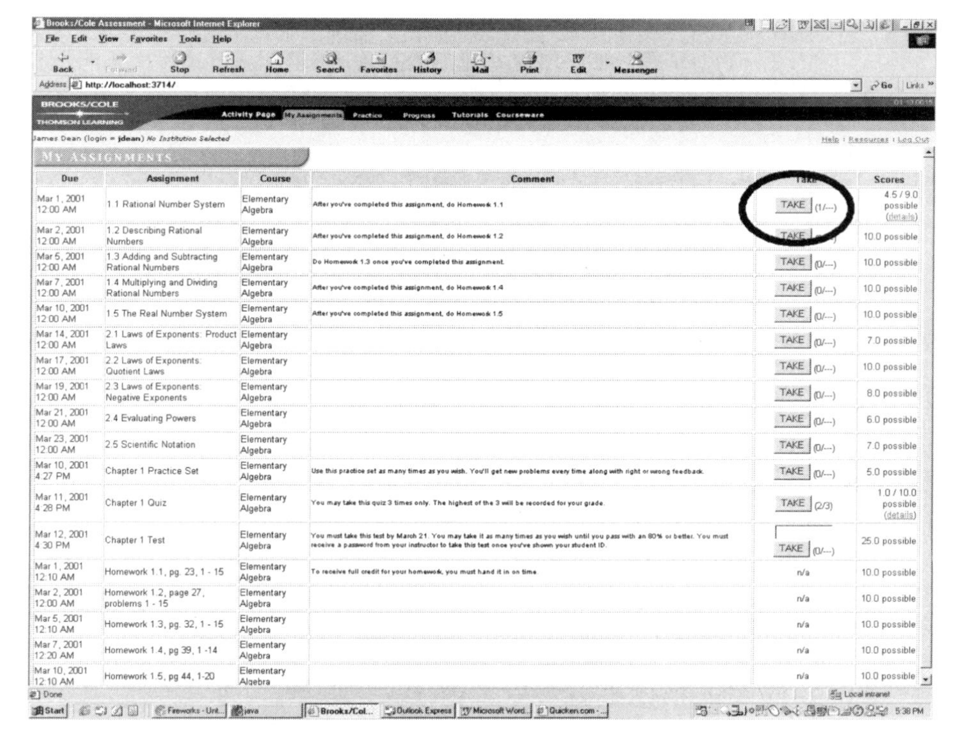

Due	Assignment	Course	Comment	Take	Scores
Mar 1, 2001 12:00 AM	1.1 Rational Number System	Elementary Algebra	After you've completed this assignment, do Homework 1.1	TAKE (1/---)	4.5 / 9.0 possible (details)
Mar 2, 2001 12:00 AM	1.2 Describing Rational Numbers	Elementary Algebra	After you've completed this assignment, do Homework 1.2	TAKE	10.0 possible
Mar 5, 2001 12:00 AM	1.3 Adding and Subtracting Rational Numbers	Elementary Algebra	Do Homework 1.3 once you've completed this assignment.	TAKE (0/---)	10.0 possible
Mar 7, 2001 12:00 AM	1.4 Multiplying and Dividing Rational Numbers	Elementary Algebra	After you've completed this assignment, do Homework 1.4	TAKE (0/---)	10.0 possible
Mar 10, 2001 12:00 AM	1.5 The Real Number System	Elementary Algebra	After you've completed this assignment, do Homework 1.5	TAKE (0/---)	10.0 possible
Mar 14, 2001 12:00 AM	2.1 Laws of Exponents: Product Laws	Elementary Algebra		TAKE (0/---)	7.0 possible
Mar 17, 2001 12:00 AM	2.2 Laws of Exponents: Quotient Laws	Elementary Algebra		TAKE (0/---)	10.0 possible
Mar 19, 2001 12:00 AM	2.3 Laws of Exponents: Negative Exponents	Elementary Algebra		TAKE (0/---)	8.0 possible
Mar 21, 2001 12:00 AM	2.4 Evaluating Powers	Elementary Algebra		TAKE (0/---)	6.0 possible
Mar 23, 2001 12:00 AM	2.5 Scientific Notation	Elementary Algebra		TAKE (0/---)	7.0 possible
Mar 10, 2001 4:27 PM	Chapter 1 Practice Set	Elementary Algebra	Use this practice set as many times as you wish. You'll get new problems every time along with right or wrong feedback.	TAKE (0/---)	5.0 possible
Mar 11, 2001 4:28 PM	Chapter 1 Quiz	Elementary Algebra	You may take this quiz 3 times only. The highest of the 3 will be recorded for your grade.	TAKE (2/3)	1.0 / 10.0 possible (details)
Mar 12, 2001 4:30 PM	Chapter 1 Test	Elementary Algebra	You must take this test by March 21. You may take it as many times as you wish until you pass with an 80% or better. You must receive a password from your instructor to take this test once you've shown your student ID.	TAKE (0/---)	25.0 possible
Mar 1, 2001 12:10 AM	Homework 1.1, pg. 23, 1 - 15	Elementary Algebra	To receive full credit for your homework, you must hand it in on time.	n/a	10.0 possible
Mar 2, 2001 12:00 AM	Homework 1.2, page 27, problems 1 - 15	Elementary Algebra		n/a	10.0 possible
Mar 5, 2001 12:10 AM	Homework 1.3, pg. 32, 1 - 15	Elementary Algebra		n/a	10.0 possible
Mar 7, 2001 12:20 AM	Homework 1.4, pg 39, 1 -14	Elementary Algebra		n/a	10.0 possible
Mar 10, 2001 12:10 AM	Homework 1.5, pg 44, 1-20	Elementary Algebra		n/a	10.0 possible

3. There are two basic kinds of assignments: computer-based assignments to be done on the computer (TLE lessons, practice/quizzes/tests), and non-computer-based assignments (such as paper- and pencil-based homework).
4. Locate your first computer-based assignment. By clicking the "TAKE" button, you will be brought to either the TLE lesson or practice/quiz/test that your instructor wants you to complete.
5. To see any of your previous work on practice, quiz, or test assignments, click the "details" link in the "score box" associated with that particular assignment.

Computer-based Assignments: TLE Lesson Assignments

TLE lesson assignments take you to the lessons that your instructor has assigned for your course. You must access TLE lessons from your Assignments page so that you receive a grade for the course. Your instructor will be able to track how much time you spent, the TLE "pages" completed correctly, and how well you do on the Self Checks for every TLE lesson assigned. When using TLE while remaining off-line, you must first go on line to update your computer. See previous sections for information.

Follow these steps to access, work through, and end a TLE lesson assignment.

1. Launch and log in to TLE if you have not already done so (see previous sections for information on how to log in to TLE).
2. Click on the "My Assignments" tab at the top of the page if you have not already done so.
3. Click the "TAKE" button for the assignment that you want to work on. If this is your first time using TLE, you will probably want to go to the first assignment. Check with your instructor.
4. If this is your first time working on a TLE lesson assignment, you will be brought to the TLE Tour. Work through the Tour to familiarize yourself with how TLE works. If this is not your first time, you will be brought to the first page of the lesson that has been assigned. Begin working through the TLE lesson. Follow any specific instructions that your instructor has given you.

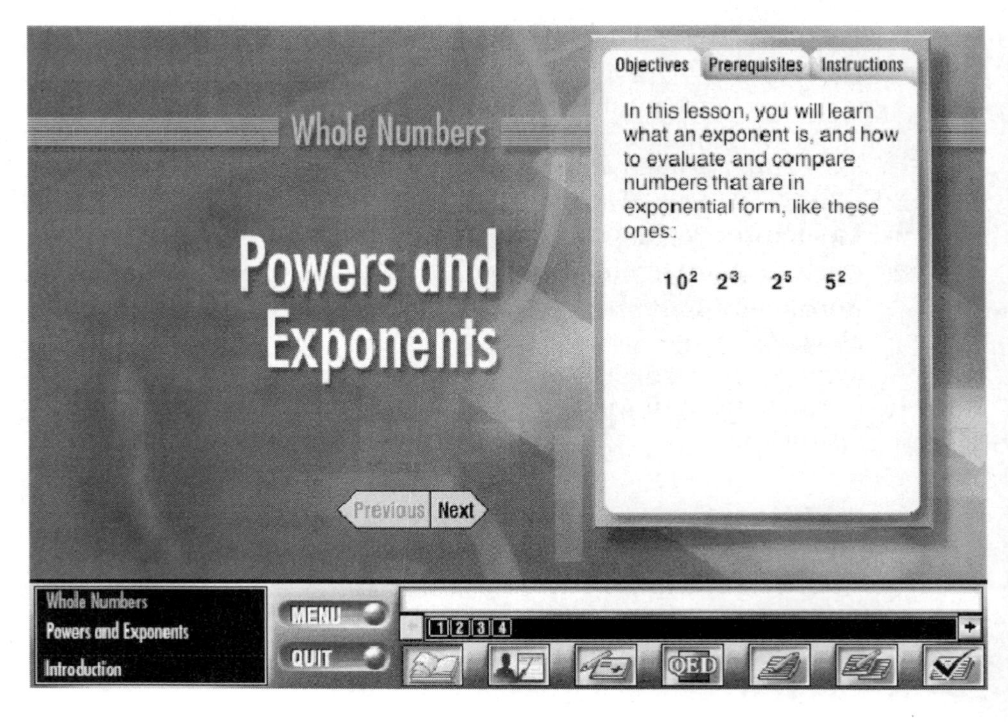

5. When you are ready to end your TLE session, click the "Quit" button at the bottom of the TLE screen.

6. When prompted to confirm the quit, either click the "Quit" button on the window that appears to end the session or "Cancel" to go back to the TLE lesson. You will then be brought back to the Assignments page. Do not click the "Menu" button as this will take you to the entire TLE table of contents.

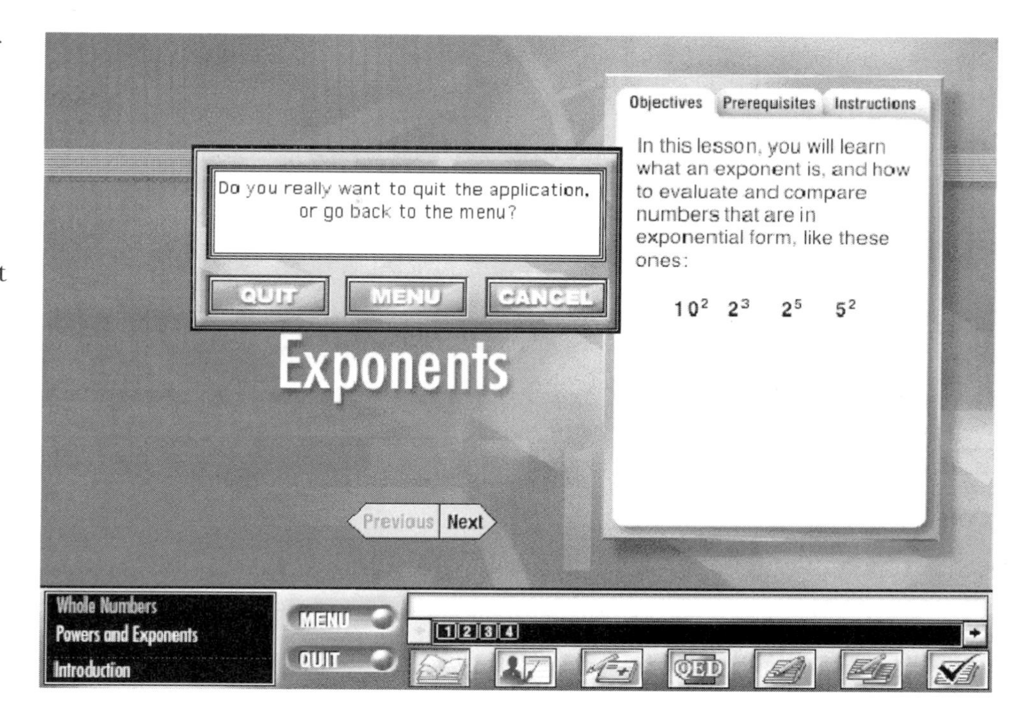

7. Note: Never click the "Menu" button, either in the "Quit" confirmation or on the taskbar at the bottom of the TLE screen. If you click the "Menu" button, you will be taken to the TLE table of contents. The program will continue to track your time spent, but you will NOT receive credit for any work you do and you will not receive a grade for any work you do. This will be time wasted. You must access TLE lessons from your Assignments page when using TLE.

Computer-based Assignments: Practice, Quiz, and Test Assignments

Your instructor may have created some on line practice, quiz, or test assignments. To take one of these assignments, follow these steps. Note: You must have an Internet connection to access these computer-based assignments.

1. Launch and log in to TLE if you have not already done so.
2. Click on the "My Assignments" tab at the top of the page if you have not already done so.
3. Click the "TAKE" button for the practice, quiz, or test assignment that you want to work on. If the assignment is a quiz or test, you may need a separate password for access. Check with your instructor.

4. You will be brought to the assignment. Follow the on screen instructions or consult your instructor.

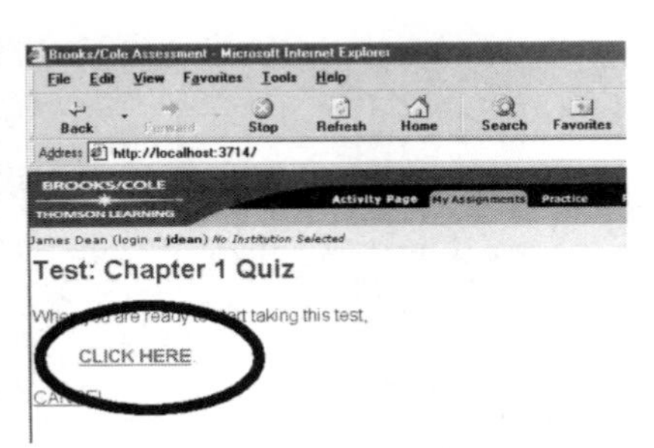

5. Enter your answer to the first question. For multiple choice questions, simply click on the choice and then click the "Go" button. For "Free-response Mathematics" type questions (where you type in mathematical answers), place the cursor in the input box, and either type your answer using calculator syntax or use the palette of mathematical operators. Once finished, click the "Go" button and you'll see "Answer Accepted" displayed. Other questions, such as completing tables, require you to directly input your numerical answer. The next problem will be automatically displayed.

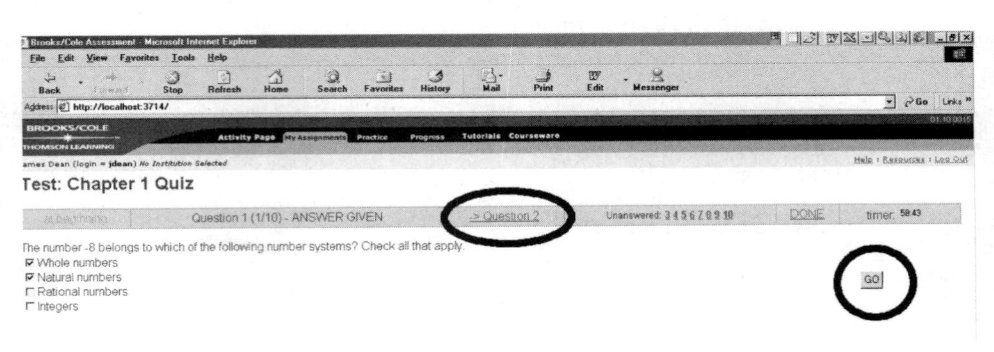

6. You may also click on any item number to go to that item directly. This will move you through the test items. Notice the question listing to the right of "Unanswered". This lists the number of the questions in the assignment (depending on your browser settings, this will appear as a "visited" link), and those questions not yet attempted. The figure below shows the palette of mathematical expressions you can use to input free response mathematics answers. For more information on how to use the palette, click on "Help".

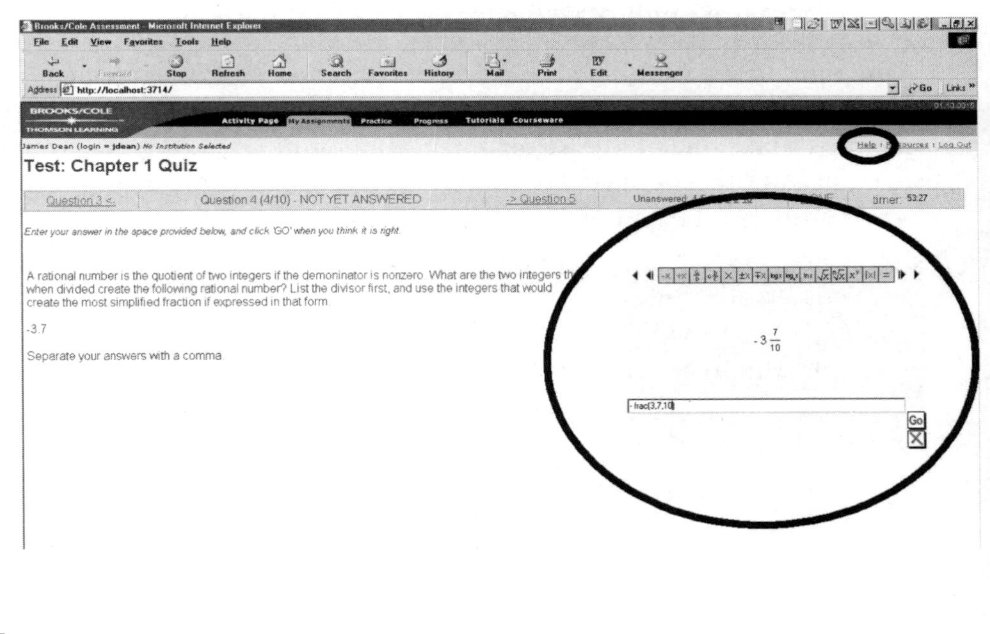

7. Once you've answered all the questions, you can go back to any of them by clicking on the question number links. Make sure you are confident of all your answers.

8. When you are confident of your answers, click the "Done" link. You will be asked to confirm whether you are done. Click "No" to go back to the test, or click "Yes" to leave the test and record your score.

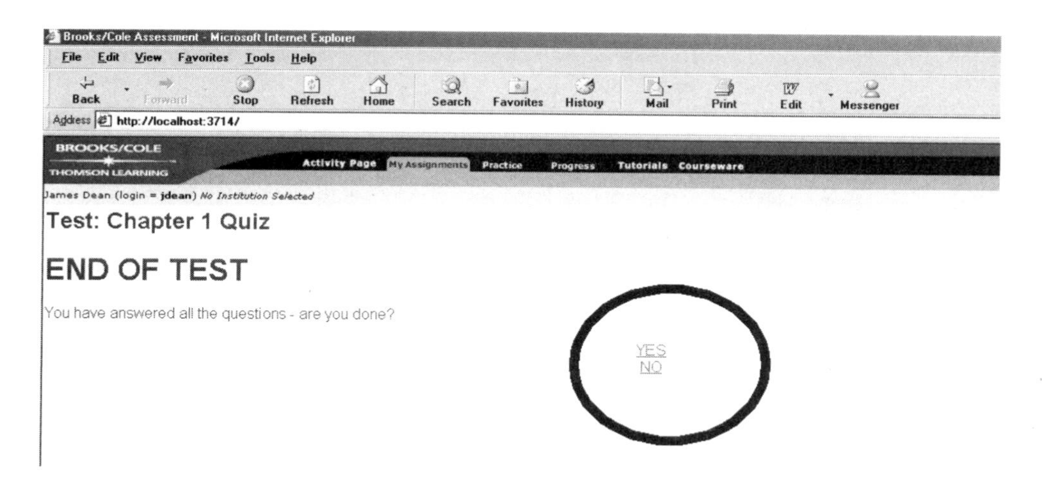

9. If your instructor has allowed for it, you may view your quiz or test results. Click the "RIGHT" or "WRONG" links to review your quiz or test item by item. Click the "return to assignments list" (below "Result Details") to return to your Assignments page.

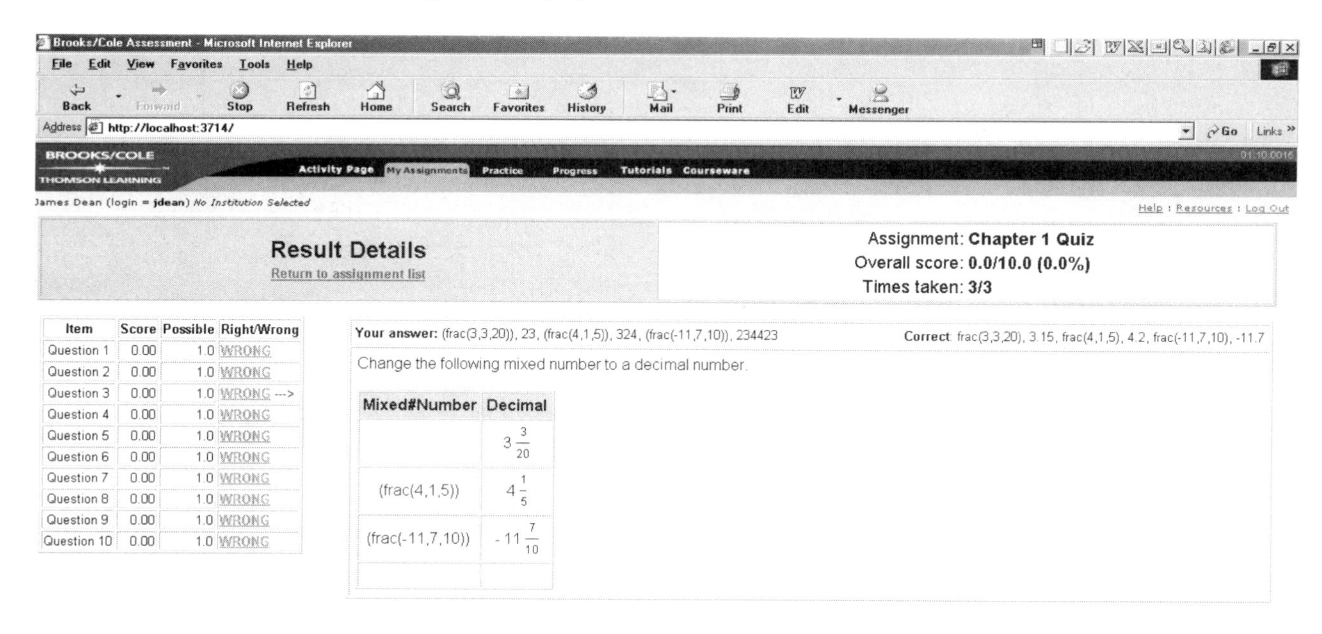

Item	Score	Possible	Right/Wrong
Question 1	0.00	1.0	WRONG
Question 2	0.00	1.0	WRONG
Question 3	0.00	1.0	WRONG --->
Question 4	0.00	1.0	WRONG
Question 5	0.00	1.0	WRONG
Question 6	0.00	1.0	WRONG
Question 7	0.00	1.0	WRONG
Question 8	0.00	1.0	WRONG
Question 9	0.00	1.0	WRONG
Question 10	0.00	1.0	WRONG

Mixed#Number	Decimal
	$3\frac{3}{20}$
(frac(4,1,5))	$4\frac{1}{5}$
(frac(-11,7,10))	$-11\frac{7}{10}$

10. To see your score recorded, click on the "Progress" tab.

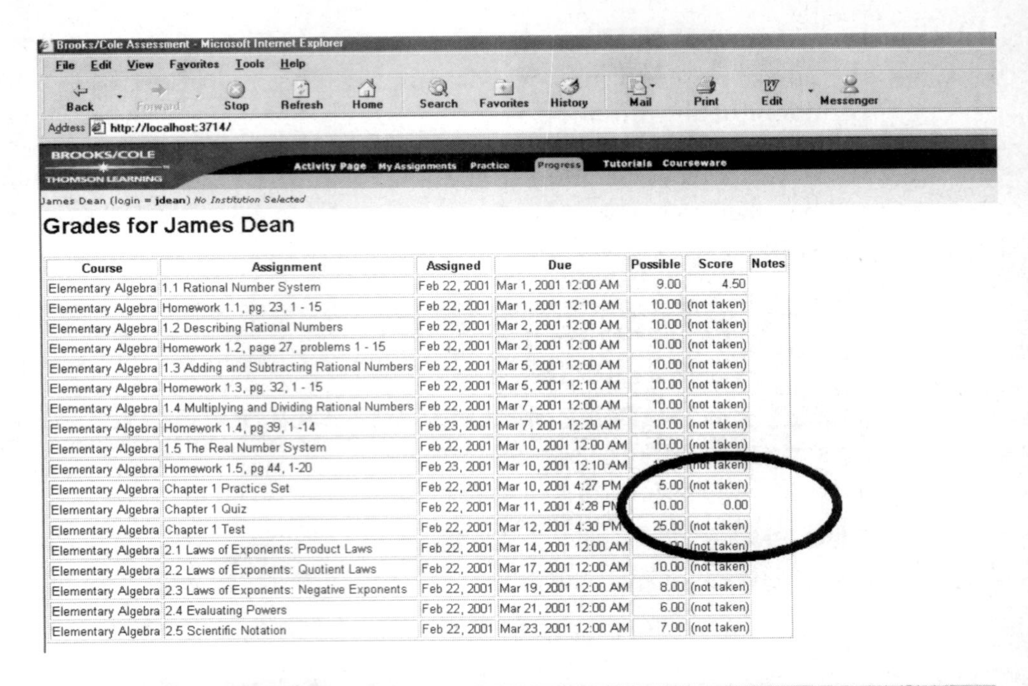

Brooks/Cole Assessment - Microsoft Internet Explorer

Address: http://localhost:3714/

Activity Page | My Assignments | Practice | Progress | Tutorials | Courseware

James Dean (login = jdean) No Institution Selected

Grades for James Dean

Course	Assignment	Assigned	Due	Possible	Score	Notes
Elementary Algebra	1.1 Rational Number System	Feb 22, 2001	Mar 1, 2001 12:00 AM	9.00	4.50	
Elementary Algebra	Homework 1.1, pg. 23, 1 - 15	Feb 22, 2001	Mar 1, 2001 12:10 AM	10.00	(not taken)	
Elementary Algebra	1.2 Describing Rational Numbers	Feb 22, 2001	Mar 2, 2001 12:00 AM	10.00	(not taken)	
Elementary Algebra	Homework 1.2, page 27, problems 1 - 15	Feb 22, 2001	Mar 2, 2001 12:00 AM	10.00	(not taken)	
Elementary Algebra	1.3 Adding and Subtracting Rational Numbers	Feb 22, 2001	Mar 5, 2001 12:00 AM	10.00	(not taken)	
Elementary Algebra	Homework 1.3, pg. 32, 1 - 15	Feb 22, 2001	Mar 5, 2001 12:10 AM	10.00	(not taken)	
Elementary Algebra	1.4 Multiplying and Dividing Rational Numbers	Feb 22, 2001	Mar 7, 2001 12:00 AM	10.00	(not taken)	
Elementary Algebra	Homework 1.4, pg 39, 1 -14	Feb 23, 2001	Mar 7, 2001 12:20 AM	10.00	(not taken)	
Elementary Algebra	1.5 The Real Number System	Feb 22, 2001	Mar 10, 2001 12:00 AM	10.00	(not taken)	
Elementary Algebra	Homework 1.5, pg 44, 1-20	Feb 23, 2001	Mar 10, 2001 12:10 AM	[illegible]	(not taken)	
Elementary Algebra	Chapter 1 Practice Set	Feb 22, 2001	Mar 10, 2001 4:27 PM	5.00	(not taken)	
Elementary Algebra	Chapter 1 Quiz	Feb 22, 2001	Mar 11, 2001 4:28 PM	10.00	0.00	
Elementary Algebra	Chapter 1 Test	Feb 22, 2001	Mar 12, 2001 4:30 PM	25.00	(not taken)	
Elementary Algebra	2.1 Laws of Exponents: Product Laws	Feb 22, 2001	Mar 14, 2001 12:00 AM	[illegible]	(not taken)	
Elementary Algebra	2.2 Laws of Exponents: Quotient Laws	Feb 22, 2001	Mar 17, 2001 12:00 AM	10.00	(not taken)	
Elementary Algebra	2.3 Laws of Exponents: Negative Exponents	Feb 22, 2001	Mar 19, 2001 12:00 AM	8.00	(not taken)	
Elementary Algebra	2.4 Evaluating Powers	Feb 22, 2001	Mar 21, 2001 12:00 AM	6.00	(not taken)	
Elementary Algebra	2.5 Scientific Notation	Feb 22, 2001	Mar 23, 2001 12:00 AM	7.00	(not taken)	

11. By clicking on "My Assignments", you can view what assignments have been completed. Note that this quiz had been taken the maximum three times (assigned by the instructor).

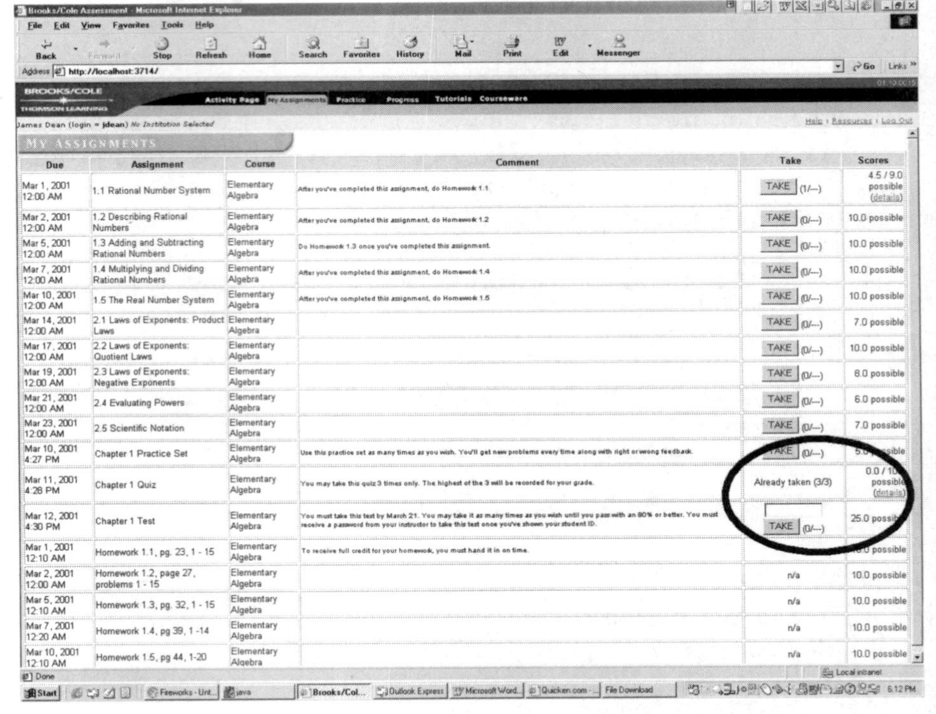

Brooks/Cole Assessment - Microsoft Internet Explorer

Address: http://localhost:3714/

Activity Page | My Assignments | Practice | Progress | Tutorials | Courseware

James Dean (login = jdean) No Institution Selected

Help : Resources : Log Out

MY ASSIGNMENTS

Due	Assignment	Course	Comment	Take	Scores
Mar 1, 2001 12:00 AM	1.1 Rational Number System	Elementary Algebra	After you've completed this assignment, do Homework 1.1	TAKE (1/---)	4.5 / 9.0 possible (details)
Mar 2, 2001 12:00 AM	1.2 Describing Rational Numbers	Elementary Algebra	After you've completed this assignment, do Homework 1.2	TAKE (0/---)	10.0 possible
Mar 5, 2001 12:00 AM	1.3 Adding and Subtracting Rational Numbers	Elementary Algebra	Do Homework 1.3 once you've completed this assignment.	TAKE (0/---)	10.0 possible
Mar 7, 2001 12:00 AM	1.4 Multiplying and Dividing Rational Numbers	Elementary Algebra	After you've completed this assignment, do Homework 1.4	TAKE (0/---)	10.0 possible
Mar 10, 2001 12:00 AM	1.5 The Real Number System	Elementary Algebra	After you've completed this assignment, do Homework 1.5	TAKE (0/---)	10.0 possible
Mar 14, 2001 12:00 AM	2.1 Laws of Exponents: Product Laws	Elementary Algebra		TAKE (0/---)	7.0 possible
Mar 17, 2001 12:00 AM	2.2 Laws of Exponents: Quotient Laws	Elementary Algebra		TAKE (0/---)	10.0 possible
Mar 19, 2001 12:00 AM	2.3 Laws of Exponents: Negative Exponents	Elementary Algebra		TAKE (0/---)	8.0 possible
Mar 21, 2001 12:00 AM	2.4 Evaluating Powers	Elementary Algebra		TAKE (0/---)	6.0 possible
Mar 23, 2001 12:00 AM	2.5 Scientific Notation	Elementary Algebra		TAKE (0/---)	7.0 possible
Mar 10, 2001 4:27 PM	Chapter 1 Practice Set	Elementary Algebra	Use this practice set as many times as you wish. You'll get new problems every time along with right or wrong feedback.	[illegible] (0/---)	5.0 [illegible]sible
Mar 11, 2001 4:28 PM	Chapter 1 Quiz	Elementary Algebra	You may take this quiz 3 times only. The highest of the 3 will be recorded for your grade.	Already taken (3/3)	0.0 / 1[illegible] possible (details)
Mar 12, 2001 4:30 PM	Chapter 1 Test	Elementary Algebra	You must take this test by March 21. You may take it as many times as you wish until you pass with an 80% or better. You must receive a password from your instructor to take this test once you've shown your student ID.	TAKE (0/---)	25.0 possi[illegible]
Mar 1, 2001 12:10 AM	Homework 1.1, pg. 23, 1 - 15	Elementary Algebra	To receive full credit for your homework, you must hand it in on time.		[illegible] possible
Mar 2, 2001 12:00 AM	Homework 1.2, page 27, problems 1 - 15	Elementary Algebra		n/a	10.0 possible
Mar 5, 2001 12:10 AM	Homework 1.3, pg. 32, 1 - 15	Elementary Algebra		n/a	10.0 possible
Mar 7, 2001 12:20 AM	Homework 1.4, pg 39, 1 -14	Elementary Algebra		n/a	10.0 possible
Mar 10, 2001 12:10 AM	Homework 1.5, pg 44, 1-20	Elementary Algebra		n/a	10.0 possible

Using the Progress Page

The Progress page allows you to check your current progress through the course. Once there, you will see which assignments you have completed and how well you have done on them. You and your instructor are the only ones who have access to this information.

To use the Progress page, follow these steps.

1. Log in to the TLE web site if you have not already done so.
2. Click on the "Progress" tab at the top of the page if you have not already done so.
3. Check your progress by reading through the list.

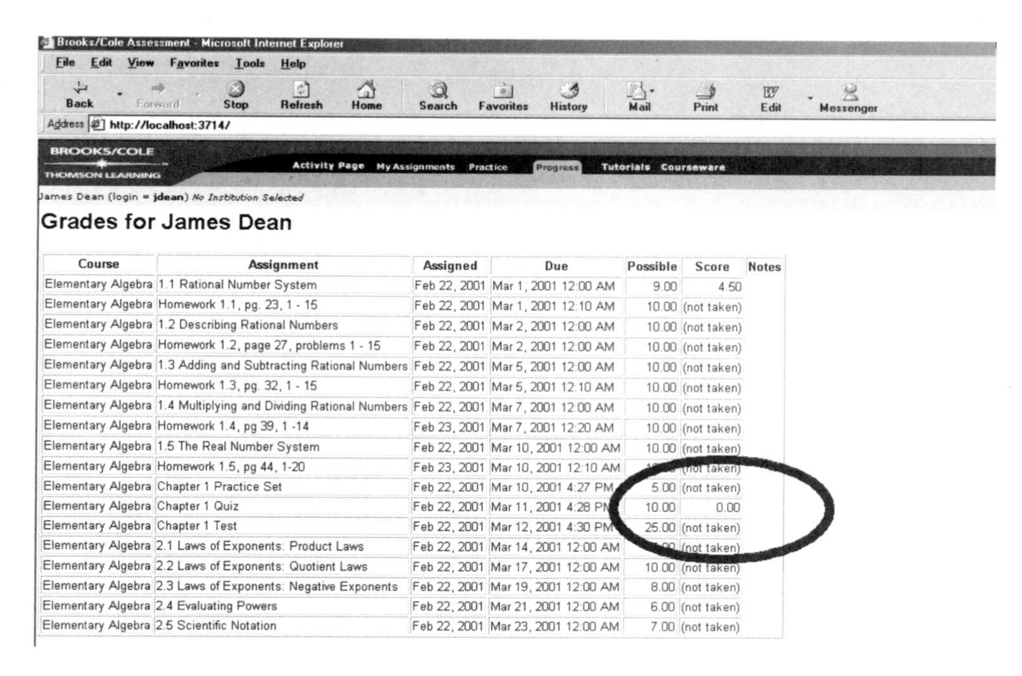

Grades for James Dean

Course	Assignment	Assigned	Due	Possible	Score	Notes
Elementary Algebra	1.1 Rational Number System	Feb 22, 2001	Mar 1, 2001 12:00 AM	9.00	4.50	
Elementary Algebra	Homework 1.1, pg. 23, 1 - 15	Feb 22, 2001	Mar 1, 2001 12:10 AM	10.00	(not taken)	
Elementary Algebra	1.2 Describing Rational Numbers	Feb 22, 2001	Mar 2, 2001 12:00 AM	10.00	(not taken)	
Elementary Algebra	Homework 1.2, page 27, problems 1 - 15	Feb 22, 2001	Mar 2, 2001 12:00 AM	10.00	(not taken)	
Elementary Algebra	1.3 Adding and Subtracting Rational Numbers	Feb 22, 2001	Mar 5, 2001 12:00 AM	10.00	(not taken)	
Elementary Algebra	Homework 1.3, pg. 32, 1 - 15	Feb 22, 2001	Mar 5, 2001 12:10 AM	10.00	(not taken)	
Elementary Algebra	1.4 Multiplying and Dividing Rational Numbers	Feb 22, 2001	Mar 7, 2001 12:00 AM	10.00	(not taken)	
Elementary Algebra	Homework 1.4, pg 39, 1 -14	Feb 23, 2001	Mar 7, 2001 12:20 AM	10.00	(not taken)	
Elementary Algebra	1.5 The Real Number System	Feb 22, 2001	Mar 10, 2001 12:00 AM	10.00	(not taken)	
Elementary Algebra	Homework 1.5, pg 44, 1-20	Feb 23, 2001	Mar 10, 2001 12:10 AM	[illegible]	(not taken)	
Elementary Algebra	Chapter 1 Practice Set	Feb 22, 2001	Mar 10, 2001 4:27 PM	5.00	(not taken)	
Elementary Algebra	Chapter 1 Quiz	Feb 22, 2001	Mar 11, 2001 4:28 PM	10.00	0.00	
Elementary Algebra	Chapter 1 Test	Feb 22, 2001	Mar 12, 2001 4:30 PM	25.00	(not taken)	
Elementary Algebra	2.1 Laws of Exponents: Product Laws	Feb 22, 2001	Mar 14, 2001 12:00 AM	[illegible]	(not taken)	
Elementary Algebra	2.2 Laws of Exponents: Quotient Laws	Feb 22, 2001	Mar 17, 2001 12:00 AM	10.00	(not taken)	
Elementary Algebra	2.3 Laws of Exponents: Negative Exponents	Feb 22, 2001	Mar 19, 2001 12:00 AM	8.00	(not taken)	
Elementary Algebra	2.4 Evaluating Powers	Feb 22, 2001	Mar 21, 2001 12:00 AM	6.00	(not taken)	
Elementary Algebra	2.5 Scientific Notation	Feb 22, 2001	Mar 23, 2001 12:00 AM	7.00	(not taken)	

4. If you have any questions, consult your instructor.
5. To see any of your previous work on practice, quiz, or test assignments, click the "details" link in the "score box" associated with that particular assignment.

Using the Practice Page

The Practice page is currently under construction. In the future, you'll be able to access further help on concepts.

Using the Courseware Page

The courseware page is designed for unlimited access to all TLE courses. This page should be used only by students NOT enrolled in a TLE-based course. If you are enrolled in a TLE-based course, you must access TLE lessons from your Assignments page. If you do not access TLE lessons from your Assignments page, you will not receive a grade for the work you do.

NOTE: The Tutorials page is designed for use with Brooks/Cole Thomson Learning textbooks and is not part of The Learning Equation.

Logging out

You must always log out of TLE to end a TLE session. Failure to log out correctly will record incorrect information and may result in a lower grade for your session. To log out, follow these steps.

1. Once you are ready to end your TLE session, click the "Log off" link in the upper right hand corner of the screen (next to "Help" and "Resources").
2. You should be brought back to the log in screen. This is an indication that your TLE session has logged out correctly and your session grades and time spent have been sent correctly to your instructor's grade book.

Technical Support

For more help, contact our technical support group by telephone or email:

800-423-0563, press option 2 (available M – F, 8:30 a.m. – 6:00 p.m. Eastern time)

or

tle.help@kdc.com

Enjoy learning! Enjoy *The Learning Equation*!

Components

The Learning Equation® Prealgebra consists of 65 lessons that cover the college prealgebra curriculum. Lessons are designed to take about 90 minutes but learners can progress at their own pace.

Every lesson consists of the following seven components.

Introduction

The opening screen briefly outlines the lesson and its objectives. Prerequisites for the lesson and instructions are available from an on-screen tabbed notebook.

To start the lesson, select the NEXT button on the opening screen.

The **Introduction** opens the lesson with a problem set in the context of a career, a real world or consumer experience, or a game. You will use knowledge you already have to solve the proposed problem. The **Introduction** is short, motivating, and highly interactive.

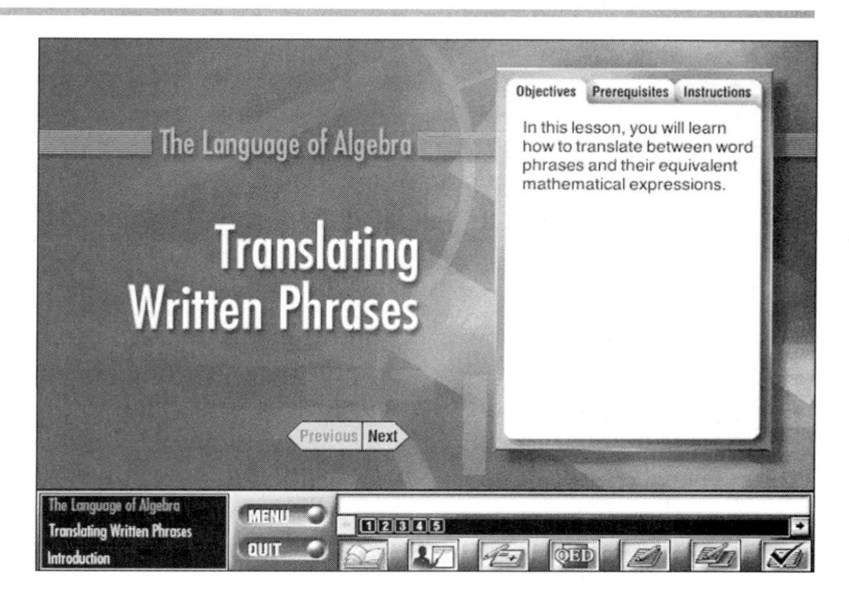

Tutorial

The *TLE* **Tutorial** offers the main instruction for the topic of the lesson. This section includes simple examples that help you understand the concepts. The **Tutorial** is intended to encourage you to learn by doing.

By providing different solutions or different routes through the same problem, *TLE* is designed to help students realize that there can be many ways to solve a problem.

TLE uses **Hints** to remind you when you can use strategies you've already learned. **Success Tips** offer ways to build confidence or to support independent study and learning.

Feedback Boxes

You can move any feedback box by selecting the green bar at the top of the box, and dragging it to the desired location. You can hide, and then reveal, the content of the feedback box by double clicking on the green bar. To close the feedback box, select the button in the upper left corner of the green bar.

Examples

The *TLE* **Examples expand** on what you learned in the Tutorial. The Examples are highly interactive and include feedback to your responses, hints, or alternative solutions.

Up to 12 examples may be presented in each lesson. A **hidden picture** is progressively revealed as you complete each example. Once the picture has been completely revealed, you may select a **Picture Information** button to learn more about it.

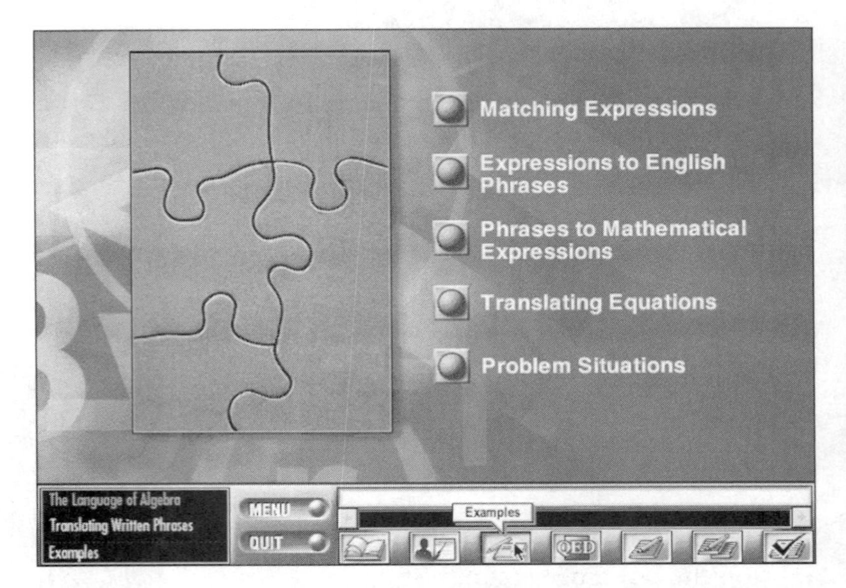

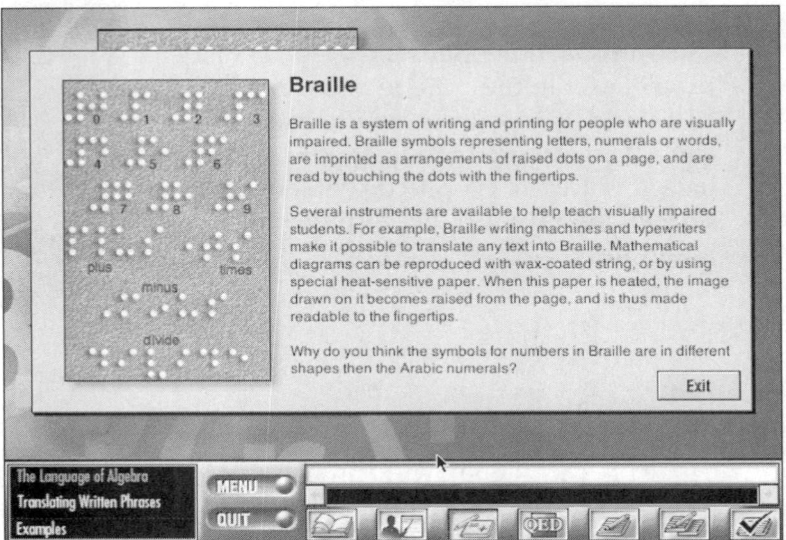

Summary

The *TLE* **Summary revisits the Introduction** and encourages you to **apply the mathematics learned** in the Tutorial and Examples to the problem in the Introduction.

Practice and Problems

The *TLE* **Practice and Problems** presents up to 25 questions. They are organized in **four or five categories** with five questions in each category. You can click on any box to try a question.

As in the **Examples,** another **hidden picture** is revealed when questions are completed correctly.

What is the first step	Solving Equations	Error Analysis	Checking Solutions	Problem Situations
1	6	11	16	21
2	7	12	17	22
3	8	13	18	23
4	9	14	19	24
5	10	15	20	25

Linear Equations
Linear Equations 1
Practice & Problems
MENU
QUIT

Extra Practice

The *TLE* **Extra Practice** presents **questions** like the ones in the Examples. After each question, you have the option of seeing a sample solution, **retrying** the question, or seeing the correct answer if they were incorrect.

Questions for each type of exercise are dynamically generated from a **mini-data bank** of up to 60 questions.

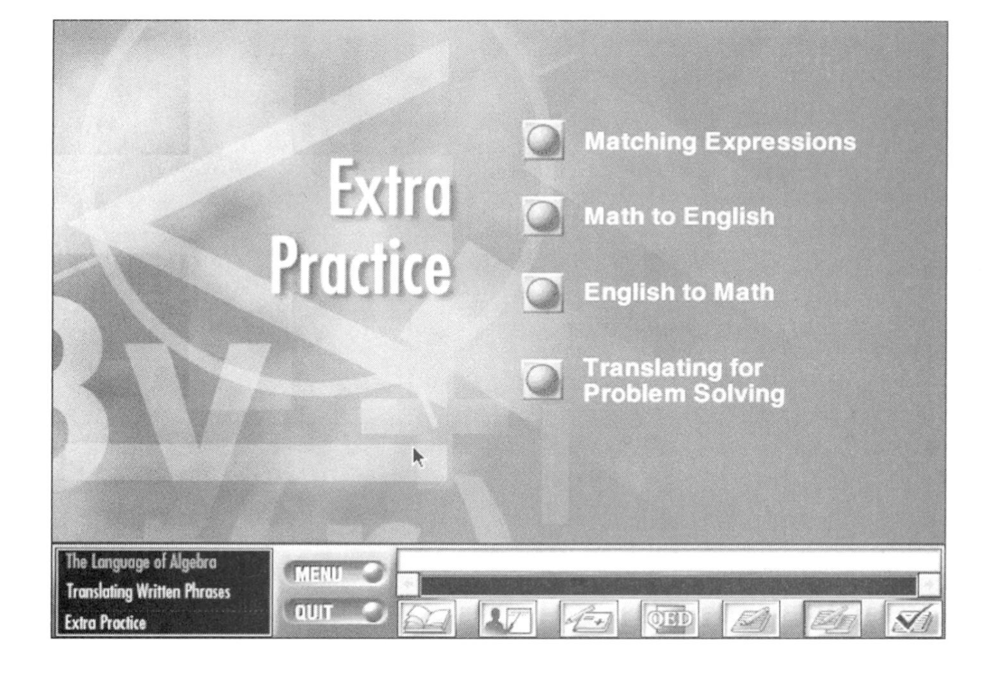

Self-Check

The *TLE* **Self-Check** presents up to **10 questions** like those in the **Extra Practice**. There are **three unique Self-Checks** for each lesson.

After you see your scores, you can **review the questions** and your **answers** one at a time, **try again**, or **see** the **correct answer** or a **sample solution**.

On achieving a minimum standard (70%), the lesson is considered to be completed. Completed lessons are indicated by a check mark that appears beside the lesson on the main menu.

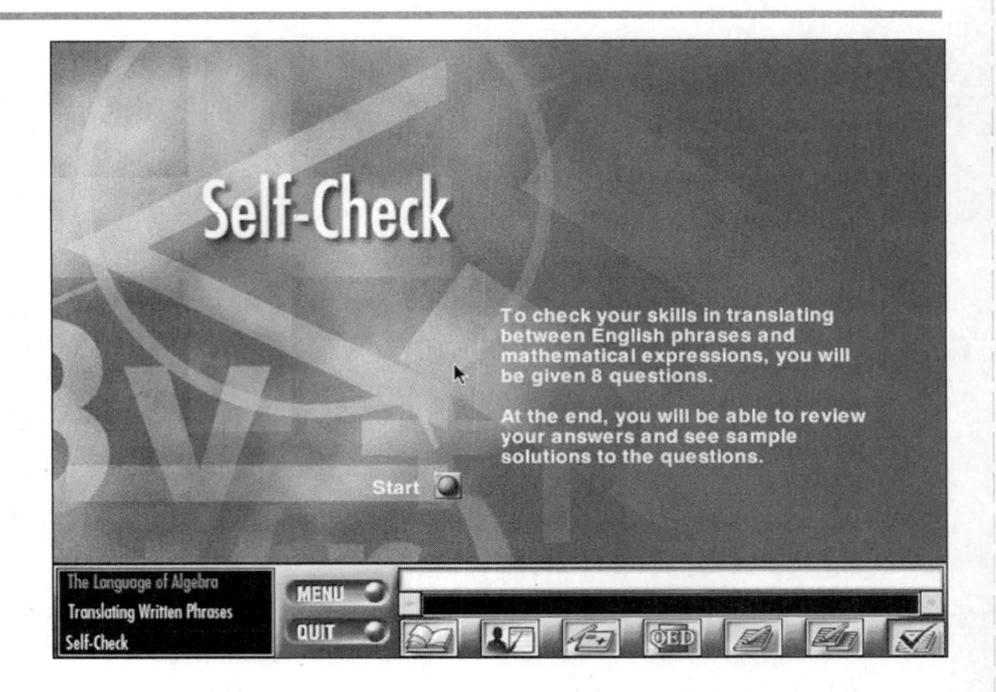

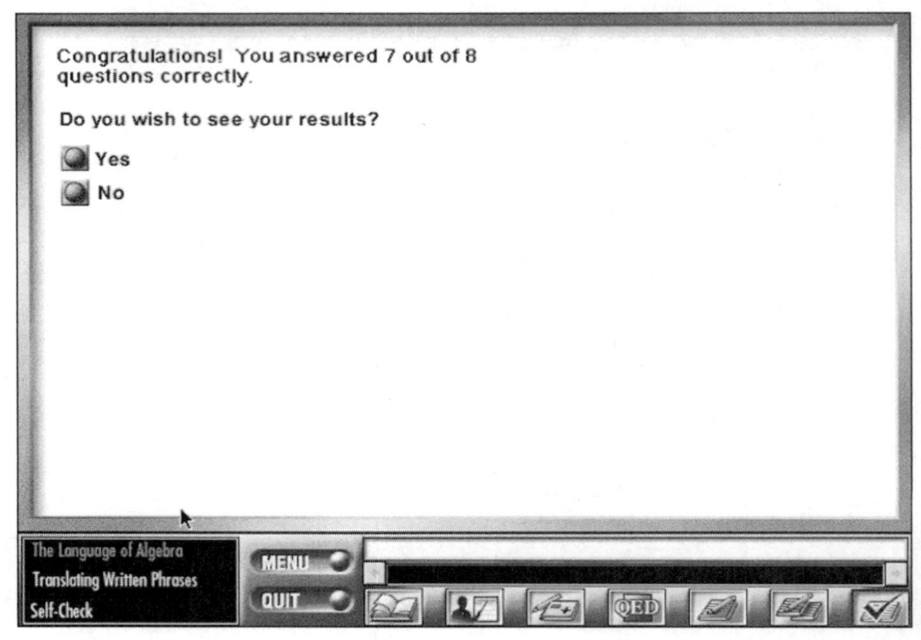

NOTES

Feedback Boxes

Please note that the Feedback Boxes can be moved by clicking on the green bar and dragging the box to the desired location on the screen. You can hide, and then reveal, the content of the Feedback Box by double-clicking on the green bar. To close the Feedback Box, click on the button in the upper left corner of the green bar.

Features

Interacting with *TLE* Courseware

You will interact with *TLE* in a variety of ways.

Interaction 1

You may input **simple text** and press RETURN after each input.

Every effort has been made to anticipate all reasonable forms in which you might enter answers. Text answers usually consist of numbers, and sometimes variables. Other alphabetic entries are rarely required, in order to avoid entering mis-spelled but mathematically correct answers.

Different numerical forms of answers may be allowed, such as 1/2, 0.5, 0.50, and so on. However, if all the information in a problem is expressed in one form, like whole numbers, the answers are usually expected to be in the same form.

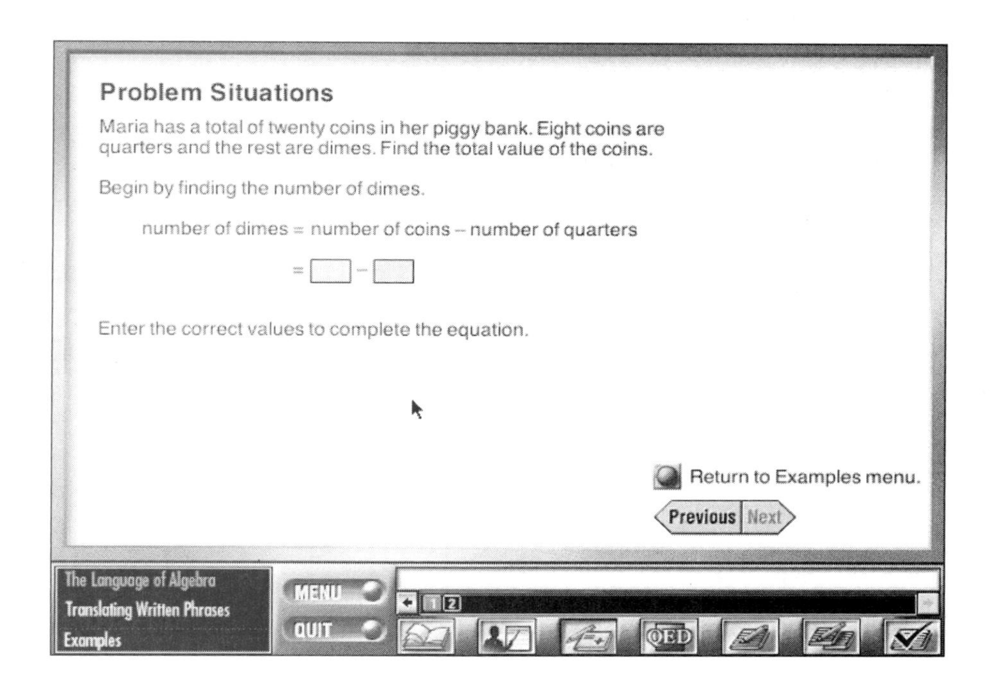

Interaction 2

You may click a button for a multiple choice response.

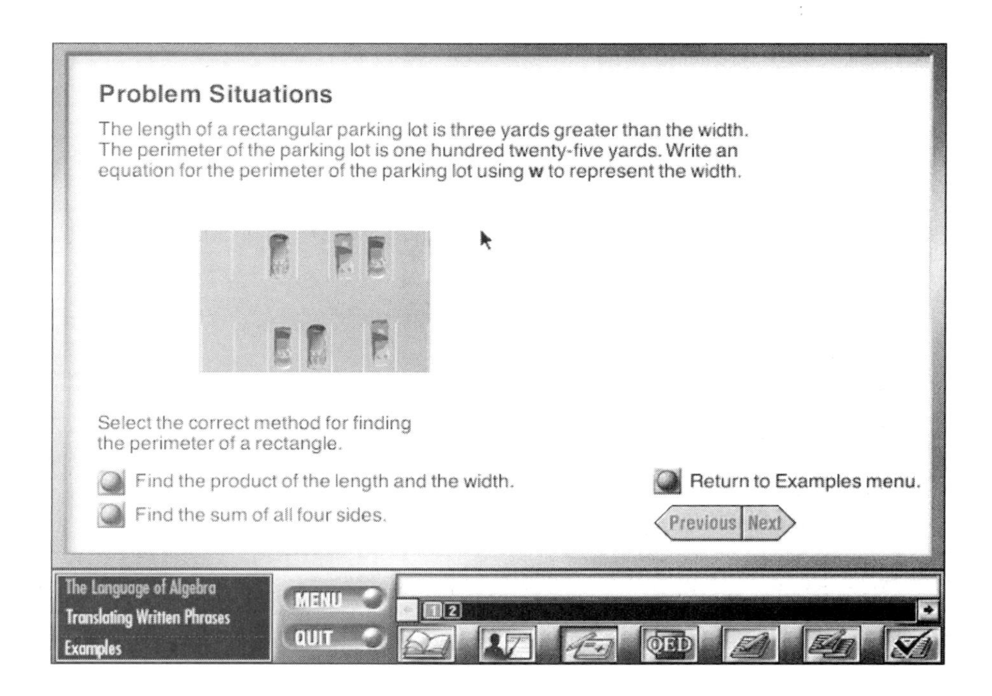

In some multiple choice questions, you may select more than one response, and then press a DONE button. You can then reflect on your choice(s).

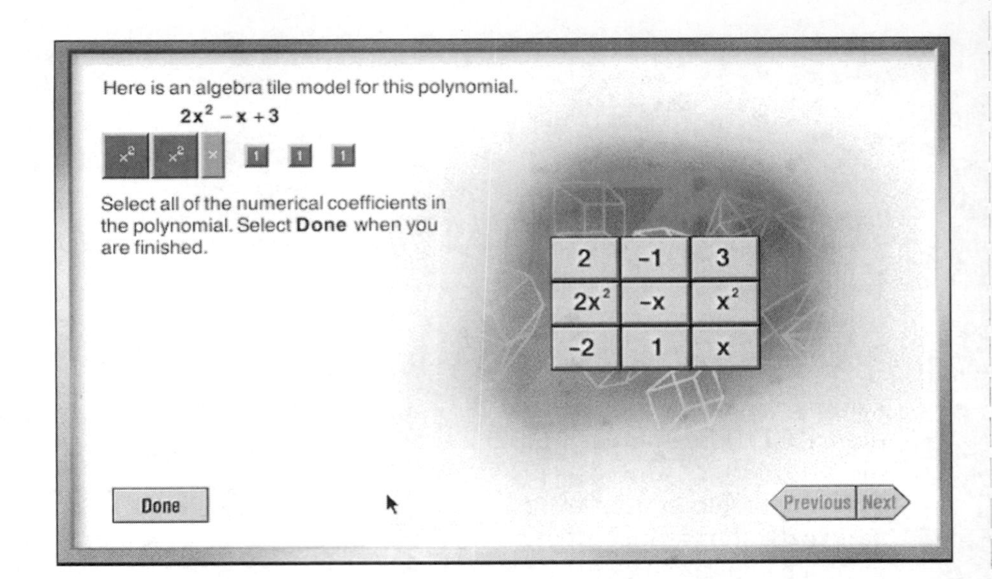

Interaction 3

You may drag one or more items to the appropriate location(s).

Click on the item and drag it with the mouse. Correctly placed items stick; incorrectly placed items bounce back.

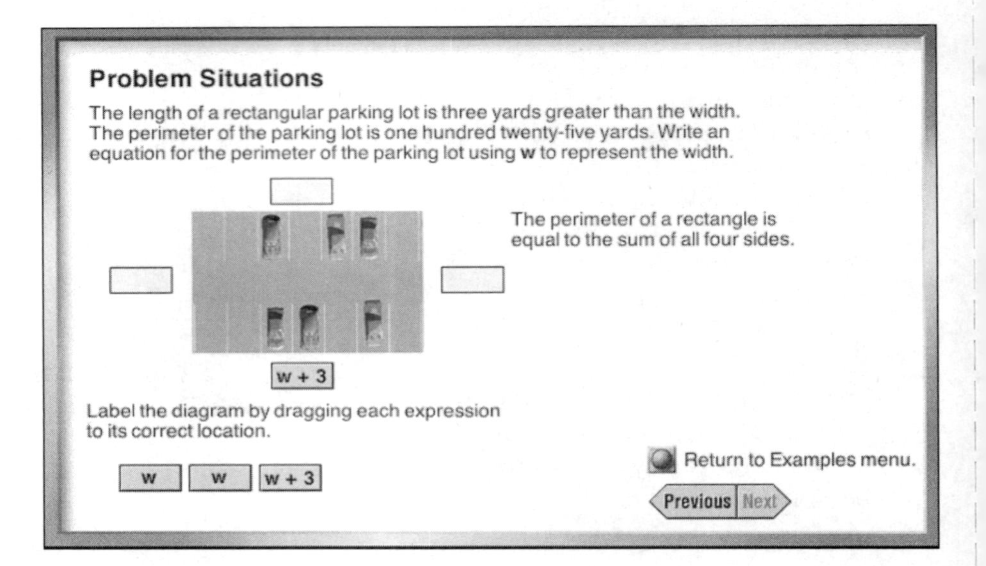

You should drag and place the items carefully with the mouse. If you are careless about the position of an item in a box, a correct item may bounce back and appear to be incorrect.

For some "drag-and-drop" questions, you may drag more than one item to the appropriate location and then press a DONE button.

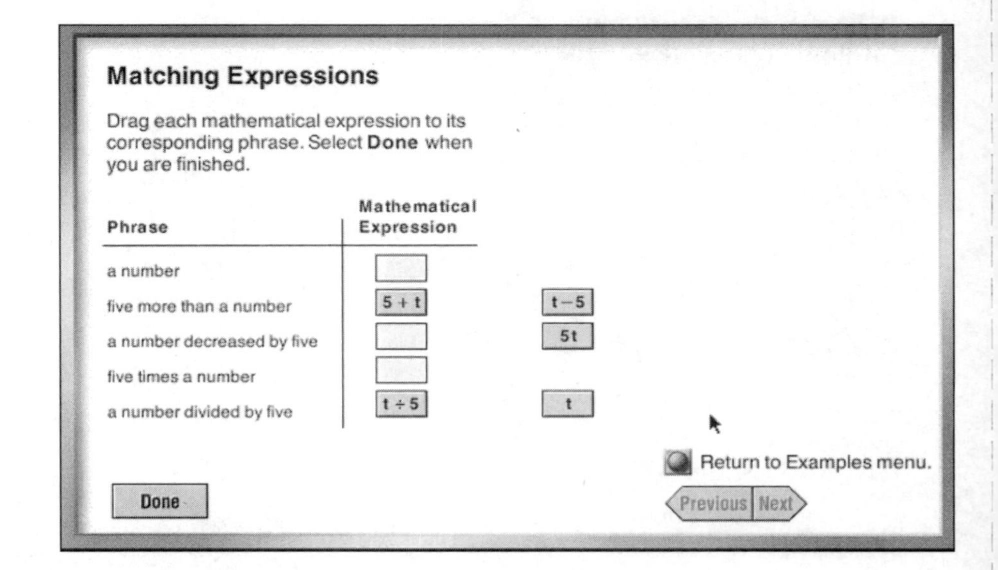

After any of the above interactions, you select NEXT to continue.

Explorers

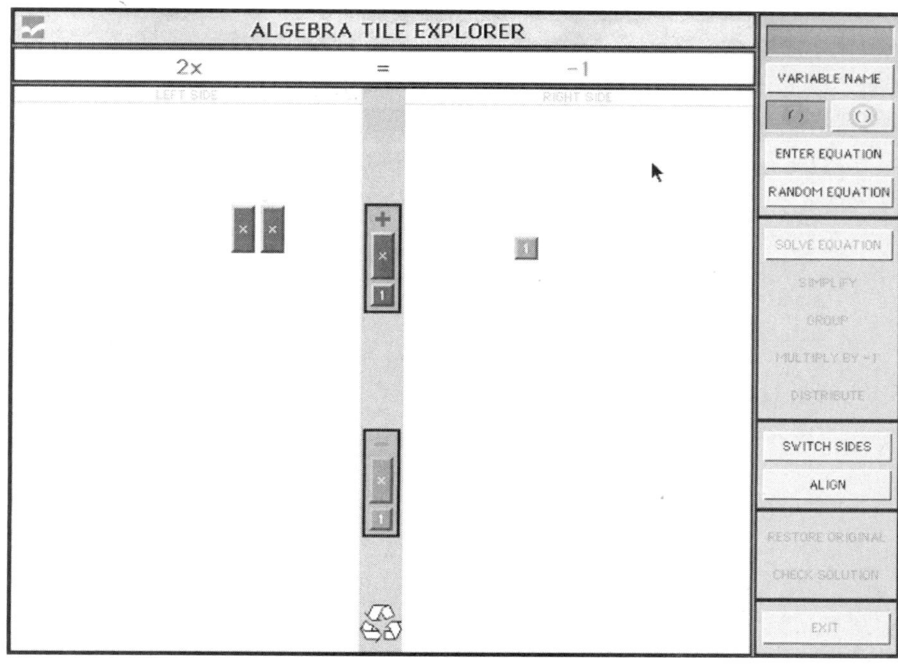

In several *TLE* lessons, you learn how to use supplementary software called **Explorers**. They are designed to **encourage** you **to explore** mathematics concepts, skills, and procedures, **test** your ideas, and **reflect** on your actions.

The **Explorers** use simulations of activities using materials like tiles or spinners, as well as programs to create graphs and geometric diagrams.

The *TLE Explorers* are available from a pull-down menu and cover the following topics:

Algebra Tiles Explorer
Volume\Surface Area Explorer

Sound

A soundtrack accompanies parts of *TLE* with a **sound control (Off or 1 to 7)**, which is available to you. Narration accompanies the Introduction, Tutorial, and Summary components.

Glossary

TLE allows you to **access** an **on-screen** Glossary using **hot text** or a pull-down **menu**. Hot text is highlighted on screen and can be clicked for direct access to the Glossary.

The Glossary provides complete definitions and examples. It helps you understand **the language of mathematics** and is just one of the many features that promote math connections within mathematics and to other disciplines.

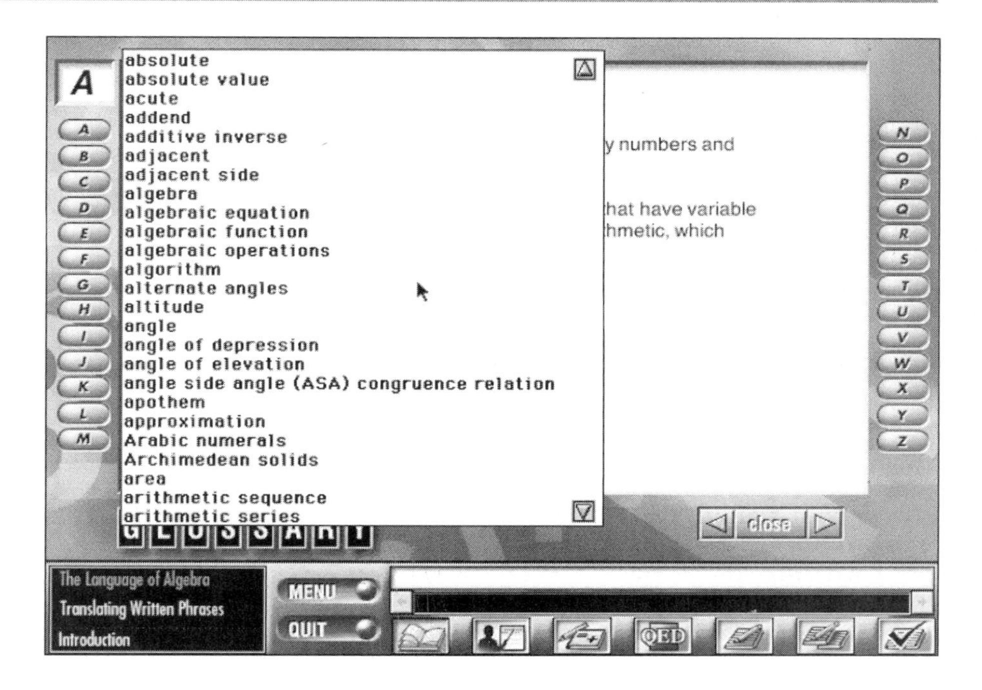

Moving Around

Lesson Title

The objectives and prerequisites for the lesson and basic navigation instructions are presented in the on-screen tabbed notebook. Click on individual heading tabs.

Select the NEXT button to start the **Introduction.**

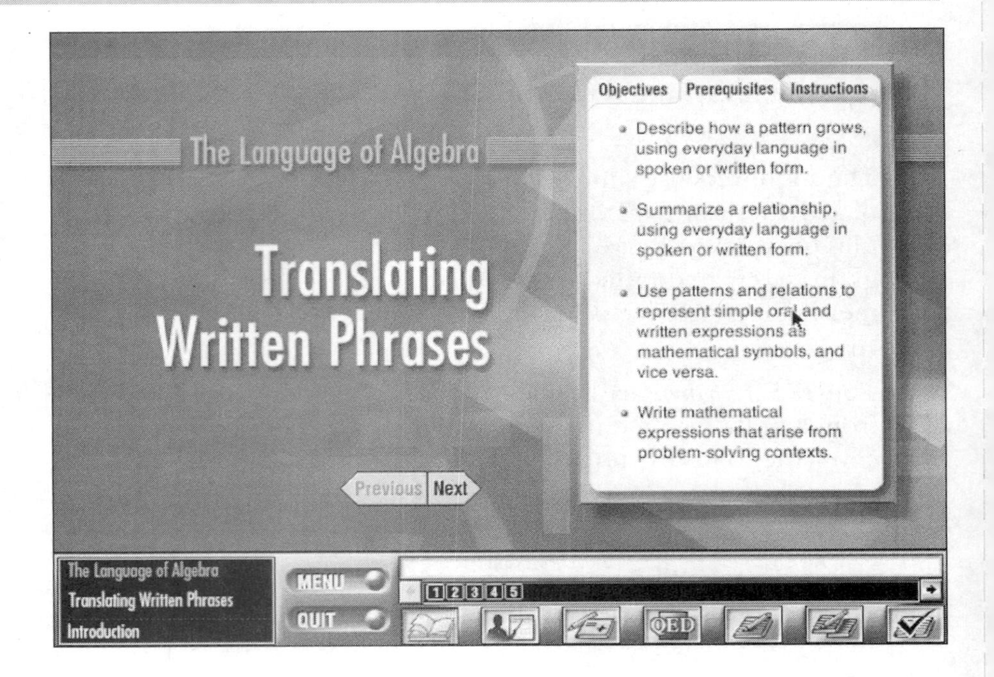

Menu Bar

The MENU bar has seven pull-down options.

Under **File**, you can **Open** your personal record file or create a **New** one. You can **Save** your current results, return to the main TLE MENU, or **Quit**. Students are advised to save their work as they go.

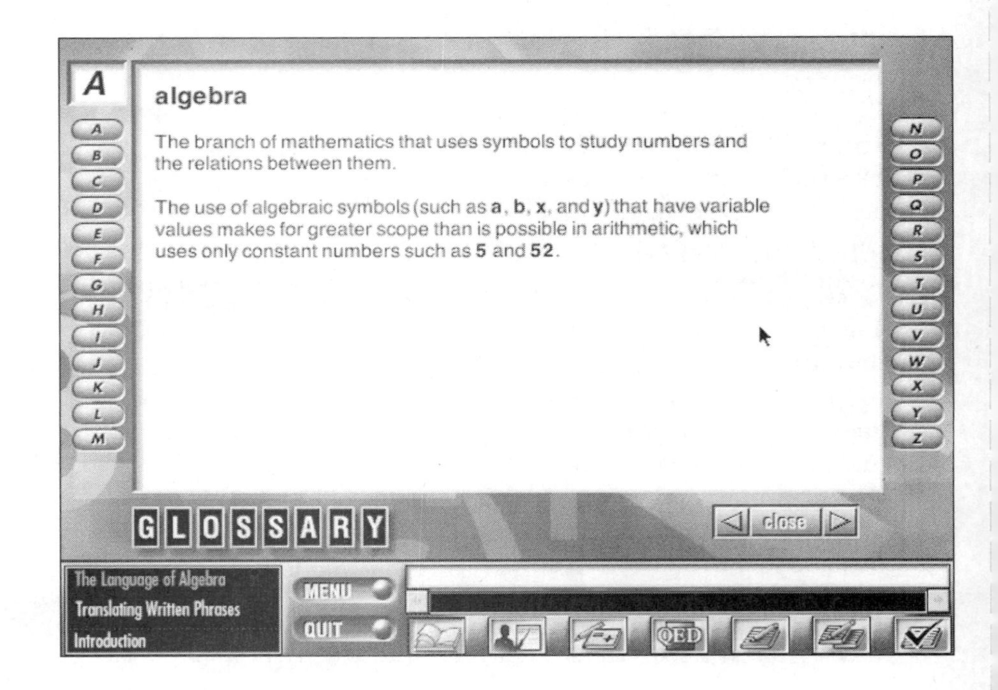

The sound can be adjusted to individual needs. Use the pull-down **Sound** menu from the menu bar to turn the sound off or to set the level from 1 to 7. Headsets may also be used with the sound feature.

Under **Explorers**, you'll find a number of programs for solving problems involving numbers, algebra, geometry, and statistics.

You can also access the **Glossary**. Within each lesson, the **Glossary** is available by double clicking words highlighted in blue ("hot text"). Once in the **Glossary,** select the word of your choice from the alphabetic menu. Return to the lesson by selecting **Close**.

Moving Around a Lesson

Navigation Bar

You always know where you are in a lesson by looking at the bottom left corner of the TLE Navigation Bar.

Chapter
Lesson
Component

The TLE Navigation Bar is designed to be easy to use, with seven icons to reference the lesson components.

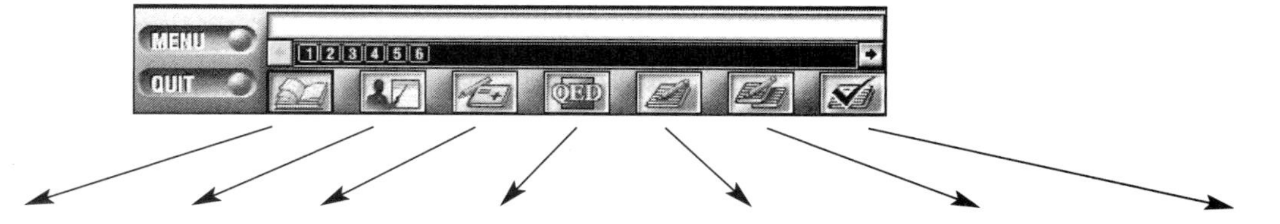

Introduction **Tutorial** **Examples** **Summary** **Practice and Problems** **Extra Practice** **Self-Check**

Click the appropriate icon to go to the lesson component of your choice.

Pages

The number of "pages" in each lesson component is also displayed in the Navigation Bar. Several screens of material may be included in a "page," which represents a "complete thought."

A yellow border highlights the current page of the lesson component. Once a page is completed, it is highlighted in red.

A user can jump to individual pages by clicking the page number on the Navigation Bar.

Previous and Next

Navigate through the lesson by selecting the NEXT button.

The PREVIOUS button allows you to go back to review earlier screens.

Success Tips for Users

1. **Take notes:** Write down an example or two that you worked out in the **Tutorial** to keep as a model for other problems throughout the lesson. Also write down any steps you find in the **Summary** or rules provided in the **Tutorial**.
2. **Use the Glossary:** Key terms are highlighted in blue throughout the lesson. You can find the meaning of these terms in the **Glossary**. To access the **Glossary** you can either click on the term itself or use the pull-down menu on the MENU bar. Keep a list of the highlighted terms, along with an example. Such a list is helpful while working through a lesson and also aids in review at the end of a lesson.
3. **Discuss:** Whenever possible, discuss what you have learned or what you do not understand with a partner.
4. **Review the Examples:** The **Examples** section usually provides different information from that covered in the **Tutorial**. Review each category in the **Examples** section.
5. **Use the calculators and other tools:** Though you can complete some examples and problems using mental math skills, you should also, where appropriate, use other tools such as pencil and paper, diagrams, and calculators, to figure out a problem.
6. **Be aware of levels of difficulty in Practice and Problems:** Recognize that the categories of questions in the **Practice and Problems** section tend to increase in difficulty from left to right across the columns, and from top to bottom within each column or category.
7. **Further enrichment:** Follow up on the Picture Information provided for each completed Hidden Screen in the **Examples** and **Practice and Problems** components.
8. **Use pencil and paper:** When completing the **Extra Practice** and **Self-Check** components take time to transfer information from screen to paper. When answers are reviewed, your work can be compared with sample answers for completeness and accuracy.
9. **Use the Self-Check:** The **Self-Check** component can be used in different ways:
   (a) as a placement tool – it can be attempted before the start of a lesson to determine whether the lesson material is new to you.
   (b) to check for understanding – it can be attempted at the end of a lesson to check that you have understood the lesson material.

   A poor mark on **Self-Check** indicates that you should go back to the **Tutorial** and **Examples** components of the lesson, or alternatively, to the **Extra Practice** section.

   Achievement of at least 70% on **Self-Check** produces a check mark indicator in the box to the left of the lesson, on the MENU.
10. **Create an instruction manual:** Create your own guide on how to use different functions and features of this program.
11. ***Explorer* screen dumps:** On Macintosh computers, it is possible to take a "screen dump" (a picture of what is on the screen) by pressing the keys "shift-command-3" simultaneously. In Windows, press alt-shift-print screen. When using the *Explorers*, you may wish to take screen dumps for later reference as well as note any observations you made about features not mentioned in the software.

The Learning Equation®

PREALGEBRA

1 Whole Numbers

1.1 Whole Numbers

This lesson reviewed whole numbers. You learned how to:

- express numbers in **standard notation, expanded notation**, and written word form.
- **round** whole numbers to indicated place values.
- add, subtract, multiply, and divide long numbers manually.
- apply calculations involving whole numbers to problem situations.

Example 1

Abraham Lincoln was first elected president in 1860 with 1,865,593 votes. Write this number in expanded notation and in words. Then round the number of votes to the nearest hundred thousand.

Solution

Step 1

Write the expanded notation.

Begin by determining the place value of the digit on the left.

There are 7 digits, so the 1 must represent 1 million.

		1	8	6	5	5	9	3
Millions			Thousands			Ones		
Hundreds	Tens	Ones	Hundreds	Tens	Ones	Hundreds	Tens	Ones

1,865,593 = 1 million + 8 hundred thousands + 6 ten thousands + 5 thousands + 5 hundreds + 9 tens + 3 ones

Step 2

Write the word form.

Written word form shows how you would read a standard-form number aloud. Write the number shown by each **period** (group of 3 digits). Separate the periods with commas and insert hyphens as necessary. Do not use the word "and."

1,865,593 = one million, eight hundred sixty-five thousand, five hundred ninety-three

Step 3

Round to the nearest hundred thousand.

Since you want to round to the nearest hundred thousand, the digit in the hundred thousands place is called the **rounding digit**. The digit immediately to its right is called the **test digit**.

rounding digit
↓
1,865,593 ≐ ________
↑
test digit

The rounding digit shows that the number is between 1,800,000 and 1,900,000. The test digit shows there are 6 ten thousands, so the number is closer to 1,900,000.

$$1{,}865{,}593 \doteq 1{,}900{,}000$$

Example 2

The table shows the natural gas reserves held by five countries in 1997.

Natural gas reserves, 1997 (in trillion cubic feet)	
United States	162
Venezuela	129
Canada	95
Mexico	70
Argentina	27

(a) Find the total amount of natural gas held by all five countries.
(b) How much more natural gas did the United States have than Canada?
(c) How many times as much natural gas did Venezuela have than Argentina?

Solution

(a)

```
  2 2
  162
  129
   95
   70
 + 27
  483
```

Write the numbers in vertical format. Align the digits according to place value. Add the ones. The sum, 23, has two digits. Record the 3 in the ones column. Carry the 2 tens into the tens column. Add the tens, including those carried from the ones. The sum is 28 tens. Carry 20 tens (2 hundreds) into the hundreds column. Then add the hundreds.
The total natural gas reserves of the five countries amounted to 483 trillion cubic feet.

(b)

```
   5 1
  162
  -95
   67
```

Write the numbers in vertical format. Align the digits according to place value. Begin with the ones column. Subtract the subtrahend from the minuend. Since the subtrahend, 5, is larger than the minuend, 2, borrow from the tens column to get enough ones to subtract. Repeat with the remaining columns.
The United States had 67 trillion cubic feet more natural gas reserves than Canada.

c)

```
      4
 27)129
    108
     21
```

Estimate the number of groups of 27 you can make from 129. Multiply to check. When you have found the greatest possible number of groups, subtract the product from 129 to see how much will be left over.
Venezuela had between 4 and 5 times as much natural gas as Argentina.

Exercises

1. Complete each statement:

(a) Multiplication means repeated ______________.

(b) Numbers that are to be multiplied are called ______________.

(c) When you divide two numbers, the result is called the ______________ .

(d) When you add two numbers, the result is called the ______________.

(e) When you subtract one number from another, the result is called the __________ .

(f) The numbers 1, 2, 3, 4, . . . form the set of ______________ numbers.

(g) The numbers 0, 1, 2, 3, 4, . . . form the set of ______________ numbers.

2. Write each number in expanded notation and in written word form.

(a) 245 **(b)** 3609 **(c)** 3200

(d) 73,009 **(e)** 104,401 **(f)** 570,003

3. Write each number in standard notation.

(a) 2 thousands + 7 hundreds + 3 tens + 6 ones

(b) 7 hundreds + 7 tens + 7 ones

(c) twenty-seven thousand five hundred ninety-eight

(d) seven million, four hundred fifty-two thousand, eight hundred sixty

4. Round 6,967,532 to each place.

(a) nearest thousand

(b) nearest ten thousand

(c) nearest hundred thousand

(d) nearest million

5. Round 69,599 to each place.

(a) nearest ten

(b) nearest hundred

(c) nearest thousand

(d) nearest hundred thousand

6. Without using a calculator, find each sum.

(a) $425 + 572$

(b) $1372 + 613$

(c) $5799 + 6889$

(d) $95 + 16 + 39$

(e) $1246 + 576 + 20$

(f) $3156 + 578 + 37$

7. Without using a calculator, find each difference.

(a) $17 - 14$

(b) $257 - 55$

(c) $423 - 305$

(d) $1521 - 729$

(e) $7357 - 3778$

(f) $15{,}700 - 15{,}397$

8. Find and correct the errors in the solution to this problem.

Problem

A tennis court is 78 ft long and 36 ft wide. Find the area of the court.

Solution

Area of a rectangle = length × width

= 78 ft × 36 ft

The area of the court is 692 ft.

$$\begin{array}{r} 78 \\ \times\ 36 \\ \hline 468 \\ 234 \\ \hline 692 \end{array}$$

9. Without using a calculator, find each product.

(a) 99×77 **(b)** 232×55 **(c)** 999×13

(d) 619×895 **(e)** 768×103 **(f)** 9234×659

10. Complete the long division.

$$\begin{array}{r} 1 \\ 68\overline{)763} \\ \underline{68\downarrow} \end{array}$$

11. Use long division to find each quotient and remainder.

(a) $949 \div 73$ **(b)** $132 \div 12$ **(c)** $1353 \div 41$

(d) $1795 \div 57$ **(e)** $3280 \div 83$ **(f)** $9876 \div 99$

12. A poster board is 24 in. wide and 36 in. long. Find its area.

13. A car with a tank that holds 14 gal of gasoline can travel 29 mi on one gallon. How far can the car travel on a full tank?

14. A radio, originally priced at $97, has been marked down to $75. By how many dollars has the radio been discounted?

15. To draw graphics on a computer screen, a computer controls each pixel (one dot on the screen). A high-resolution graphics image is 800 pixels wide and 600 pixels high. How many pixels does the computer control?

16. A man went on a business trip. His airplane ticket cost $415, his rental car cost $197, his hotel cost $612, and his food cost $140. Find his total expenses.

17. Create a problem that could be solved using addition, subtraction, multiplication, or division. Show how to solve your problem without using a calculator.

1.2 Powers and Exponents

In this lesson, you learned that a **power** is a short way to write an expression where the same factor is multiplied repeatedly. Recall:

- The expression 3^4 means $3 \times 3 \times 3 \times 3$.
- The number that is multiplied repeatedly is called the **base**. In 3^4, the base is 3.
- The **exponent** shows how many times the base occurs.
- You can read 3^4 as "3 to the exponent 4" or "3 to the fourth."
- There are special ways to read the exponents 2 and 3. The power 3^2 can be read as "3 squared," because it can represent the area of a square. The power 3^3 can be read as "3 cubed," because it can represent the volume of a cube.

Example 1

The number 32 is written in standard form. Write this number as a repeated multiplication and as a power.

Solution

To write a number as a repeated multiplication, start by writing the factors you know. Then write the factors of the factors. Keep going until you can see a way to write the repeated multiplication. (Don't use any factor pairs with $1 \times$.)

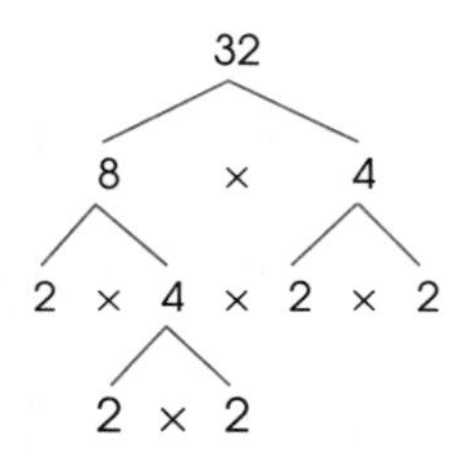

In repeated multiplication form, $32 = 2 \times 2 \times 2 \times 2 \times 2$.

To write 32 as a power, count the number of times the factor occurs in repeated multiplication form. The factor 2 occurs five times, so the power is 2^5.

Example 2

Write the area of the square in standard form, in repeated multiplication form, and as a power.

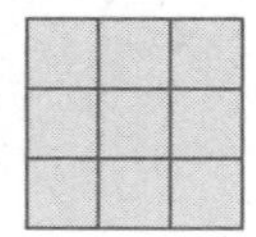

Solution

The square is 3 units long and 3 units wide, so the area is 3×3 square units.

Repeated Multiplication: 3×3

Power: 3^2

Standard Form: 9

Exercises

1. Describe what the base and the exponent of a power tell you. Use an example or a diagram to help with your explanation.

2. Describe an example situation where it is helpful to know how to write a number as a power.

3. Draw a diagram to show why 2^3 is not equal to 3^2.

4. Identify the base.
   (a) 5^4

   (b) 4^5

5. Identify the exponent.
   (a) 3^4

   (b) 4^3

6. Another term for repeated multiplication is *factored form.* Write each power in factored form and in standard form.
   (a) 1^2

   (b) 2^3

   (c) 3^1

   (d) 5^4

   (e) 7^2

   (f) 6^3

7. Write each number as a power and in standard form.
   (a) four squared

   (b) three cubed

   (c) two to the exponent five

   (d) five to the exponent three

8. Complete the chart.

Name	Power	Base	Exponent	Standard Form
four squared				
	5^2			
				8

9. Write each number as a power.
   (a) 8

   (b) 27

   (c) 64

   (d) 100

   (e) 1000

   (f) 144

10. Shira was asked to write 216 as a power. She made this diagram and concluded that 216 could not be written in this form. Was she correct? Explain.

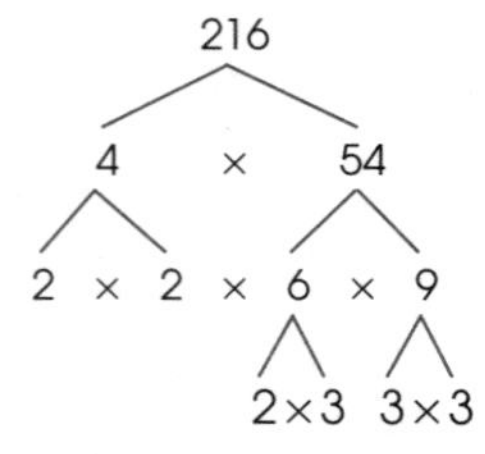

$216 = 2 \times 2 \times 2 \times 3 \times 3 \times 3$

11. Use >, <, or = to make each statement true.

(a) $2^3 \square\ 3^2$

(b) $3^4 \square\ 4^3$

(c) $10^2 \square\ 5^3$

(d) $2^4 \square\ 4^2$

(e) $10^3 \square\ 5^4$

12. A square has a side length of 9 cm.

(a) Write a multiplication expression for the area of the square.

(b) Calculate the area of the square.

13. Kelsey's backyard is a square with sides 24 ft long.

(a) Draw a diagram to show the area in square units.

(b) Express the area of the yard in factored form, as a power, and in standard form.

14. A cube-shaped block of ice has a volume of 64 in.^{3}.

(a) What is the side length of the cube? How do you know?

(b) Express the volume of the ice in factored form and as a power.

15. Find five numbers between 0 and 1000 that can be written as powers. (Choose numbers that aren't already in these exercises.) Write each number in standard form, in factored form, and as a power.

1.3 Divisibility Rules

In this lesson, you learned how to determine whether one number can be divided by another number with no remainder, without actually completing the division. Recall:

- A number is **divisible by 2** if it ends in 0, 2, 4, 6, or 8.
- A number is **divisible by 3** if the sum of the digits is 3 or a multiple of 3.
- A number is **divisible by 4** if the last two digits form a multiple of 4.
- A number is **divisible by 5** if it ends in 5 or 0.
- A number is **divisible by 6** if it is divisible by both 2 and 3.
- A number is **divisible by 9** if the sum of the digits is 9 or a multiple of 9.
- A number is **divisible by 10** if it ends in 0.

Example 1

Is 740 divisible by 4?

Solution

A number is divisible by 4 if the last two digits form a multiple of 4.

The number formed by the last two digits of 740 is 40. Since 40 is divisible by 4, then 740 is divisible by 4.

Example 2

Which number is divisible by both 5 and 10?

735 850 505

Solution

If a number is divisible by 5, the ones digit must be 5 or 0.

All three of these numbers are divisible by 5.

If a number is divisible by 10, the ones digit must be 0.

Only 850 ends in 0, so this is the only multiple of 10.

Of these numbers, only 850 is divisible by both 5 and 10.

Example 3

Is 9 a factor of 6572?

Solution

The sum of the digits of 6572 is $6 + 5 + 7 + 2 = 20$.

Since 20 is not divisible by 9, 6572 is not divisible by 9.

9 is not a factor of 6572.

Example 4

Is 23,946 a multiple of 6?

Solution

A multiple of 6 must be an even number whose digits add to 3 or a multiple of 3.

23,946 is even, because the ones digit, 6, is even.

The sum of the digits of 23,946 is 24.

$$24 \div 3 = 8 \text{ R } 0$$

Since the sum of the digits is divisible by 3, 23,946 is also divisible by 3.

23,946 is an even multiple of 3, so 23,946 is also a multiple of 6.

Example 5

A rectangle with an area of 5312 ft^2 measures exactly 3 ft along one side. Can the other side length be a whole number of feet?

Solution

The area of a rectangle is equal to *length* × *width.*

If the width is 3 ft, then the length can only be a whole number of feet if 5312 is divisible by 3.

A number is divisible by 3 if the sum of the digits is divisible by 3.

$$5 + 3 + 1 + 2 = 11$$

Since 11 is not a multiple of 3, then 5312 is not divisible by 3.

Therefore, the length of the rectangle cannot be a whole number of feet.

You can use a calculator to check this conclusion:

[5] [3] [1] [2] [÷] [3] [=] 1770.6666

This means that the length of the rectangle is a bit more than halfway between 1770 ft and 1771 ft.

Exercises

1. If one number is divisible by another, what remainder would you get if you divided the two numbers?

2. How can you use a calculator to determine if one number is divisible by another number? Explain what it means if the quotient is a whole number and what it means if the quotient is a decimal number.

3. Complete.
   (a) Use the divisibility rule for 3 to determine if 3479 is divisible by 3. Show your work.

   (b) Use a calculator to divide 3479 by 3. Does the result verify your conclusion from part (a)? Explain.

4. Which numbers are divisible by 6?
   (a) 4530 **(b)** 873
   (c) 20,556 **(d)** 7564

5. Which numbers are multiples of 4?
   (a) 8042 **(b)** 616
   (c) 9528 **(d)** 7144

6. For which numbers is 3 a factor?
   (a) 16,491 **(b)** 9612
   (c) 33,520 **(d)** 5720

7. Paige said,

   "There's a shortcut you can use to find out if a number is divisible by 3. When you're adding the digits, you don't need to include digits that are 3s, 6s, or 9s."

   Is Paige right? Explain.

8. Which numbers are factors of 132? Show how you know.

(a) 2

(b) 3

(c) 4

(d) 5

(e) 6

(f) 9

(g) 10

9. Which numbers are factors of 30,645? Explain your reasoning.

(a) 2

(b) 3

(c) 4

(d) 5

(e) 6

(f) 9

(g) 10

10. Write a number greater than 1000 that is divisible by each number. Explain your reasoning.

(a) 2

(b) 3

(c) 4

(d) 5

(e) 6

(f) 9

(g) 10

11. Start with 5286.

(a) What is the sum of the digits?

(b) Is 5286 divisible by 3?

(c) Is 5286 divisible by 9?

(d) Is 5286 divisible by 6?

12. Complete the table. Check the boxes to show which factors divide evenly into each starting number.

		Divisible By						
	Name	2	3	4	5	6	9	10
(a)	282							
(b)	8396							
(c)	20,736							
(d)	111,111,111							
(e)	536,240							
(f)	237,912							
(g)	937,125							
(h)	274,680							

13. Is the statement true or false? Why?

(a) If a number is divisible by 10, it must also be divisible by 5.

(b) If a number is divisible by 5, it must also be divisible by 10.

14. What is the smallest number that is divisible by 2, 3, 4, 5, and 6?

15. Fill in the missing tens digit so that 19,9☐4 is divisible by 3, 4, and 6. List every possibility.

16. Is it possible to complete the number in Problem 15 to make it divisible by 5? Explain.

17. If 324 people attend a concert, can they fill rows of 12 with no empty seats? Explain how you know.

(Hint: You used the rule for 3 to help you find a rule for 6. How can you use the rule for 6 to help you find a rule for 12?)

1.4 Prime Factors and Exponents

This lesson introduced **prime** and **composite** numbers. It also showed how to use **prime factorization** to find the **greatest common factor** (GCF) and the **least common multiple** (LCM) of two or more numbers. Recall:

- A prime number has two factors — 1 and itself. A composite number has two or more factors.
- Any natural number can be expressed as a product of prime factors. Make a factor tree or use step factorization.
- To find the GCF of two numbers, multiply the factors that appear in both prime factorizations.
- To find the LCM of two numbers, find the shortest possible string of prime factors that can be multiplied to produce both numbers.

Example 1

State the greatest common factor of 36 and 54.

Solution

Write 36 and 54 as products of prime factors.

```
2 |36            2 |54
   2 |18            3 |27
      3 |9             3 |9
         3                3
```

$36 = 2 \times 2 \times 3 \times 3$ $\qquad 54 = 2 \times 3 \times 3 \times 3$

Method 1

Use the common prime factors.

The GCF is the product of the prime factors common to 36 and 54.

$36 = 2 \times \mathbf{2} \times \mathbf{3} \times \mathbf{3}$ $\qquad 54 = \mathbf{2} \times \mathbf{3} \times \mathbf{3} \times 3$

$\text{GCF} = 2 \times 3 \times 3$
$= 18$

The GCF of 36 and 54 is 18.

Method 2

Use a diagram.

Classify the factors in this circle diagram. Place the prime factors common to both numbers in the intersection area.

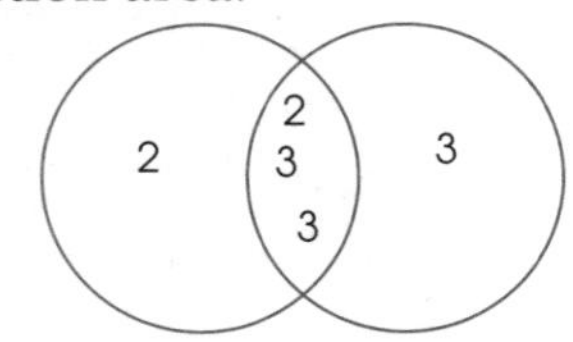

Prime Factors of 36 Prime Factors of 54

The GCF of 36 and 54 is the product of the numbers in the intersection area.

$$2 \times 3 \times 3 = 18$$

The GCF of 36 and 54 is 18.

Example 2

At a school cafeteria, hot dogs arrive in bags of 18, and buns in bags of 24. What is the least number of each that must be purchased so no buns will be left over?

Solution

Find the least number you can make with sets of 18 and 24. This is the LCM.

Write 18 and 24 as products of prime factors.

```
2 |18            2 |24
   3 |9             2 |12
      3                2 |6
                          3
```

$18 = 2 \times 3 \times 3$ $\quad 24 = 2 \times 2 \times 2 \times 3$

Method 1

Use the common prime factors.

$18 = \mathbf{2} \times \mathbf{3} \times 3$ $\quad 24 = 2 \times 2 \times \mathbf{2} \times \mathbf{3}$

Since 2×3 is part of both prime factorizations, you can find the LCM by eliminating one set of 2×3.

$\text{LCM} = 2 \times 3 \times 3 \times 2 \times 2$
$= 72$

Method 2

Use a diagram.

Classify the factors in this circle diagram. Place the prime factors common to both numbers in the intersection area.

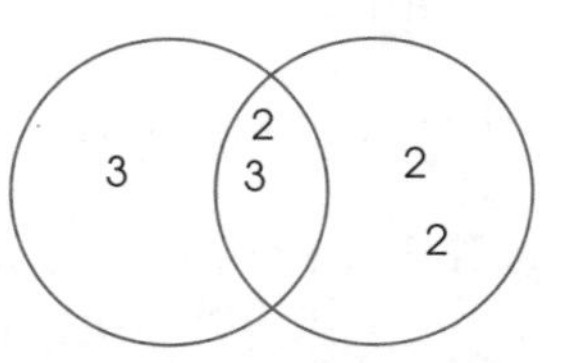

Prime Factors of 18 Prime Factors of 24

Since the factors shared by both numbers are only written once in the diagram, you can find the LCM by multiplying all the numbers shown in the circles.

$$2 \times 2 \times 2 \times 3 \times 3 = 72$$

The cafeteria manager must purchase at least 72 hot dogs and 72 hot dog buns.

Exercises

1. Explain how you can tell if a number is prime.

2. Why is 1 not considered a prime number?

3. Choose any composite number. Show two different ways to express your number as a product of prime factors.

4. Is the number prime, composite, or neither?
   (a) 6
   (b) 5
   (c) 27
   (d) 1
   (e) 29
   (f) 31
   (g) 0
   (h) 33

5. Use the diagram to put the numbers in their proper places: 16, 22, 24, 28, 30.

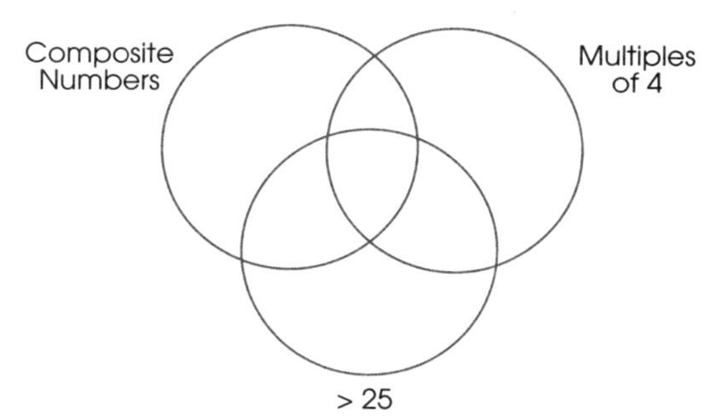

6. Write the prime factorization.
   (a) 36
   (b) 78
   (c) 45
   (d) 126
   (e) 234
   (f) 366
   (g) 405
   (h) 567

7. Use exponents to rewrite each statement.
   (a) $24 = \square^{\square} \times \square$
   (b) $18 = \square \times \square^{\square}$
   (c) $252 = \square^{\square} \times \square^{\square} \times \square$

8. Complete. Write the prime factorization.

(a)

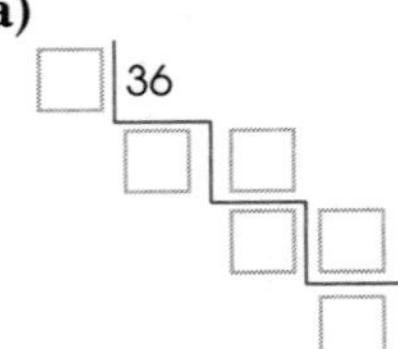

(b)

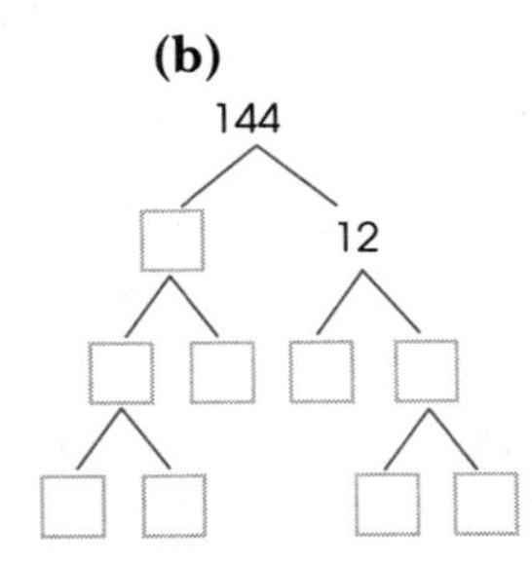

9. Use prime factorization to find:

(a) the prime factors of 104.

(b) the GCF of 36 and 42.

(c) the LCM of 15 and 25.

(d) the GCF of 72 and 48.

(e) the LCM of 42 and 56.

10. If the digits of 13 are reversed to create 31, both numbers are prime numbers. Find three other pairs of prime numbers that have this relationship.

11. Find a path through this maze so you can multiply all the numbers along the path to get a product of 510. You can move horizontally and vertically but not diagonally.

	7	3	1
	5	5	17
Enter →	2	3	7

12. These posters all have the same height, but they have different areas. Determine the greatest possible height they could have.

180 ft^2

450 ft^2

13. Jayden spent $78 on football tickets while Morgan spent $65. If all the tickets were the same price:

(a) What was the price of one ticket?

(b) How many tickets did Jayden buy?

(c) How many tickets did Morgan buy?

14. The Fun Zone, an indoor playground, wants to cover its 51 ft × 135 ft floor with alternating yellow and red squares. The owner wants to use the largest possible squares. What will be the side length of each square?

15. Use prime factorization to create a puzzle similar to the one in Problem 11. Exchange puzzles with a classmate.

16. Write the prime factorizations of any five composite numbers.

(a) How many different ways are there to pair the five numbers?

(b) Find the GCF and the LCM for each pair.

2 FRACTIONS AND DECIMALS

2.1 Fraction Conversions

You have learned to express quantities as **improper fractions** and **mixed numbers**. You have also learned to write equivalent fractions and to reduce fractions to simplest terms. Recall:

- A mixed number includes a whole number and a fraction. An example is $3\frac{1}{2}$.
- An improper fraction has a numerator that is more than the denominator. An example is $\frac{7}{2}$.
- To find an equivalent fraction with a different denominator, multiply or divide the numerator and denominator of the fraction by the same number. This is like multiplying or dividing by a fraction equal to 1.
- To reduce a fraction to simplest terms, divide the numerator and the denominator by their greatest common factor (GCF).

Example 1

Write an improper fraction and a mixed number to describe the shaded part.

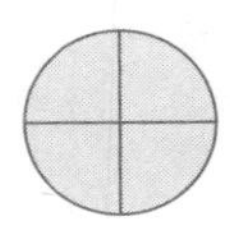 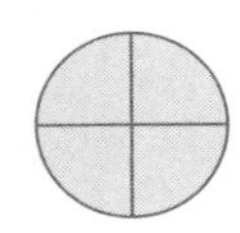 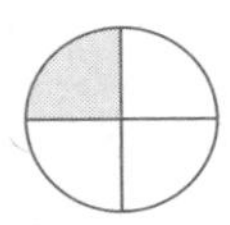

Solution

Each circle is divided into fourths. Nine parts are shaded.

The shaded region represents 9 fourths or $\frac{9}{4}$.

Two whole circles are shaded, as well as 1 fourth of another circle. The mixed number is $2\frac{1}{4}$.

Since the diagram is the same for both numbers, $\frac{9}{4}$ must equal $2\frac{1}{4}$.

Example 2

Write the improper fraction and the mixed number in simplest terms.

 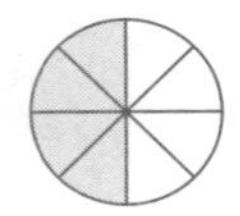

Solution

Find the Improper Fraction

Each circle is divided into eighths.

Twelve parts are shaded.

The shaded region represents 12 eighths or $\frac{12}{8}$.

You can tell that this fraction is not in simplest terms because there is at least one number, 2, that is a factor of both the numerator and the denominator. List the factors of 12 and 8 to see if 2 is the greatest common factor.

Factors of 12: 1, 2, 3, **4**, 6, 12

Factors of 8: 1, 2, **4**, 8

The greatest factor common to 12 and 8 is 4.

To express $\frac{12}{8}$ in simplest terms, divide the numerator and denominator by 4.

$$\frac{12 \div 4}{8 \div 4} = \frac{3}{2}$$

Find the Mixed Number

One whole circle is shaded, as well as 4 eighths of the second circle.

The mixed number is $1\frac{4}{8}$.

The fraction part of this number, $\frac{4}{8}$, is not yet in simplest terms.

You can see from the diagram that $\frac{4}{8}$ and $\frac{1}{2}$ are different names for the same quantity. In simplest terms, the mixed number is $1\frac{1}{2}$.

Exercises

1. Fractions such as $\frac{1}{4}$ and $\frac{2}{3}$ are called proper fractions.

(a) What is a proper fraction?

(b) Why are numbers such as $2\frac{1}{4}$ and $3\frac{2}{5}$ called mixed numbers?

(c) How can you tell that numbers such as $\frac{7}{2}$ and $\frac{9}{4}$ are improper fractions?

2. Write the improper fraction and mixed number for each diagram. Make sure each number is in simplest terms.

(a)

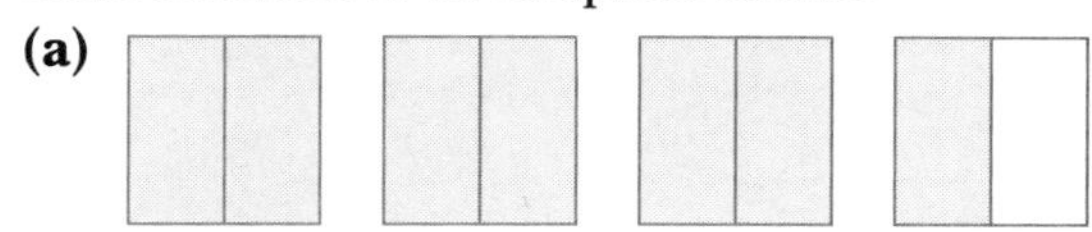

(b)

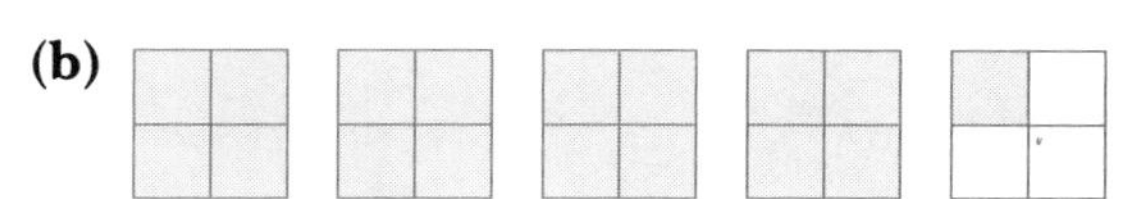

(c)

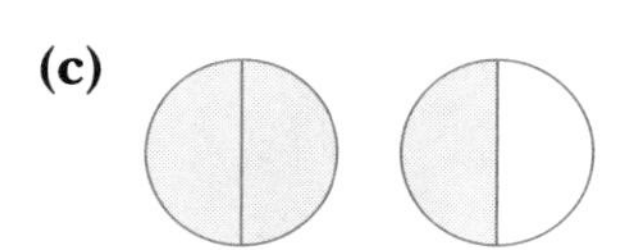

(d) 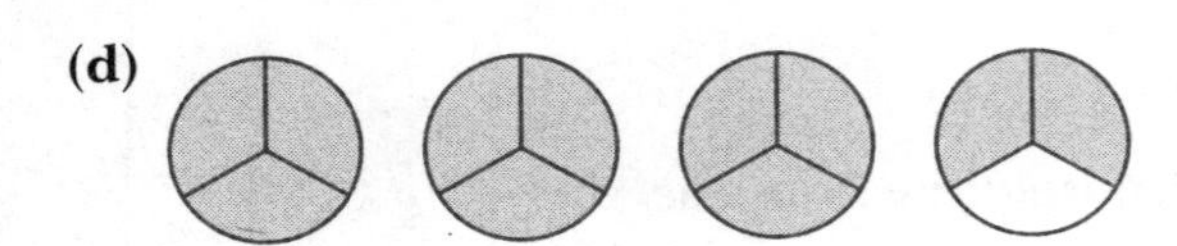

(e) 

(f) 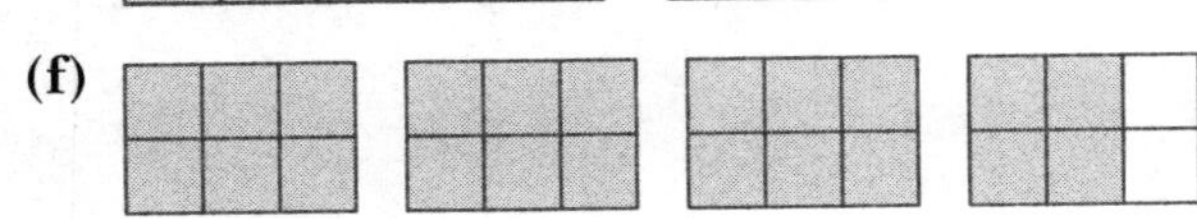

3. Find the GCF for the numerator and denominator, then write the improper fraction in simplest terms.

(a) $\frac{15}{12}$

(b) $\frac{24}{20}$

(c) $\frac{16}{10}$

(d) $\frac{30}{18}$

4. Alex used long division to convert $\frac{36}{8}$ to a mixed number.

$$\begin{array}{r} 4 \\ 8\overline{)36} \\ \underline{32} \\ 4 \end{array}$$

He concluded that you could make 4 wholes, with $\frac{4}{36}$ left over. Since $\frac{4}{36} = \frac{1}{9}$, he decided that $\frac{36}{8} = 4\frac{1}{9}$. Was he correct? Explain.

5. Write each improper fraction as a mixed number in simplest terms.

(a) $\frac{5}{3}$ **(b)** $\frac{13}{7}$ **(c)** $\frac{18}{12}$ **(d)** $\frac{6}{4}$

(e) $\frac{24}{5}$ **(f)** $\frac{21}{2}$ **(g)** $\frac{14}{10}$ **(h)** $\frac{75}{50}$

(i) $\frac{11}{4}$ **(j)** $\frac{31}{8}$

6. Romesh used this method to convert $3\frac{5}{8}$ to an improper fraction:

Each whole is equal to 8 eighths, so $3\frac{5}{8}$ must be equal to 8 eighths + 8 eighths + 8 eighths + 5 eighths.
$8 + 8 + 8 + 5 = 29$, so $3\frac{5}{8}$ is equal to 29 eighths or $\frac{29}{8}$.

Did Romesh convert the mixed number correctly?

7. Show another way to convert $3\frac{5}{8}$ to an improper fraction.

8. Write each mixed number as an improper fraction in simplest terms.

(a) $2\frac{3}{5}$ (b) $1\frac{5}{6}$ (c) $3\frac{2}{3}$ (d) $4\frac{4}{14}$

(e) $7\frac{6}{12}$ (f) $4\frac{5}{6}$ (g) $1\frac{7}{8}$ (h) $5\frac{3}{9}$

(i) $2\frac{3}{6}$ (j) $6\frac{1}{4}$

9. A baker sells pies that have been pre-sliced into equal-sized pieces. Without making any additional cuts, he can serve his customers portions of 1 whole pie, $\frac{1}{2}$ pie, $\frac{1}{4}$ pie, and $\frac{1}{6}$ pie. Into how many slices is each pie cut?

10. A type of chocolate bar has four equal sections. Shauna has already eaten some of one bar, but she has $2\frac{1}{4}$ bars left.

(a) Draw a diagram of this situation.

(b) Write $2\frac{1}{4}$ as an improper fraction.

(c) Shauna wants to give $\frac{1}{4}$ of a bar to each of her friends. If she keeps $\frac{1}{4}$ of a bar for herself, how many friends can share the rest of the chocolate?

11. Write five mixed numbers and five improper fractions. On another piece of paper, show how you would convert each number to the other form. Include some examples that are not in simplest form.

Exchange lists with a classmate and convert the numbers. Then check your answers.

2.2 Decimal Conversions

In this lesson you learned three different ways to **convert a fraction to decimal form**:
- Write an equivalent fraction with a denominator that is a power of 10.
- Use long division to divide the numerator by the denominator.
- Use a calculator to divide the numerator by the denominator.

Example 1

Nancy wants to know which of these six test marks was her best: $\frac{7}{10}$, $\frac{3}{5}$, $\frac{21}{25}$, $\frac{5}{8}$, or $\frac{10}{11}$.

To compare the marks, convert each fraction to decimal form.

Solution

The easiest fraction to convert is $\frac{7}{10}$, since 7 tenths can be written as either $\frac{7}{10}$ or 0.7.

The fractions $\frac{3}{5}$, $\frac{21}{25}$, and $\frac{5}{8}$ all have denominators that are compatible with powers of ten.

$$\begin{aligned}\frac{3}{5} &= \frac{3\times 2}{5\times 2}\\ &= \frac{6}{10}\\ &= 0.6\end{aligned} \qquad \begin{aligned}\frac{21}{25} &= \frac{21\times 4}{25\times 4}\\ &= \frac{84}{100}\\ &= 0.84\end{aligned} \qquad \begin{aligned}\frac{5}{8} &= \frac{5\times 125}{8\times 125}\\ &= \frac{625}{1000}\\ &= 0.625\end{aligned}$$

The fraction $\frac{10}{11}$ cannot easily be converted using equivalent fractions. Use division instead.

$$\begin{array}{r} 0.9090 \\ 11\overline{)10.000} \\ \underline{9.9} \\ 0.10 \\ \underline{0.00} \\ 0.100 \\ \underline{0.099} \\ 0.001 \end{array}$$

Add zeros after the decimal point. Then divide as with whole numbers. The decimal form will repeat, since the remainders will continue in the same pattern.

Check by dividing with a calculator: [1] [0] [÷] [1] [1] [=] 0.909090.

In order from least to greatest, the decimals are 0.6, 0.625, 0.7, 0.84, and $0.\overline{90}$.

Nancy's best mark is $0.\overline{90}$, or $\frac{10}{11}$.

Example 2

Use a calculator to write $2\frac{3}{4}$ as a decimal number.

Solution

Think "$2\frac{3}{4}$ means $\frac{3}{4}$ more than 2."

Enter [3] [÷] [4] [+] [2] [=].

The display shows 2.75, so $2\frac{3}{4} = 2.75$.

Exercises

1. Describe how you can use long division to find out if the decimal form of a fraction terminates or repeats.

2. Why is it often helpful to write a fraction in simplest terms before you convert it to decimal form? Give an example.

3. Does *equivalent* mean "the same"? Give reasons for your answer.

4. Write an equivalent fraction with a denominator that is a power of 10. Then write the decimal.

   (a) $\frac{12}{25}$ **(b)** $\frac{9}{15}$ **(c)** $\frac{7}{8}$

   (d) $\frac{4}{5}$ **(e)** $\frac{15}{4}$ **(f)** $\frac{3}{2}$

5. Use long division to write the decimal form.

   (a) $\frac{3}{8}$ **(b)** $\frac{5}{11}$ **(c)** $\frac{3}{16}$

   (d) $\frac{2}{15}$ **(e)** $\frac{2}{3}$ **(f)** $\frac{10}{9}$

6. Divide with a calculator. Write the decimal form.

   (a) $\frac{1}{6}$ **(b)** $\frac{12}{11}$ **(c)** $\frac{1}{15}$

   (d) $\frac{1}{12}$ **(e)** $\frac{15}{11}$ **(f)** $\frac{8}{9}$

7. Write the decimal form.

(a) $2\frac{1}{4}$ **(b)** $4\frac{1}{3}$ **(c)** $4\frac{1}{2}$

(d) $1\frac{1}{16}$ **(e)** $2\frac{7}{12}$ **(f)** $3\frac{5}{6}$

8. Use estimation to write each set of fractions in order. Then check by writing each number in decimal form.

(a) $\frac{3}{5}, \frac{5}{8}, \frac{3}{4}, \frac{7}{10}$ **(b)** $\frac{2}{7}, \frac{4}{9}, \frac{1}{4}, \frac{4}{12}$ **(c)** $\frac{11}{15}, \frac{12}{16}, \frac{15}{21}, \frac{33}{50}$

9. Use a calculator to convert each fraction to decimal form. Match the equivalent pairs.

(a) $\frac{144}{180}$ **(b)** $\frac{159}{477}$ **(c)** $\frac{104}{364}$ **(d)** $\frac{162}{486}$

(e) $\frac{536}{670}$ **(f)** $\frac{310}{496}$ **(g)** $\frac{455}{728}$ **(h)** $\frac{62}{217}$

10. (a) Write decimal equivalents for the fractions in this pattern: $\frac{1}{2}, \frac{1}{3}, \frac{1}{4}, \ldots, \frac{1}{11}$.

(b) How many decimals are repeating? terminating?

(c) Estimate the decimal form for $\frac{1}{12}$ and then check with a calculator.

11. (a) Write the decimal equivalents for $\frac{1}{2}, \frac{1}{4}$, and $\frac{1}{8}$.

(b) Will $\frac{1}{16}$ and $\frac{1}{32}$ be repeating or terminating decimals? Why?

(c) Write the decimal equivalents for $\frac{1}{3}, \frac{1}{6}$, and $\frac{1}{9}$. Will $\frac{1}{12}$ and $\frac{1}{15}$ be repeating or terminating decimals? Why?

12. Amal conducted a survey to find out whether students at his school would prefer to hold a school dance on a Thursday night, a Friday night, or a Saturday night. The table shows his results.

Thursday	8
Friday	7
Saturday	4

(a) If each student surveyed made one choice, what fraction of all the students surveyed chose each night?

(b) Write each fraction from part (a) in decimal form. Round to three decimal places.

13. Tannis works part-time and earns about $200 per month. The table shows Tannis's monthly spending.

Entertainment	$25
Clothes	$50
Snacks	$20

(a) Write a fraction and a decimal number to represent Tannis's spending on entertainment, clothes, and snacks.

(b) Tannis saves the rest of her money. Write a fraction and a decimal number to represent the amount she saves.

14. Write five fractions different from the ones used in previous problems. Show two different ways to convert each fraction to decimal form. (Make sure you use each of the three methods at least once.)

2.3 Ordering Numbers

You have learned how to **compare** and **order fractions, mixed numbers**, and **decimals**. Recall:

- When two fractions have the same denominator, the greater fraction is the one with the greater numerator.
- When two fractions have the same numerator, the greater fraction is the one with the lesser denominator.
- When two fractions have different numerators and denominators, you can rewrite each fraction with a common denominator, or divide the numerator by the denominator to find the decimal equivalent.

Example

Which fraction is greater?

Solution

(a) $\frac{10}{11}$ or $\frac{9}{11}$ When the parts are the same size, you can compare the number of parts.

10 is greater than 9, so $\frac{10}{11} > \frac{9}{11}$.

(b) $\frac{3}{8}$ or $\frac{3}{7}$ When there are the same number of parts, you can compare the size of the parts.

Eighths are smaller than sevenths, so $\frac{3}{7} > \frac{3}{8}$.

(c) $\frac{2}{7}$ or $\frac{4}{11}$ These fractions cannot be compared directly, but you can change the first fraction to an equivalent with 4 in the numerator.

$\frac{2}{7} \times \frac{2}{2} = \frac{4}{14}$

Fourteenths are smaller than elevenths, so $\frac{4}{11} > \frac{2}{7}$.

(d) $\frac{3}{5}$ or $\frac{4}{9}$ Compare both fractions to $\frac{1}{2}$.

$\frac{3}{5} > \frac{1}{2}$

$\frac{4}{9} < \frac{1}{2}$

$\frac{3}{5}$ must be greater than $\frac{4}{9}$.

(e) $\frac{11}{6}$ or $\frac{15}{8}$ Rewrite each fraction with the same denominator.

Multiples of 6: 6, 12, 18, **24**, 30, ...

Multiples of 8: 8, 16, **24**, 32, ...

The least common denominator of 6 and 8 is 24.

$\frac{11}{6} = \frac{11 \times 4}{6 \times 4} = \frac{44}{24}$

$\frac{15}{8} = \frac{15 \times 3}{8 \times 3} = \frac{45}{24}$

$\frac{45}{24} > \frac{44}{24}$, so $\frac{15}{8} > \frac{11}{6}$

Exercises

1. Explain why it is easier to compare and order numbers when they are in the same form.

2. List five hints you can use to compare fractions and mixed numbers.

3. Explain how you know each statement is true. Use diagrams if you like.

(a) $\frac{3}{4} > \frac{1}{2}$

(b) $\frac{2}{5} < \frac{1}{2}$

(c) $\frac{7}{9} > \frac{5}{9}$

(d) $\frac{2}{5} > \frac{2}{15}$

(e) $\frac{1}{4} = 0.25$

(f) $1\frac{7}{8} < 1.901$

(g) $1\frac{3}{7} = \frac{10}{7}$

(h) $2.658 > 2.568$

4. Order the numbers in each set from least to greatest.

(a) 0.787, 0.8, 0.71, 0.792, 0.778

(b) 2.045, 2.004, 2.504, 2.444, 2.5, 2.4

(c) 4.999, 5.012, 4.099, 5.1, 4.99, 5.12

5. Show where you would place the numbers along this number line. Draw a new number line for each set.

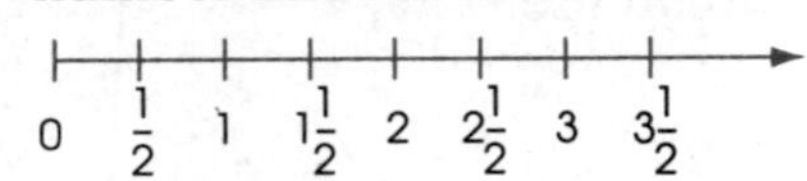

(a) $1\frac{1}{5}, \frac{2}{3}, \frac{1}{4}, \frac{3}{4}, \frac{4}{5}$

(b) $\frac{4}{7}, \frac{15}{8}, \frac{11}{7}, \frac{1}{8}, \frac{14}{7}, \frac{7}{8}$

(c) $\frac{3}{7}, 1\frac{1}{3}, \frac{5}{9}, \frac{13}{12}, \frac{11}{5}$

6. Which fraction is greater? How do you know?

(a) $\frac{2}{5}, \frac{4}{5}$ **(b)** $\frac{4}{7}, \frac{3}{7}$ **(c)** $1\frac{1}{3}, 1\frac{2}{3}$

(d) $\frac{2}{5}, \frac{2}{15}$ **(e)** $2\frac{5}{8}, 2\frac{5}{9}$ **(f)** $\frac{9}{7}, \frac{9}{5}$

(g) $\frac{7}{12}, \frac{3}{4}$ **(h)** $1\frac{1}{6}, 1\frac{5}{18}$

7. Use <, >, or = to make the statement true.

(a) $\frac{3}{8}$ ☐ 0.4 **(b)** 1.6 ☐ $\frac{12}{3}$ **(c)** $2\frac{1}{3}$ ☐ $\frac{5}{2}$

(d) 3.222 ☐ $\frac{17}{5}$ **(e)** $\frac{5}{4}$ ☐ 1.2 **(f)** $1\frac{1}{8}$ ☐ 1.13

8. Leora drew this number line. Redraw the line to correct her errors.

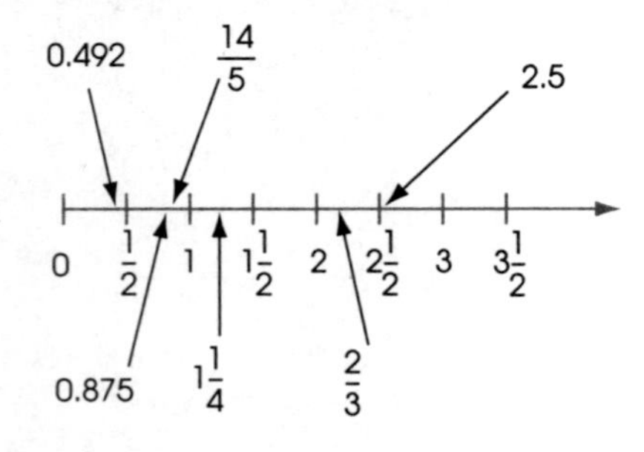

9. Write a fraction and a decimal that fall between each pair of fractions.

(a) $\frac{1}{2}, \frac{3}{4}$ **(b)** $\frac{2}{3}, \frac{7}{9}$ **(c)** $\frac{5}{12}, \frac{1}{2}$

10. Draw a diagram to determine whether this statement is true or false. Test your conclusion with another example.

It's easy to compare two fractions when both of them are one part less than a whole. For example, $\frac{3}{4}$ is less than $\frac{7}{8}$ because fourths are bigger than eighths. This means $\frac{3}{4}$ is farther away from one whole, so $\frac{3}{4}$ is less than $\frac{7}{8}$.

11. Use $>$ or $<$ to make the statement true.

(a) 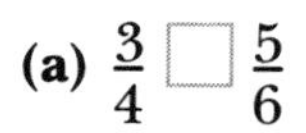 $\frac{3}{4} \square \frac{5}{6}$

(b) $\frac{11}{12} \square \frac{9}{10}$

(c) $1\frac{4}{5} \square 1\frac{2}{3}$

(d) $3.9 \square 3\frac{19}{20}$

12. Rico finished $\frac{2}{5}$ of a walkathon and Lucie finished $\frac{2}{3}$. Who stopped closer to the finish line? Explain.

13. Can the following situation happen? Explain.

Nick and Angelina share a submarine sandwich. Nick eats $\frac{2}{3}$ and Angelina eats $\frac{5}{8}$.

14. Estimate a numerator that will make each statement true. Use a calculator to check.

(a) $0.4 < \frac{\square}{8} < 0.6$

(b) $0.3 < \frac{\square}{3} < 0.5$

(c) $0.6 < \frac{\square}{4} < 0.8$

15. Create three word problems about comparing numbers. Write a solution to each problem on a separate sheet of paper. Then exchange problems with a partner, solve, and compare solutions.

2.4 Terminating and Repeating Decimals

In this lesson you learned to **convert** terminating decimals and decimals with a single repeating digit to **equivalent fractions**. Recall:

- Use the place value of the final digit to help you write the fraction. For example, 0.348 means 348 thousandths or $\frac{348}{1000}$.
- If the fraction is not in simplest form, divide the numerator and the denominator by their greatest common factor (GCF).

Example 1

Convert 0.24 to a fraction in simplest terms.

Solution

0.24 means 24 hundredths.

$0.24 = \frac{24}{100}$ This fraction is not in simplest terms.

Find the greatest common factor of 24 and 100.

Factors of 24: 1, 2, 3, **4**, 6, 8, 12, 24

Factors of 100: 1, 2, **4**, 5, 10, 20, 25, 50, 100

The greatest common factor of 24 and 100 is 4.

To write $\frac{24}{100}$ in simplest terms, divide the numerator and the denominator by 4.

$$\frac{24}{100} = \frac{24 \div 4}{100 \div 4} = \frac{6}{25}$$

In simplest terms, 0.24 equals $\frac{6}{25}$.

Example 2

Convert 1.004 to a mixed number in simplest terms.

Solution

1.004 means 1 and 4 thousandths.

When 1.004 is converted to a fraction, the whole number part will remain the same. The fraction part will be the fraction form of 0.004.

0.004 means 4 thousandths.

$$0.004 = \frac{4}{1000} \quad \text{Divide by the GCF.}$$
$$= \frac{4 \div 4}{1000 \div 4}$$
$$= \frac{1}{250}$$

In simplest terms, 1.004 equals $1\frac{1}{250}$.

Example 3

Convert $1.\overline{7}$ to a mixed number in simplest terms.

Solution

Any decimal number with a single repeating digit after the decimal point can be converted to a fraction in ninths. Since the repeating digit is 7, the fraction is $\frac{7}{9}$.

In simplest terms, $1.\overline{7}$ equals $1\frac{7}{9}$.

Check by dividing with a calculator to restore the fraction to decimal form.

$1\frac{7}{9} = \frac{16}{9}$

Enter [1] [6] [÷] [9] [=].

The display shows 1.7777778. (The final digit is 8 because the decimal was rounded up for the calculator display. Only some calculators do this.)

Therefore, the fraction $1\frac{7}{9}$ is the correct equivalent for $1.\overline{7}$.

Exercises

1. List the decimals whose fraction forms you have memorized. Write the fraction equivalent for each one.

2. Describe how you use place value to convert a terminating decimal to fraction form. Give an example.

3. Why can you not use place value to convert a repeating decimal to fraction form?

4. How can you tell whether a fraction is written in simplest terms?

5. Let one flat square represent 1 whole. Write the decimal and the fraction or mixed number in simplest terms.

   (a) 

   (b) 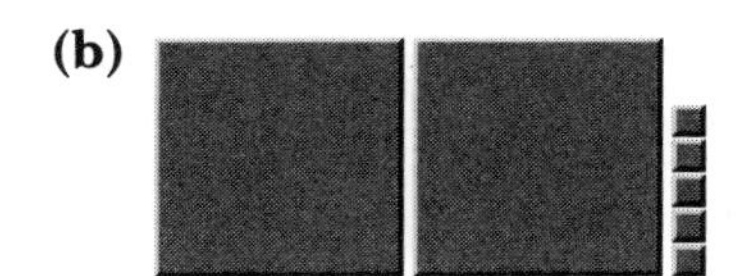

   (c) 

   (d)

   (e) 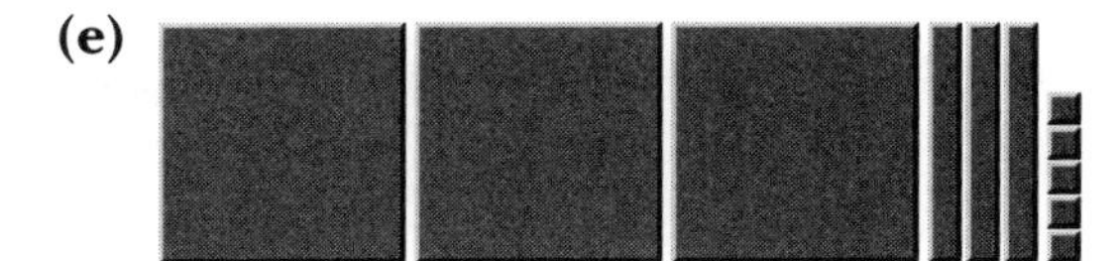

   (f)

6. Write each decimal as a fraction or mixed number in simplest terms.

(a) 0.3 **(b)** 0.5 **(c)** 0.2 **(d)** 0.1

(e) 1.8 **(f)** 2.9 **(g)** 1.4 **(h)** 6.7

7. Write each decimal as a fraction or mixed number in simplest terms.

(a) 0.07 **(b)** 0.45 **(c)** 0.12 **(d)** 1.17

(e) 2.75 **(f)** 5.85 **(g)** 3.22 **(h)** 4.02

8. Write each decimal as a fraction or mixed number in simplest terms.

(a) 0.042 **(b)** 0.111 **(c)** 2.232

(d) 0.125 **(e)** 0.875 **(f)** 5.09

9. Explain what happens when you convert a fraction in ninths to decimal form.

10. Write each decimal as a fraction or mixed number in simplest terms.

(a) $0.\overline{4}$ **(b)** $0.\overline{2}$ **(c)** $1.\overline{1}$

(d) $2.\overline{8}$ **(e)** $3.\overline{6}$ **(f)** $7.\overline{7}$

11. Write the decimal.

(a) $\frac{19}{20}$ **(b)** $\frac{5}{8}$ **(c)** $3\frac{1}{5}$

(d) $2\frac{3}{4}$ **(e)** $5\frac{2}{9}$ **(f)** $4\frac{1}{3}$

12. Check the solution and correct any errors.

Problem
Convert 8.175 to a fraction in simplest terms.

Solution

$$8.175 = 8 + \frac{175}{1000}$$

$$\frac{175}{1000} = \frac{175 \div 5}{1000 \div 5}$$

$$= \frac{35}{200}$$

In simplest terms, 8.175 is $\frac{35}{200}$.

13. Search newspaper and magazine articles (or other available sources) to find decimal numbers. Write each number as a fraction in simplest terms.

14. **(a)** Create an exercise similar to one from this lesson.

(b) Write a step-by-step solution and analyze each step to determine where errors might occur.

(c) On a clean sheet of paper, rewrite the solution so that it contains an error. Exchange with a classmate and identify the errors.

2.5 Multiplication with Decimals

This lesson demonstrated how to calculate the product of **two decimal numbers** by:
- making an area model with base 10 blocks.
- multiplying the numbers as whole numbers, then placing the decimal point in the product so the number of decimal places matches the number of decimal places in both factors.

You also learned some strategies for **estimating and calculating mentally**:
- Think of the × sign as the word "of." The multiplication 0.3×2.4 can mean 3 tenths of 2 and 4 tenths.
- Round the decimals to friendly whole numbers.
- Convert one decimal to a fraction equivalent. For example, 0.728 is about 0.75 or $\frac{3}{4}$.
- To multiply by 0.1, 0.01, or 0.001, move the decimal point one, two, or three places to the left.
- When you halve one factor and double the other, the product does not change. This is especially useful when you are multiplying by a decimal ending in .5, since doubling this number results in a whole number.

Example

Multiply 2.4×1.5.

Solution

Step 1

Estimate the product.

2.4 rounds down to 2.
1.5 rounds up to 2.
2.4×1.5 becomes 2×2 or 4.

Step 2

Multiply the numbers as whole numbers.

$$\begin{array}{r} 24 \\ \underline{\times 15} \\ 120 \\ \underline{240} \\ 360 \end{array}$$

The whole-number product is 360.

Step 3

Place the decimal point in the product.

The sum of the number of decimal places in the factors equals the number of decimal places in the product.

$$\begin{array}{rl} 24 & \text{1 decimal place} \\ \underline{\times 15} & \text{1 decimal place} \\ 360 & \text{2 decimal places} \end{array}$$

Since the 0 in the hundredths place does not affect the value of this number, you can shorten the product to 3.6. This is close to the estimated product, 4.

Step 4

Check.

Since one factor ends in .5, you can use halving and doubling to check the product.

Half of 2.4 is 1.2.
Double 1.5 is 3.
So $2.4 \times 1.5 = 1.2 \times 3$
$= 3.6$

The product, 3.6, was calculated correctly .

Exercises

1. Explain how multiplication and addition are related. Illustrate your explanation with an example.

2. Explain how this block model represents 3.2×1.4.

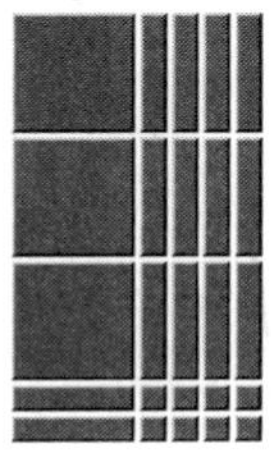

3. State the product.

(a) 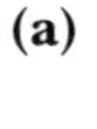 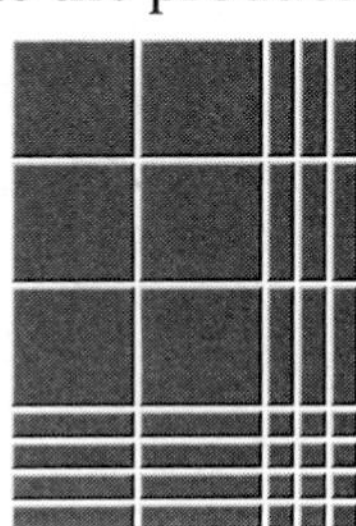

(b) 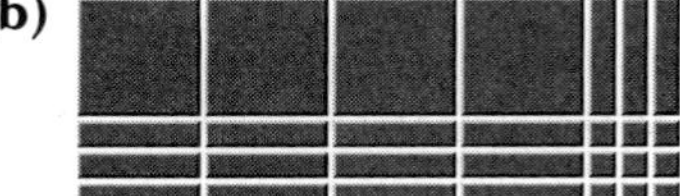

4. State the factors and the product.

(a) 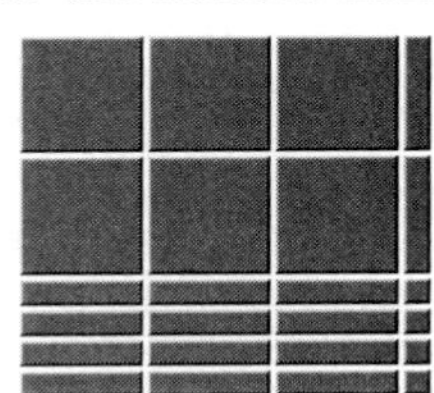

(b) 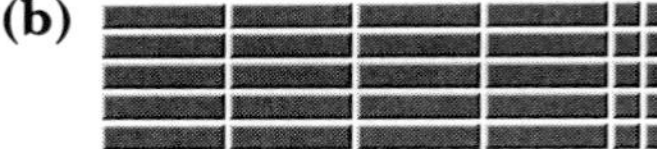

5. Use blocks or draw a block diagram to show how you would find 3.2×4.5.

6. Estimate and then multiply.

(a) 2.5×0.3 **(b)** 0.44×2 **(c)** 0.23×0.4

(d) 8.3×5.2 **(e)** 1.2×4.5 **(f)** 0.75×4.5

7. Choose the best estimate.

(a) 49×8.2 40 400 130

(b) 9.9×74 7 74 730

(c) 0.75×81 60   75   560

(d) 0.3×2.4 0.15   0.6   6.2

8. Round each product.
(a) 7.3×5.9 to the nearest tenth

(b) 65×4.3 to the nearest tenth

(c) 2.8×6.5 to the nearest hundredth

(d) 17.3×0.6 to the nearest hundredth

(e) 0.83×4 to the nearest whole number

9. Calculate mentally.
(a) 1000×5.4 **(b)** 16.89×0.1 **(c)** 4.23×200

(d) 0.013×100 **(e)** 186.9×0.01 **(f)** 2.001×4000

10. Complete each pattern.

(a) 2.8×10
2.8×100
2.8×1000

(b) 3.67×0.1
3.67×0.01
3.67×0.001

11. When a number is multiplied by 10, 100, 1000, and so on, is the product less or greater than the original number?

12. When a number is multiplied by 0.1, 0.01, 0.001 and so on, is the product less or greater than the original number?

13. To place a direct long-distance call from a pay phone, a customer pays $0.95 for the first three minutes, and $0.10 for each additional minute. If Paula spoke with a friend for 15 min, what was the cost of the call?

14. A hummingbird is 2.8 in. long. A photograph shows the hummingbird at 1.8 times its actual size.

(a) Is the photograph smaller or larger than the actual hummingbird? Explain.

(b) Calculate the length of the hummingbird in the photograph.

15. At a bulk food store, caramels cost $2.95 per pound. Albert has $15 and wants to purchase 3.75 lb of caramels.

(a) Estimate to see whether Albert has enough money.

(b) Calculate the exact cost of 3.75 lb of caramels.

16. Doreen and Bill met for lunch at a highway restaurant. After they left the restaurant, Bill drove at a steady speed of 60 mph and Doreen drove steadily at 55 mph. They drove for 3.5 h. How far apart are they if:

(a) they drove in opposite directions?

(b) they drove in the same direction?

17. Create a word problem that could be solved using this block model. Have a partner verify your work.

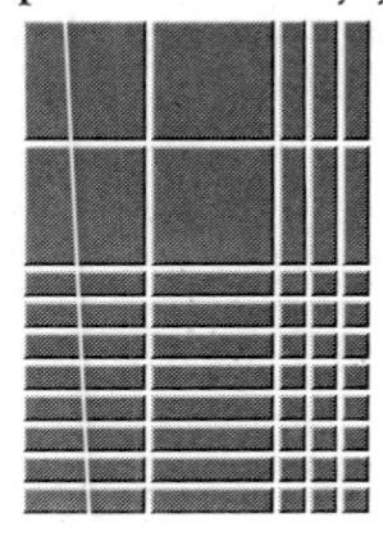

18. Create a multiplication problem where the factors are even whole numbers. Use halving and doubling to create a decimal multiplication problem that has the same product.

19. Create three decimal multiplication problems of each type.

(a) The product is greater than both factors.

(b) The product is less than one factor and greater than the other.

2.6 Division with Decimals

This lesson demonstrated how to find the **quotient of two numbers** where the divisor is a decimal. You learned how to:

- build a **base 10 model** and then divide the blocks into the required number of groups (partitioning) or groups of the required size (measurement).
- multiply the dividend and the divisor by the same power of 10 to create a whole-number divisor and then use long division.

You also learned some strategies for estimating and calculating mentally:

- Interpret the situation in words. For example, $2.32 \div 0.4$ can mean "How many sets of 0.4 can you make from 2.32?"
- Round the decimals to friendly whole numbers.
- Multiply the dividend and the divisor by the same number to convert the divisor to a whole number. Doing this will not change the quotient.

Example

Calculate $8.82 \div 4.2$.

Solution

Step 1

Estimate the quotient.

$8.82 \div 4.2$ is close to $8.8 \div 4$ or 2.2

Step 2

Write the long division.

$4.2\overline{)8.82}$

Step 3

Rewrite the division with a whole-number divisor.

The divisor is 4.2. Multiply by 10 to eliminate one decimal place.

Then multiply the dividend by 10 to keep the relationship the same.

$4.2\overline{)8.82}$ becomes $42\overline{)88.2}$

×10 ×10

Step 4

Divide.

$$
\begin{array}{r}
2.1 \\
42\overline{)88.2} \\
\underline{84} \\
4\,2 \\
\underline{4\,2} \\
0
\end{array}
$$

Step 5

Check the location of the decimal point in the quotient.

The estimate was 2.2, so 2.1 is a reasonable quotient.

Step 6

Verify the solution by multiplying.

If $8.82 \div 4.2 = 2.1$, then 2.1×4.2 should equal 8.82.

Use a calculator: $2.1 \times 4.2 = 8.82$

The quotient is verified.

Exercises

1. Explain the relationship between division and subtraction. Use an example to illustrate your explanation.

2. What can you predict about the quotient if:
   (a) the divisor is greater than the dividend?

   (b) the divisor is less than the dividend?

3. Write the long division form for each operation.
   (a) $15.64 \div 6.8$ **(b)** $\frac{27.52}{3.2}$

4. Rewrite each operation with a whole-number divisor.
   (a) $0.56\overline{)1.792}$ **(b)** $4.2\overline{)15.96}$

5. Calculate mentally. Explain your thinking.
   (a) $12 \div 0.25$ **(b)** $78 \div 0.5$ **(c)** $30 \div 0.2$ **(d)** $168.94 \div 0.1$

6. Which expressions have quotients greater than 15? Explain how you can tell this without calculating.
   (a) $15 \div 0.3$ **(b)** $15 \div 9.4$ **(c)** $15 \div 1.2$

   (d) $15 \div 0.005$ **(e)** $15 \div 0.98$ **(f)** $15 \div 1.07$

7. Use estimation to locate the decimal point in each quotient.
   (a) $8.32 \div 1.6 = 52$ **(b)** $39.15 \div 0.9 = 435$

   (c) $1.876 \div 2.8 = 67$ **(d)** $48.6 \div 0.48 = 10{,}125$

8. Estimate, then use long division.

(a) 50 ÷ 0.005 **(b)** 6.5 ÷ 0.13 **(c)** 472.8 ÷ 0.02

(d) 0.7296 ÷ 0.64 **(e)** 0.0294 ÷ 6 **(f)** 50.53 ÷ 31

9. Use multiplication to verify each quotient from Problem 8.

10. Dane said,

"A shortcut way to find decimal quotients is to think of the numbers in word form, using the same unit for both numbers. For example, to find 72.9 ÷ 0.9, I think of 729 tenths ÷ 9 tenths. Then I just divide 729 by 9 to get the answer."

Test Dane's method on several example exercises. Evaluate its advantages and disadvantages.

11. Find and correct any errors in the solution.

Problem
If Bart's car travels about 27.5 mi on each gallon of gas, how many gallons of gas will be needed for a 335.5 mi trip?

Solution

$27.5\overline{)335.5}$ becomes $275\overline{)335.5}$

$$\begin{array}{r}
1.22 \\
275\overline{)\,335.50} \\
\underline{275} \\
605 \\
\underline{550} \\
550 \\
\underline{550} \\
0
\end{array}$$

Bart will need 1.22 gal of gas.

12. In a Name-That-Song contest on a local radio station, a 3.45 min song was played in 0.5 min intervals throughout the day. How many intervals were needed to play the whole song? (Round your answer to the nearest integer.)

13. Apples cost $0.49 per pound. Nathan has $6 and wants to purchase a 9.75 lb bag of apples.

    (a) Estimate to see whether Nathan has enough money. Explain how you made your estimate.

    (b) Calculate the cost, to the nearest cent, of the bag of apples.

14. Laura purchased 3 equally priced CDs and a tape priced at $11.95. She paid $71.20 in total before taxes.

    (a) Did each CD cost more or less than $24? Explain how you know.

    (b) Find the exact cost of a CD.

15. A dime is about 1.18 mm thick. What is the value of a stack of dimes 2.95 cm tall?

16. Create two division word problems that involve some of the following data. Write solutions to your problems on a separate sheet of paper. Then exchange problems with a classmate, solve again, and compare solutions.

    (Hint: You may need to round some quotients to the nearest hundredth.)

2.14 gal	39.2 mi	48.5 mi
$1.89	$0.85	4.5 h
12.6 mi	6.8 gal	49.9¢

17. Create a division problem of each type. Write the solutions.

    (a) The quotient is greater than the divisor and the dividend.

    (b) The quotient is less than either the divisor or the dividend.

2.7 Order of Operations

You have learned how to simplify expressions that include more than one operation and you learned these **rules for ordering operations**:

- Do operations in **parentheses** first. If one set of parentheses is inside another, start with the inside set and work out.
- Then do **multiplication and division** in the order in which they occur from left to right.
- Then do **addition and subtraction** in the order in which they occur from left to right.

Example 1

Find $30 - 4 \times (12 \div 2)$.

Solution

$30 - 4 \times (12 \div 2) = 30 - 4 \times 6$	Find $12 \div 2$, because this operation is in parentheses.
$= 30 - 24$	Multiply 4×6, because you multiply before you subtract.
$= 6$	Subtract.

Example 2

Find $\dfrac{6 \times 5 - 15}{(3+9) \div 4}$.

Solution

$\dfrac{6 \times 5 - 15}{(3+9) \div 4} = \dfrac{30 - 15}{(3+9) \div 4}$	Start with the numerator. Multiply before subtracting.
	Finish calculating the numerator by subtracting.
$= \dfrac{15}{(3+9) \div 4}$	Calculate the denominator. Start with the operation in parentheses.
$= \dfrac{15}{12 \div 4}$	Finish calculating the denominator.
$= \dfrac{15}{3}$	Divide the numerator by the denominator.
$= 5$	

Example 3

Heather's quiz grades in science are 72, 90, 86, 69, and 75. Calculate her average.

Solution

Add all Heather's grades, then divide by the number of quizzes. This represents her average — the grade she would have on every test if all her grades were the same.

$(72 + 90 + 86 + 69 + 75) \div 5$	Use parentheses to show that the numbers must be added first.
$= 392 \div 5$	Add the numbers in parentheses.
$= 78.4$	Use a calculator to complete the division.

Heather's average in science is 78.4.

Check to make sure this answer is reasonable:

Heather's lowest grade is 69 and her highest is 90, so the average is between these numbers. The sum of her grades is about 400, so the average is around $400 \div 5$, or 80. An average of 78.4 seems reasonable.

Exercises

1. Explain why it is important to follow the rules for ordering operations. Give an example to illustrate your answer.

2. Calculate.

   (a) $10 + 9 \times 2$

   (b) $14 \div 7 - 1$

   (c) $11 \times (8 - 3)$

   (d) $6 + 9 \div (5 - 2)$

   (e) $(4 + 8) \div (8 - 4)$

   (f) $20 - (14 - 5) \times 2$

3. Calculate.

   (a) $6.3 \times 4.5 + 2.7$

   (b) $15.2 \times (6.3 + 2.9)$

   (c) $4.2 \times 2.9 \div 3.5$

   (d) $2.1 + 4.6 \times 4.8 - 3.7$

   (e) $(9.1 \times 3.6) - (4.2 + 1.7)$

   (f) $[(11.2 - 6.7) \times 1.9] \div 2.0$

4. Evaluate.

   (a) $\dfrac{8-5}{18 \div 6}$

   (b) $\dfrac{2 \times 2 \times 2}{16 \div 4}$

   (c) $\dfrac{(11+4) \div 3}{16 \div 8 - 1}$

   (d) $\dfrac{(4+5) \times 2}{3 \times (6-4)}$

   (e) $\dfrac{(22-2) \div 5}{4 \div (8 \div 4)}$

5. Replace each box with an operation to make the statement true.

(a) $4 \square 7 \square 2 = 14$

(b) $9 \square 6 \square 1 = 3$

(c) $8 \square 2 \square 4 \square 2 = 6$

(d) $7 \square 5 \square 3 \square 1 = 5$

6. Insert parentheses to make the statement true.

(a) $8 \div 2 + 2 = 2$

(b) $7 - 6 \times 5 = 5$

(c) $4 \times 3 - 1 + 2 = 9$

(d) $12 - 4 \times 3 \div 6 = 4$

7. Insert parentheses in different places to get as many different answers as you can.

$4 + 3 \times 6 - 4 \div 2$

8. Terri measures the temperature every day at noon for seven days. Calculate the average temperature for the week if the temperatures were 27°F, 30°F, 28°F, 23°F, 25°F, 23°F, and 26°F.

9. If your calculator does not do order operations automatically, you can still find the correct answer by using parentheses or the calculator memory keys.

Record keystroke sequences you could use to complete each calculation using parentheses and/or memory keys.

(a) $4 + 3 \times 6 - 8$

(b) $833 - 84 \div 2 \times 12$

(c) $14.2 - 7.8 + 3.5 \times 6.1$

10. Create a problem similar to one in Problem 6. Exchange problems with a partner and solve.

11. Imagine that you are holding a contest. Create an entry form that includes a skill-testing question. Exchange entry forms with a classmate and complete the forms.

12. **(a)** Create an expression that involves several operations.

(b) Write a step-by-step solution and analyze each step to determine where errors might occur.

(c) On a clean sheet of paper, rewrite the solution so that it contains an error. Exchange with a classmate and identify the errors.

3 INTEGERS

3.1 An Introduction to Integers

This lesson introduced the meaning of integers and showed some of the ways we use these numbers in everyday life. Recall:

- Integers are used to represent opposite ideas, such as profit and loss.
- Integers include all the **whole numbers** and their **opposites**. A + sign means the integer is positive (greater than zero) and a – sign means that the integer is negative (less than zero).
- Zero is the only integer that is neither positive nor negative.
- The set of integers extends infinitely in both directions. You can represent it like this:
  {..., –3, –2, –1, 0, +1, +2, +3, ...}
- To represent integers, you can use colored disks (black for positive and red for negative) or a number line.
- Opposite integers, such as –3 and +3, are the same distance from zero on a number line, but in different directions.
- If you add a pair of opposite integers, the sum is always zero.

Example

Write the integers in order from least to greatest:

–1, +3, +1, –2, –4, 0

Solution

Step 1

Identify the least and greatest integers.

Imagine an integer number line.

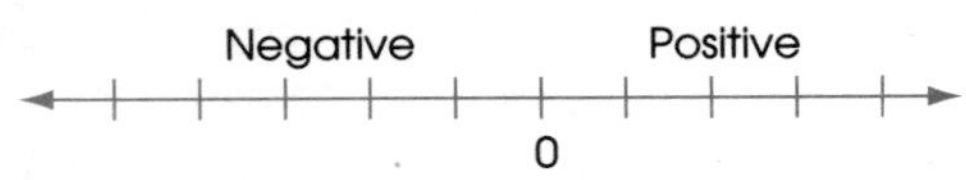

The greatest integer will be the positive integer that is farthest from 0.

In this set, the positive integers are +1 and +3, so +3 is the greatest.

The least integer will be the negative integer that is farthest from 0.

In this set, the negative integers are –1, –2, and –4, so –4 is the least.

Negative Positive

–4 0 +3

Step 2

Arrange the remaining integers along the number line.

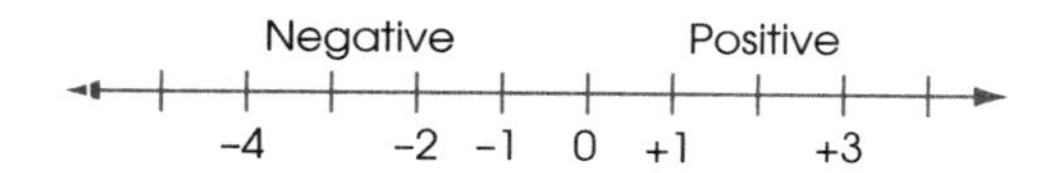

In order from least to greatest, the numbers are –4, –2, –1, 0, +1, and +3.

Exercises

1. Which integer is less: –8 or –12? Why?

2. How is comparing integers like comparing whole numbers? How is it different?

3. Which integer would you use to represent 6 steps taken down a ladder? Why?

4. In the following situation, is the second statement true or false? Give an example to support your answer.

   Integer A is greater than Integer B. Therefore, the opposite of Integer A must be greater than the opposite of Integer B.

5. Locate each integer on a number line from –10 to +10.
   (a) three more than zero
   (b) four less than zero
   (c) two more than positive four
   (d) three more than negative three
   (e) five less than negative two

6. Write the integer and draw the colored disks you would use to model it.
   (a) 6°C above freezing
   (b) 5 ft below sea level
   (c) a loss of 7 lb
   (d) a deposit of $5

(e) backing up 3 ft

(f) moving 4 jumps to the right along a number line

7. Write the opposite of each integer in Problem 6.

8. Continue the pattern.

(a) +3, 0, –3, ☐, ☐, …

(b) –6, –8, –10, ☐, ☐, …

9. Write two integers that fall between each pair.

(a) –1 and +3

(b) –9 and –4

(c) +2 and –4

10. Use < or > to create a true statement.

(a) +7 ☐ +2

(b) +8 ☐ – 10

(c) –2 ☐ –3

(d) –4 ☐ –3

(e) –8 ☐ –5

(f) +10 ☐ –12

11. Write each set in order from least to greatest:

(a) +9, –6, +10, –3, 0

(b) 0, –25, +2, –15, –29

(c) –3, 0, –6, –15, –9

(d) –2, –3, +2, +5, –4

12. Which integers are less than +2 but greater than –3? Use a number line to check.

–2   +1   +3   –4   0   +4

13. Use + or – to make a true statement. Give as many different answers as you can.

(a) $\square 5 < \square 4$

(b) $\square 3 > \square 4$

(c) $0 > \square 7$

14. The chart shows the approximate daily maximum and minimum temperatures in Barrow, Alaska, for each month.

Month	Maximum (°F)	Minimum (°F)
January	-7	-19
February	-12	-24
March	-9	-21
April	+5	-9
May	+24	+14
June	+38	+30
July	+45	+34
August	+42	+33
September	+34	+27
October	+18	+9
November	+4	-7
December	-5	-17

(a) What is the lowest high temperature?

(b) What is the highest low temperature?

(c) In which months is the maximum temperature colder than 10°F?

(d) In which months is the minimum temperature warmer than 10°F?

15. Solve the code to discover Barry's favorite food.

Move from left to right across each row.

Start at G.	+1	-7	+12
-11	+19	-3	-11
-2	+13	+1	End

16. Create your own integer code. Exchange messages with a partner.

3.2 Adding Integers

In this lesson, you learned to **add positive and negative numbers** by using algebra tiles, number lines, and pencil-and-paper calculations. Recall:

- A black tile represents +1 and a red tile represents –1. When you combine one red tile with one black one, the **resulting value is 0**.
- On a number line, move **right** to add a positive number, and **left** to add a negative one.
- When you add **two integers with the same sign**, find the sum of the values and keep the sign. When you add **two integers with different signs**, find the difference between the values and use the sign of the one that is farther from 0.

Example 1

Draw a model and write a number sentence to illustrate the sum of 4 black tiles and 7 red tiles.

Solution

Set out 4 black tiles and 7 red ones. Match pairs of different-colored tiles to make zeros.

The 4 black tiles represent +4.

The 7 red tiles represent –7.

4 positive tiles and 4 negative tiles form 4 zeros, leaving 3 negative tiles.

There were 3 red tiles left over, so the sum is –3.

The number sentence is (+4) + (–7) = –3

Example 2

Write the number sentence.

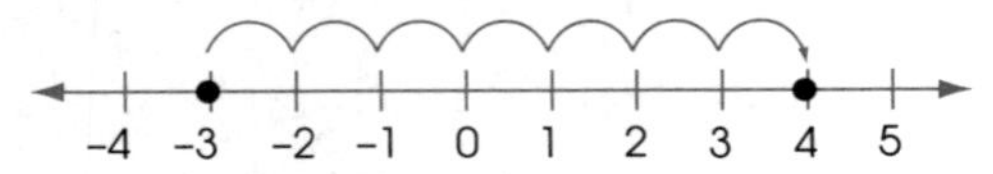

Solution

The arrow begins at –3, then travels 7 jumps in a positive direction to +4.

The number sentence is (–3) + (+7) = +4

Example 3

Calculate (+7) + (–4).

Solution

The addends +7 and –4 have different signs. When the signs are different, adding the numbers will bring the sum closer to 0, not farther away.

To add two integers with different signs, find the difference between the values and use the sign of the one that is farther from 0.

7 – 4 = 3 — The difference between the values is 3, so the sum is either +3 or –3.

+7 is farther from 0 than –4. — The sum is positive.

(+7) + (–4) = (+3)

Exercises

1. Draw the integer model and the number line you would use to model each situation. Then complete the number sentence.

   (a) $(+2) + (+5) = \square$

   (b) $(+5) + (-3) = \square$

   (c) $(-6) + (-1) = \square$

   (d) $(-3) + (+5) = \square$

2. Define each term and give an example.

   (a) addend

   (b) zero model

   (c) sum

3. Draw a number line to find each sum.

   (a) $(+5) + (+4)$

   (b) $(+3) + (+1)$

   (c) $(-7) + (+2)$

   (d) $(-3) + (-6)$

   (e) $(+5) + (-9)$

   (f) $(-3) + (-2)$

4. Use or draw an integer tile model to find each sum.

   (a) $(+2) + (+3)$

   (b) $(-4) + (+5)$

   (c) $(+3) + (-4)$

   (d) $(-5) + (-6)$

   (e) $(+4) + (-2)$

   (f) $(+7) + (-10)$

5. Which expressions have the same sum?

(a) $(+8) + (-5)$

(b) $(-2) + (-1)$

(c) $(-5) + (+2)$

(d) $(+7) + (-4)$

(e) $(-4) + (+1)$

(f) $(-3) + 0$

6. Find the sums in each set. Explain the pattern.

(a) $(+6) + (+4)$
$(+5) + (+4)$
$(+4) + (+4)$
$(+3) + (+4)$

(b) $(-5) + (+3)$
$(-4) + (+3)$
$(-3) + (+3)$
$(-2) + (+3)$

7. Which has the least sum? the greatest sum?

(a) $(-3) + (+5)$

(b) $(+3) + (-5)$

(c) $(+12) + (-9)$

(d) $(-5) + (-5)$

(e) $(-4) + (-8) + (+3)$

(f) $(-21) + (+7)$

(g) $(+3) + (+5) + (-6)$

(h) $(+6) + (+4) + (-9)$

8. Use a calculator to find each sum.

(a) $(-89) + (+56)$

(b) $(-92) + (-67)$

(c) $(-43) + (+58)$

(d) $(-35) + (+78) + (-29)$

(e) $(+52) + (-47) + (+63) + (-82)$

9. The temperature in Boston starts at –2°F, rises 19°F, and then falls 12°F. What is the final temperature?

10. A submarine is located at a depth of –88 ft. Find its final position after it completes the following set of vertical maneuvers: +15 ft, –23 ft, and + 29 ft.

11. In a magic square, each row, column, and diagonal has the same sum. Find the integers that complete this magic square.

-4		
	-1	
-2		+2

12. Use a black die and a red die. The numbers on the black die are positive and the numbers on the red die are negative.
   (a) Make a table to show all the sums you could get by tossing the two dice. Which sum do you think you will get most often? least often? Why?

   (b) Conduct an experiment to test the prediction you made for part (a). Roll the dice 50 times and record the sum each time. Compare the results with your prediction, and with those obtained by other groups.

3.3 Subtracting Integers

In this lesson, you learned to subtract positive and negative numbers by using **algebra tiles**, **number lines**, and **pencil-and-paper calculations**. Recall:

- If you do not have enough tiles to subtract, you can **add zero pairs** until subtraction becomes possible.
- On a number line, **use a balloon image** to help you decide whether to move up or down. Subtracting a positive number is like subtracting hot air, so you move down. Subtracting a negative number is like subtracting a sandbag, so you move up.
- Subtracting an integer gives the same result as adding its opposite. For example, $(-5) - (-3) = (-5) + (+3)$.

Example

Carl is scuba diving 5 ft below the surface of a lake while Jaclyn is sitting on a diving tower 2 ft above the water. What is the difference in height between Carl's position and Jaclyn's position?

Demonstrate four different ways to use integers to solve this problem.

Solution

Use a Tile Model.

Start by assigning integers to each position.

Carl is 5 ft below the water, so his position is –5 ft.

Jaclyn is 2 ft above the water, so her position is +2 ft.

Use integer tiles to model the difference between Carl's position and Jaclyn's. Start with Carl's position.

Model $(-5) - (+2)$.

Use 5 red tiles to model –5. Since there are no black tiles, you must add before you can subtract +2.

When you subtract 2 black tiles, the result is 7 red tiles.
$(-5) - (+2) = (-7)$

To reach Jaclyn's position, you would have to subtract –7 ft from Carl's position.

Use a Balloon Model.

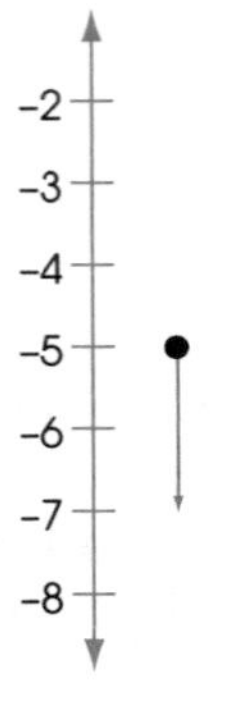

The balloon starts at –5.
Taking away (+2) is like subtracting 2 hot air bursts. The balloon moves down 2 to –7.

Add the Opposite.

Subtracting +2 is the same as adding –2.

$$(-5) - (+2) = (-5) + (-2)$$
$$= -7$$

Think of the related addition.

In the subtraction $(-5) - (+2) = $ ▲, the starting amount is –5. If you take away +2, you will have ▲ left. This means that if you add +2 and ▲, the result must be –5.

To find $(-5) - (+2)$, think:

"What would I have to add to +2 to get –5?"

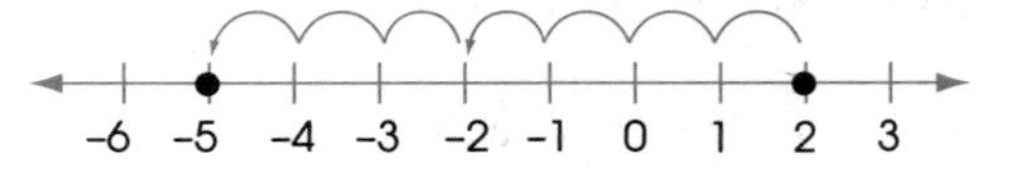

Start at +2.
Jump –2 to get to 0.
Jump –5 more to reach –5.
You jumped –7 in all.

You would have to add –7 to +2 to get –5, so ▲ must equal –7.

Exercises

1. Explain why you need to add zeros to model (+5) – (–8) with tiles.

2. Describe four different ways to subtract a negative integer from a positive one. Use a specific example.

3. Draw tiles to show why (–2) – (–3) has the same value as (–2) + (+3).

4. Model each operation. Write the number sentence.
   (a) Start with 5 red tiles. Remove 2 of them.

   (b) Start with 3 red tiles. Add enough zeros so that 6 red tiles can be removed.

   (c) Start with 4 black tiles. Add enough zeros so that 5 red tiles can be removed.

   (d) Start with 3 red tiles. Add enough zeros so that 7 black tiles can be removed.

5. Use the additive inverse to complete each statement. What statement comes next in the pattern?
   (–5) – (–2) is the same as (–5) + (+2) or –3.

   (–5) – (–1) is the same as (–5) + ☐ or ☐.

   (–5) – 0 is the same as (–5) + ☐ or ☐.

   (–5) – (+1) is the same as (–5) + ☐ or ☐.

   (–5) – (+2) is the same as (–5) + ☐ or ☐.

6. Use tiles to find each difference.

(a) $(+4) - (+3)$ **(b)** $(+4) - (-2)$ **(c)** $(+6) - (+7)$

(d) $(-5) - (-6)$ **(e)** $(+6) - (+8)$ **(f)** $(-3) - (+5)$

7. Which expressions have the same value?

(a) $(+9) - (+5)$ **(b)** $(-4) - (-9)$ **(c)** $(+7) - (+2)$

(d) $(+2) - (-3)$ **(e)** $(-4) - (+1)$ **(f)** $(-3) - (-2)$

8. Complete each pattern.

(a) $(+7) - (+4) = 3$

$(+7) - (+3) = \square$

$(+7) - (+2) = \square$

$(+7) - (+1) = \square$

(b) $(-5) - (-9) = 4$

$(-5) - (-8) = \square$

$(-5) - (-7) = \square$

$(-5) - (-6) = \square$

9. Simplify.

(a) $(+5) - (+3)$ **(b)** $(-2) - (+2)$

(c) $(-4) - (-5)$ **(d)** $(-5) - (+8)$

(e) $(+2) - (+11)$ **(f)** $(+3) - (-7)$

(g) $(-10) - (-20)$ **(h)** $(-4) - (+5) - (+6) + (-4)$

(i) $(+6) - (-2) - (+1)$ **(j)** $(-5) - (+8) + (+7) + (-3) - (+4)$

10. The table shows the approximate daily minimum temperatures for a few places in the U.S.

Location	January (°F)	February (°F)	March (°F)
McGrath, AK	-18	-14	-3
Mt. Washington, NH	-5	-3	5
Alamosa, CO	-4	5	16
Williston, ND	-2	5	17

(a) What is the difference between the January minimum temperatures in Mt. Washington and McGrath?

(b) In Bettles, Alaska, the minimum temperature in February is 4°F less than in McGrath. What is the February minimum temperature in Bettles?

(c) Which location has the greatest change in temperature between January and February? between February and March?

(d) Find two temperatures that have a sum of –7°F and a difference of –1°F.

(e) Find two temperatures that have a sum of 2°F and a difference of 8°F.

11. The melting point of krypton is –157°C. The melting point of platinum is +1772°C. What would you need to subtract from the melting point of platinum to reach the melting point of krypton?

12. The air temperature is 15°F. With the wind blowing at a speed of 15 mph, this temperature feels like –11°F. How many degrees does the temperature change because of the wind chill?

13. During a football game, Jake carried the ball +6 yards and then was tackled and lost +2 yards. What was Jake's actual gain on the play?

14. In 1960, a U.S. Navy deep submergence vessel called the *Trieste* set a world record by diving to a depth of –35,813 ft. The floor of the Caspian Sea is located at a depth of about –560 ft. How much deeper than the Caspian Sea was the *Trieste*'s dive?

15. Create a word problem that involves integer subtraction. Exchange problems with a classmate.

3.4 Multiplying and Dividing Integers

In this lesson, you learned to multiply and divide positive and negative numbers by using algebra tiles, number patterns, and pencil-and-paper calculations.
Recall:

- If you multiply or divide **two positive** numbers, the result is **positive**.
- If you multiply or divide **two negative** numbers, the result is **positive**.
- If you multiply or divide **one positive** number and **one negative** number, the result is **negative**.

Example 1

Use tiles to find $(+3) \times (-2)$.

Solution

The expression $(+3) \times (-2)$ means 3 groups of -2.

There are 6 red tiles in all, so $(+3) \times (-2) = -6$.

Example 2

Use tiles to model $(-4) \div (-2)$.

Solution

The expression $(-4) \div (-2)$ can mean -4 shared into groups of -2.

There are 2 groups of tiles, so $(-4) \div (-2) = (+2)$.

Example 3

Use opposites to find $(+12) \div (-3)$.

Solution

$(+12) \div (-3)$ is the opposite of $(+12) \div (+3)$.

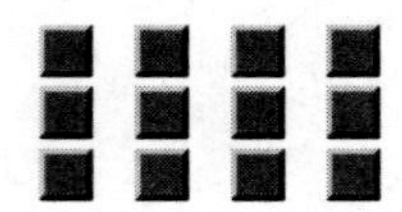

$(+12) \div (+3) = (+4)$

The expression $(+12) \div (-3)$ has the opposite quotient, so $(+12) \div (-3) = (-4)$.

Exercises

1. Define each term and give an example.

   (a) factor

   (b) product

   (c) quotient

   (d) divisor

   (e) dividend

2. Describe three multiplication or division situations that can't be modeled directly with tiles. Explain another method you could use to find each result.

3. Rewrite each addition statement as multiplication. Use tiles to show why the products are the same.

   (a) $(-3) + (-3) + (-3) + (-3)$

   (b) $(-4) + (-4) + (-4)$

4. Use a tile model or a number pattern to find each product.

   (a) $(+2) \times (+9)$

   (b) $(+4) \times (+8)$

   (c) $(-2) \times (-6)$

   (d) $(-20) \times (-4)$

   (e) $(-3) \times (+5)$

   (f) $(-13) \times (+1)$

   (g) $(-7) \times (-3)$

   (h) $(+5) \times (-7)$

   (i) $(+4) \times (-6)$

   (j) $(-44) \times (-11)$

5. Divide.

(a) $(-12) \div (-3)$

(b) $(+9) \div (-3)$

(c) $(-6) \div (+3)$

(d) $(+35) \div (-7)$

(e) $(-45) \div (-3)$

(f) $(+14) \div (-1)$

6. Calculate. Use the rules for order of operations.

(a) $(+5) + (-18) \div (+6)$

(b) $(+12) \times (-4) \div (-6)$

(c) $(+14) \div (-7) + 2$

(d) $(+8) + (-9) \div (-9) + (-9)$

(e) $[(+27) + (3)] \div (-5) - (-40)$

(f) $(-12) \times [(+4) - (+2)] \div (+3)$

7. Explain how each multiplication or division pattern is formed. Then write the next two numbers.

(a) +3, +9, +27, ...

(b) –160, –80, –40, ...

(c) +81, –27, +9, ...

(d) –625, +125, –25, ...

8. –2 is multiplied by itself 11 times. Will the result be positive or negative? Why?

9. The product of two integers is –10. One integer is 7 more than the other. Find two pairs of integers that fit this description.

10. Write a multiplication expression and a division expression that would have each result.

(a) -6

(b) $+8$

(c) -10

(d) -18

11. Write an expression with two or more operations that has each result. Remember to use parentheses where they are necessary.

(a) $+24$

(b) -15

(c) $+36$

(d) 0

(e) -100

(f) -17

12. Determine the operation signs that would make each statement true.

(a) $(+6) \square (-3) \square (-4) = -12$

(b) $(24) \square (-3) \square (-4) = -4$

13. Insert parentheses to make each expression equal 0.

(a) $(+4) + (-5) \times (-3) - (-3)$

(b) $(-8) \div (-2) + (+6) - (-2)$

14. Create a word problem that involves multiplication or division with integers. Write a full solution for your problem on another sheet of paper, then exchange problems with a classmate.

4 Rational Numbers and Square Roots

4.1 Calculating Square Roots

In this lesson you learned to **represent square roots** with square models, number lines, and mathematical symbols. Recall:

- A **perfect square** is the product of two equal natural numbers. For example, 25 is a perfect square with the square root 5. We write $5^2 = 25$ and $\sqrt{25} = 5$.
- To **find the square root** of a number with a calculator, use the [√] key.
- To find the square of a number, use the [x^2] key.

Example 1

Use square tiles to find the square root of 25.

Solution

Step 1

Arrange 25 small squares to make a larger square.

5

5 $\quad 5 \times 5 = 25$

The square has a side length of 5 and an area of 25.

Therefore, $\sqrt{25} = 5$.

Step 2

Check with a calculator.

Enter [2][5][√]. (Some calculators require you to use the [=] key, or use a different order.)

The calculator shows that $\sqrt{25} = 5$, so the square root is correct.

Example 2

Use a square model to estimate the square root of 56.

Solution

First, use 56 small squares to make the largest possible square.

Cut the remaining 7 squares in half and fit them around the edges, making a square 7.5 units long and 7.5 units high.

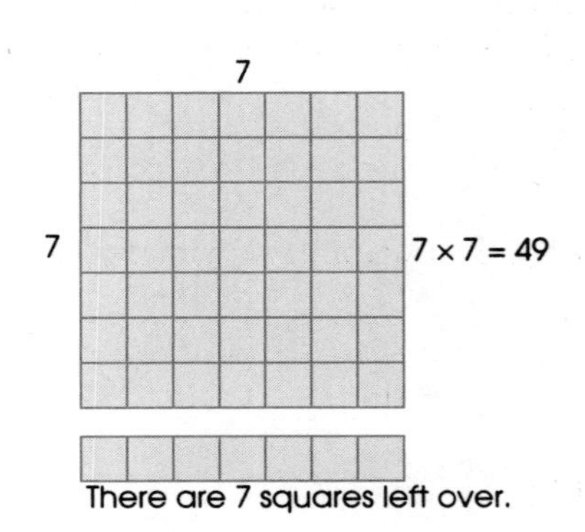

Use a calculator to check.

Press [7] [.] [5] [x²] [=].

The calculator shows 56.25, so 7.5 is a good estimate for $\sqrt{56}$.

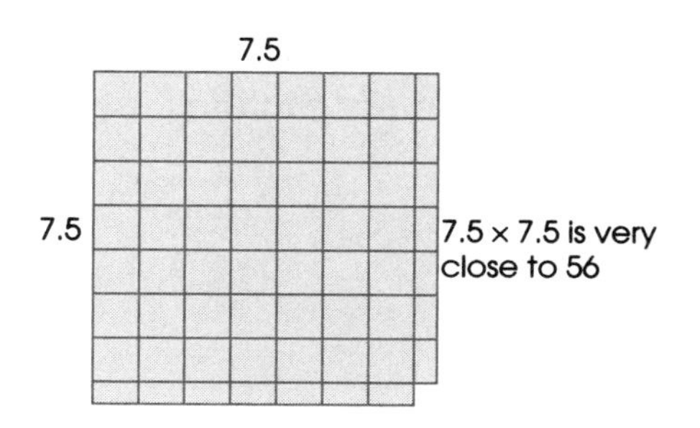

Example 3

Estimate $\sqrt{12}$.

Solution

Step 1

Use a number line.

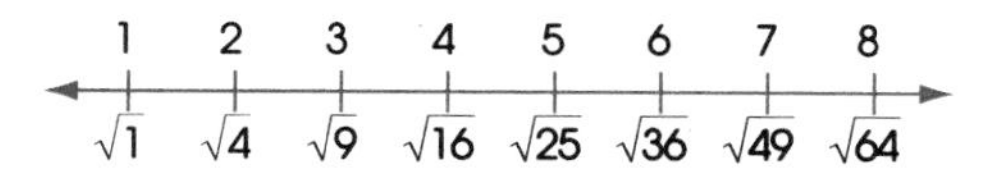

The square root of 12 must be between 3 and 4, since $\sqrt{12}$ is between $\sqrt{9}$ and $\sqrt{16}$. Since 12 is a bit less than halfway between 9 and 16, the root should be about 3.4.

Step 2

Check with a calculator.

$3.4^2 = 11.56$

This is close to 12, but a better estimate may be possible.

Check 3.5^2, since this will be a bit greater than 3.4^2.

$3.5^2 = 12.25$ This is closer to 12 than 11.56.

Therefore, $\sqrt{12}$ is about 3.5.

Exercises

1. What is a square root?

2. Do all natural numbers have square roots? Explain.

3. Explain how to calculate the side length of a square if you know its area.

4. Draw each square on grid paper. Write the side length.
   (a) 16 cm^2 **(b)** 36 cm^2
   (c) 100 cm^2 **(d)** 64 cm^2

5. Write each number as a product of two equal factors.
   (a) 25 **(b)** 49
   (c) 144 **(d)** 121

6. Write the perfect square of every number from 1 to 15. Fill in as many as you can from memory and then use a calculator to find the rest.

7. Try to write each square root from memory. Refer to your list from Problem 6 for the ones you don't know.

(a) $\sqrt{9}$ **(b)** $\sqrt{144}$

(c) $\sqrt{169}$ **(d)** $\sqrt{225}$

(e) $\sqrt{16}$ **(f)** $\sqrt{121}$

(g) $\sqrt{36}$ **(h)** $\sqrt{100}$

(i) $\sqrt{49}$ **(j)** $\sqrt{196}$

(k) $\sqrt{81}$ **(l)** $\sqrt{64}$

8. Example 2 showed how to use squares to estimate the square root of 56. Use a similar method to estimate the square root of 20. You can draw your model on grid paper if you like.

9. Use the number line from Example 3 to estimate each square root.

(a) $\sqrt{40}$

(b) $\sqrt{30}$

(c) $\sqrt{80}$

10. Estimate each square root. Explain your method. Compare your estimate with the calculator result.

(a) $\sqrt{57}$ **(b)** $\sqrt{43}$

(c) $\sqrt{21}$ **(d)** $\sqrt{133}$

(e) $\sqrt{72}$ **(f)** $\sqrt{98}$

11. Use the square root key on a calculator. Express each root to the nearest tenth.

(a) $\sqrt{17}$ **(b)** $\sqrt{28}$

(c) $\sqrt{117}$ **(d)** $\sqrt{350}$

(e) $\sqrt{219}$ **(f)** $\sqrt{399}$

12. Josh used a calculator to find $\sqrt{5}$ and rounded the result to 2.24. To check, he entered 2.24×2.24.

(a) What product did Josh get?

(b) Why is the product not 5?

13. Josh wanted to express $\sqrt{5}$ as accurately as possible. This time, he copied the entire decimal number from the calculator display: 2.236067.

To check, he entered 2.236067×2.236067. To his surprise, the calculator still did not show 5 as a product.

(a) What product did the calculator show?

(b) Why is the product not exactly 5?

14. Jenny also used a calculator to find $\sqrt{5}$. To check her calculation, she pressed x^2 while the square root, 2.236067977, was still showing in the display.

Try Jenny's method with your calculator.

(a) What product do you get this time?

(b) Why do you think that Jenny's product is different from the one Josh got in Problem 13?

15. The square root of 81 is 9, because $9 \times 9 = 81$.

(a) What negative number multiplied by itself gives a product of 81?

(b) How many square roots does a perfect square usually have?

(c) What number has exactly one square root?
(Hint: The $\sqrt{}$ sign usually indicates the positive square root. If the negative square root is needed, the sign is $-\sqrt{}$.)

16. This rectangle is half of a square.
If the square has an area of 7.84 in.^{2},
what are the side lengths of the rectangle?

4.2 Adding and Subtracting Fractions

You have learned how to add and subtract fractions by using pattern block models, fraction strips, and number sentences. Recall:

- To add or subtract **fractions with the same denominator**, add or subtract the numerators and leave the denominator the same. Always write the result in simplest terms.
- If the numerator of a fraction is larger than the denominator, then the fraction is greater than 1 and is called an **improper fraction**. Add and subtract improper fractions using the same methods you use for other fractions.
- If a fraction is combined with a whole number, the combination is called a **mixed number**. To add and subtract mixed numbers, you can either separate the wholes from the parts, or rewrite the mixed numbers as improper fractions.

Example 1

Use pattern blocks to show $\frac{1}{6} + \frac{3}{6}$.

Solution

It takes 6 blue rhombuses to fill 1 whole, so each blue rhombus represents $\frac{1}{6}$.

3 blue rhombuses + 1 blue rhombus = 4 blue rhombuses

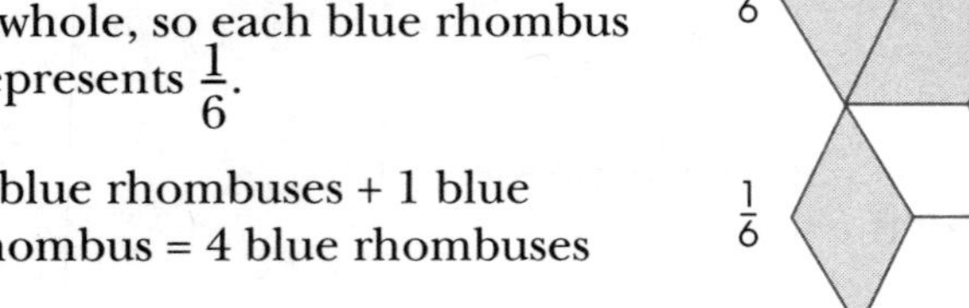

$\frac{3}{6}+\frac{1}{6}=\frac{4}{6}$ — $\frac{4}{6}$ is not in simplest terms. The GCF of 4 and 6 is 2. $\frac{3}{6}+\frac{1}{6}=\frac{4\div 2}{6\div 2}$. Divide by $\frac{2}{2}$ to write the sum in simplest terms.

$\frac{3}{6}+\frac{1}{6}=\frac{2}{3}$ — $\frac{2}{3}$ of the whole is filled, so the sum of $\frac{1}{6}$ and $\frac{3}{6}$ is $\frac{2}{3}$.

Example 2

Use fraction strips to show $\frac{7}{10} - \frac{2}{10}$.

Solution

$\frac{7}{10}$ of this strip is shaded.

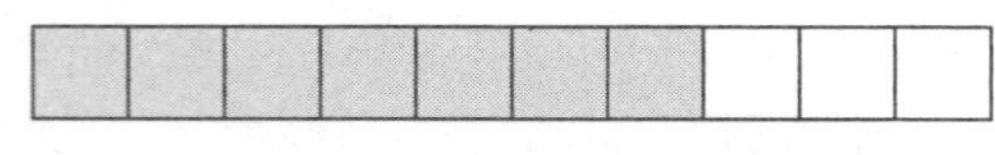

$\frac{2}{10}$ of this strip is shaded.

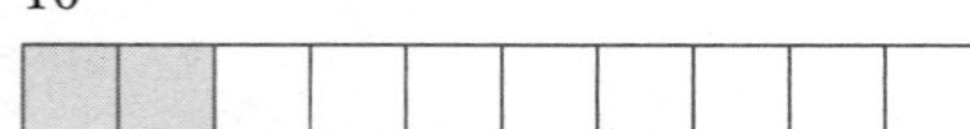

The difference between $\frac{7}{10}$ and $\frac{2}{10}$ is $\frac{5}{10}$.

Therefore, $\frac{7}{10}-\frac{2}{10}=\frac{5}{10}=\frac{1}{2}$.

Example 3

Tony worked for $5\frac{3}{4}$ h, and Kathleen worked for $7\frac{1}{4}$ h. How much longer did Kathleen work than Tony?

Solution

Subtract $5\frac{3}{4}$ from $7\frac{1}{4}$ to find the difference.

Method 1: **Separate the wholes from the parts.**

You can subtract 5 wholes from 7 wholes, but you can't subtract 3 fourths from 1 fourth. To get more fourths in the first number, regroup one of the wholes to fourths.

$$7\frac{1}{4}=6+\frac{4}{4}+\frac{1}{4}$$
$$=6\frac{5}{4}$$

Now you can subtract.

$$6\frac{5}{4}-5\frac{3}{4}=1 \text{ and } \frac{2}{4}$$
$$=1\frac{2}{4}$$
$$=1\frac{1}{2}$$

Kathleen worked for $1\frac{1}{2}$ h more than Tony.

Method 2: **Change to improper fractions.**

$7\frac{1}{4}-5\frac{3}{4}$

$=\frac{29}{4}-5\frac{3}{4}$ — 7 is $\frac{28}{4}$, so $7\frac{1}{4}$ is $\frac{29}{4}$. 5 is $\frac{20}{4}$, so $5\frac{3}{4}$ is $\frac{23}{4}$.

$=\frac{29}{4}-\frac{23}{4}$ — Subtract the numerators.

$=\frac{6}{4}=\frac{3}{2}$ — Write the difference in simplest terms.

$=1\frac{1}{2}$ — Simplify further by changing $\frac{3}{2}$ to a mixed number.

Exercises

1. With a partner, brainstorm some everyday situations where people add or subtract fractions.

2. Describe the method you use to:
   (a) write a fraction in simplest terms.

   (b) add or subtract fractions with like denominators.

   (c) add or subtract mixed numbers.

3. Express each fraction in simplest terms.
   (a) $\frac{8}{20}$ **(b)** $\frac{9}{15}$ **(c)** $\frac{6}{16}$
   (d) $\frac{15}{48}$ **(e)** $\frac{20}{24}$ **(f)** $\frac{75}{100}$

4. Write as a mixed number in simplest terms.
   (a) $\frac{17}{2}$ **(b)** $\frac{7}{5}$ **(c)** $\frac{6}{4}$
   (d) $\frac{11}{4}$ **(e)** $\frac{14}{8}$ **(f)** $\frac{27}{8}$

5. Write the number sentence. Express the sum in simplest terms.
   (a)
   (b)
   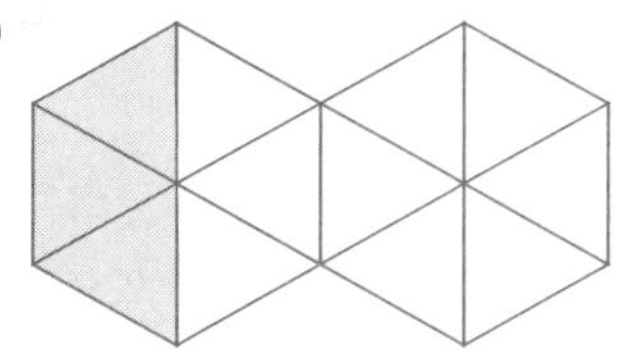
   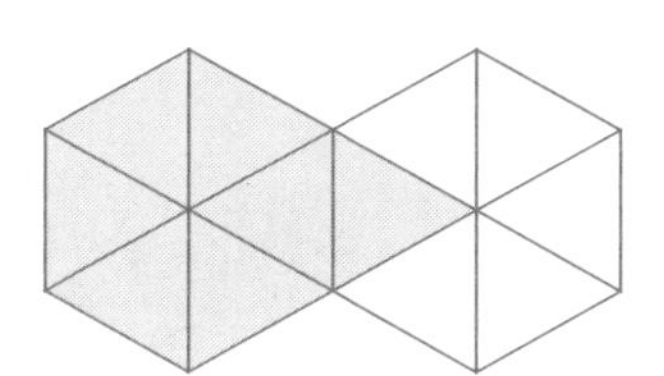

6. Write the number sentence. Express the difference in simplest terms.
   (a)
   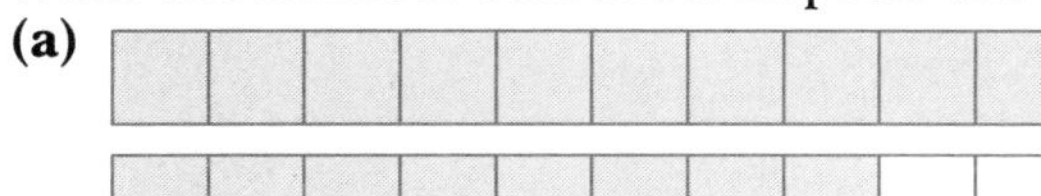
   (b)
   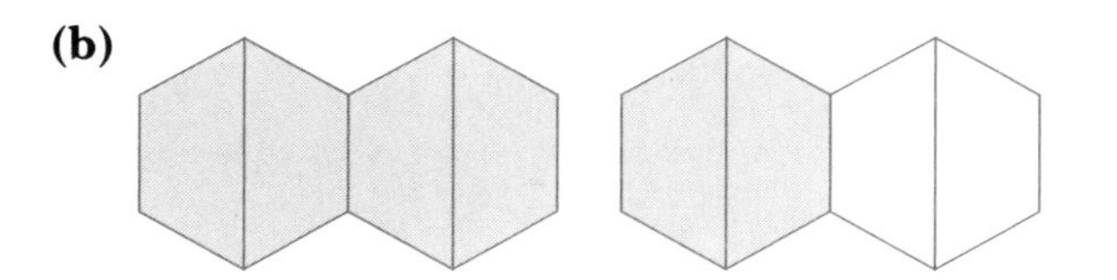

7. Express the sum or difference in simplest terms.

(a) $\frac{7}{8}-\frac{1}{8}$

(b) $\frac{9}{13}+\frac{2}{13}$

(c) $\frac{15}{17}-\frac{11}{17}$

(d) $\frac{9}{11}+\frac{4}{11}$

8. Choose the best estimate.

(a) $2\frac{7}{10}+1\frac{2}{10}$ **(i)** 2 **(ii)** 3 **(iii)** 4 **(iv)** 5

(b) $4\frac{9}{12}-1\frac{7}{12}$ **(i)** 1 **(ii)** 3 **(iii)** 4 **(iv)** 6

(c) $6\frac{7}{8}+2$ **(i)** 8 **(ii)** 9 **(iii)** 10 **(iv)** 12

(d) $8-5\frac{1}{10}$ **(i)** 1 **(ii)** 2 **(iii)** 3 **(iv)** 5

9. Estimate, then find the sum or difference.

(a) $3\frac{2}{10}-2\frac{1}{5}$

(b) $5\frac{3}{5}+2\frac{2}{5}$

(c) $6\frac{3}{4}-2\frac{1}{4}$

(d) $1\frac{3}{4}+1\frac{2}{3}+2$

(e) $5\frac{1}{2}-2\frac{4}{5}$

(f) $10-9\frac{1}{4}$

(g) $2\frac{1}{3}-1\frac{2}{3}$

(h) $4\frac{1}{6}-2\frac{5}{6}$

10. Example 3 shows two ways to subtract mixed numbers. Discuss with a partner the advantages and disadvantages of each method. In writing, explain which method you prefer and why.

11. The Dunlops are paying their daughters to work on their farm in the summertime. The time sheet for the first two days shows the number of hours each girl spent working.

Daughter	Day 1 (h)	Day 2 (h)
Linda	$2\frac{1}{4}$	4
Shannon	$5\frac{3}{4}$	2
Jessica	$3\frac{1}{4}$	$3\frac{3}{4}$

(a) On which day were more hours worked? How many more?

(b) Which daughter worked longest over these two days? How many more hours did this girl work than each of her sisters?

12. Examine the maze. Draw a pathway through the side openings of triangles that have a sum of 2. The path must begin at a vertex of the maze.

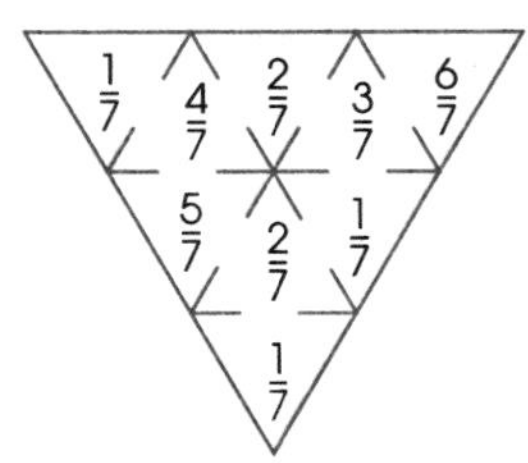

Create your own maze and exchange with a partner.

4.3 Adding Fractions (Unlike Denominators)

You have learned how to add fractions with unlike denominators by **building models** and by finding **equivalent fractions with a common denominator**. Recall:

- To add two fractions, change the fractions so they have the same denominator. The **least common denominator** is the least number that is a multiple of both denominators.
- To change a fraction to an equivalent with a different denominator, multiply or divide the numerator and denominator by the same number. This is like multiplying or dividing the fraction by 1, so the value does not change.
- Once two fractions have the same denominator, you can add the numerators and leave the denominator the same. Always write the result in simplest terms.

Example 1

Add: $\frac{1}{3}+\frac{3}{5}$

Solution

First, estimate the sum:

Since $\frac{1}{3}$ is a bit less than one half and $\frac{3}{5}$ is a bit more, the sum should be about 1 whole.

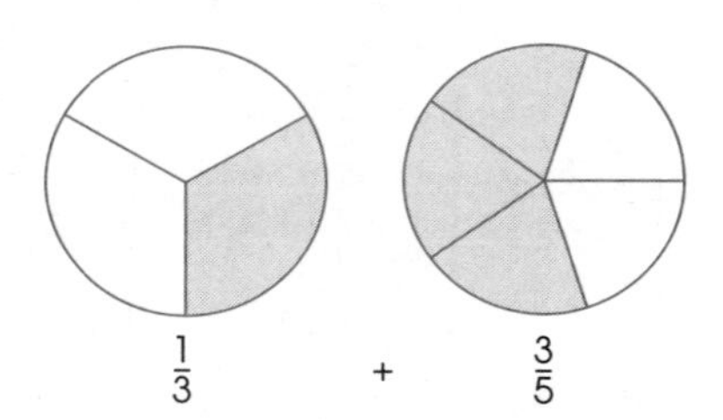

$\frac{1}{3}$ + $\frac{3}{5}$

In order to find the sum, all of the parts must be the same size.

You can make both thirds and fifths out of fifteenths.

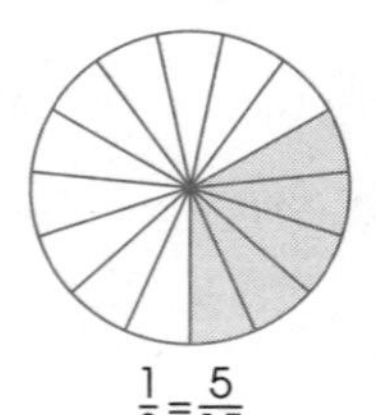

$\frac{1}{3}=\frac{5}{15}$

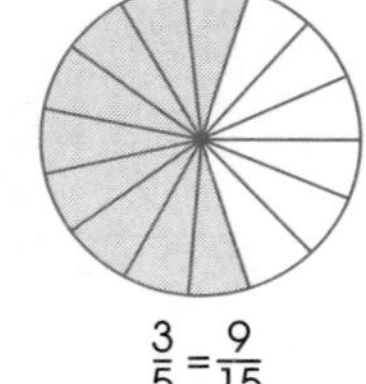

$\frac{3}{5}=\frac{9}{15}$

$$\frac{1}{3}+\frac{3}{5}=\frac{5}{15}+\frac{9}{15}$$
$$=\frac{14}{15}$$

This result is close to the estimated sum, 1 whole, so the answer seems reasonable.

Example 2

Gina made several large cakes for a bake sale. By the end of the sale, $1\frac{1}{2}$ chocolate cakes and $\frac{2}{3}$ of a vanilla cake were left over. How much cake was left over?

Solution

Step 1

Estimate.

$\frac{2}{3}$ is slightly more than $\frac{1}{2}$, so the sum of $1\frac{1}{2}$ and $\frac{2}{3}$ should be a bit more than 2 wholes.

Step 2

Find the least common denominator (LCD).

List multiples of the denominators and identify the least common multiple.

Multiples of 2: 2, 4, **6**, …
Multiples of 3: 3, **6**, 9, …

The first number common to both lists is 6, so 6 is the LCD for $\frac{1}{2}$ and $\frac{2}{3}$.

Step 3

Express each fraction in sixths.

$\frac{1}{2}=\frac{1\times 3}{2\times 3}=\frac{3}{6}$ Therefore, $1\frac{1}{2}=1\frac{3}{6}$.

$\frac{2}{3}=\frac{2\times 2}{3\times 2}=\frac{4}{6}$

Step 4

Add.

$$1\frac{1}{2}+\frac{2}{3}=1\frac{3}{6}+\frac{4}{6}$$
$$=1\frac{7}{6}$$

It is still possible to simplify this sum.

$$=1+\frac{6}{6}+\frac{1}{6}$$
$$=1+1+\frac{1}{6}=2\frac{1}{6}$$

There were $2\frac{1}{6}$ cakes left.

This result is close to the estimated sum, 2, so the answer seems reasonable.

Exercises

1. Complete.
   (a) What is a least common denominator?

   (b) Write the steps you use to find the least common denominator for two fractions. Can you describe more than one method?

2. Estimate. Explain your reasoning.
   (a) $\frac{1}{4}+\frac{2}{3}$
   (b) $\frac{8}{9}+\frac{3}{5}$
   (c) $\frac{7}{25}+\frac{1}{4}$
   (d) $1\frac{3}{12}+\frac{8}{7}$
   (e) $\frac{13}{10}+1\frac{1}{5}$

3. Find the LCD.
   (a) $\frac{7}{8}$ and $\frac{3}{4}$
   (b) $\frac{15}{10}$ and $\frac{7}{8}$
   (c) $\frac{1}{3}$ and $\frac{3}{10}$
   (d) $\frac{1}{9}$ and $\frac{1}{6}$
   (e) $\frac{2}{3}$ and $\frac{1}{7}$

4. Use the LCD to write equivalent fractions.

(a) $\frac{1}{4}$ and $\frac{1}{3}$

(b) $\frac{3}{4}$ and $\frac{7}{10}$

(c) $\frac{5}{8}$ and $\frac{2}{3}$

(d) $\frac{4}{5}$ and $\frac{9}{15}$

(e) $1\frac{1}{7}$ and $\frac{3}{2}$

5. With a partner or small group, discuss the process you use to add fractions with unlike denominators. Make a detailed step-by-step list that describes your method.

6. Draw a diagram to illustrate the addends and the sum. You can use pattern blocks, fraction strips, fraction circles, or a different model of your choice.

(a) $\frac{5}{9}+\frac{1}{3}$

(b) $\frac{1}{7}+\frac{2}{5}$

(c) $\frac{1}{6}+\frac{3}{4}$

7. Estimate and then find the sum.

(a) $\frac{1}{3}+\frac{1}{6}$

(b) $\frac{3}{5}+\frac{1}{6}$

(c) $\frac{3}{4}+\frac{7}{8}$

(d) $1\frac{1}{2}+2\frac{1}{4}$

(e) $3\frac{5}{8}+1\frac{1}{2}$

(f) $2\frac{1}{3}+\frac{6}{5}$

8. Find any errors and correct this solution.

Problem

Add: $\frac{7}{8}+\frac{3}{12}$

Solution

$$\frac{7}{8}+\frac{3}{12}=\frac{84}{96}+\frac{24}{96} \quad \text{LCD for 8 and 12 is 96.}$$
$$=\frac{108}{96}$$
$$=1\frac{12}{96}$$

9. **(a)** Write two different fractions that have a sum of $\frac{1}{2}$.

(b) Write two different fractions that have a sum of $1\frac{1}{2}$.

(c) Exchange your answers for parts (a) and (b) with a classmate. Did you choose the same addends? How many different solutions do you think there are in all?

10. In her first two playoff hockey games, Kimberly played for $1\frac{1}{2}$ periods and $1\frac{3}{4}$ periods. About how many periods in total did she play?

11. Create a word problem about the addition of fractions or mixed numbers with different denominators. Exchange problems with a partner and solve.

4.4 Subtracting Fractions (Unlike Denominators)

In this lesson, you have learned how to subtract fractions with unlike denominators by building models and by finding equivalent fractions with a common denominator. Recall:

- To subtract two fractions, **change the fractions** so they have the same denominator. The least common denominator is the least number that is a multiple of both denominators.
- To change a fraction to an equivalent with a different denominator, **multiply or divide the numerator and denominator by the same number**. This is like multiplying or dividing the fraction by 1, so the value does not change.
- Once two fractions have the same denominator, you can **subtract the numerators** and leave the denominator the same. Always write the result in simplest terms.

Example 1

Subtract: $\frac{2}{3} - \frac{1}{2}$

Solution

The fraction strips show the difference between $\frac{2}{3}$ and $\frac{1}{2}$. The third strip shows the smallest parts, sixths, that can be used to make both thirds and halves. This means that 6 is the least common denominator.

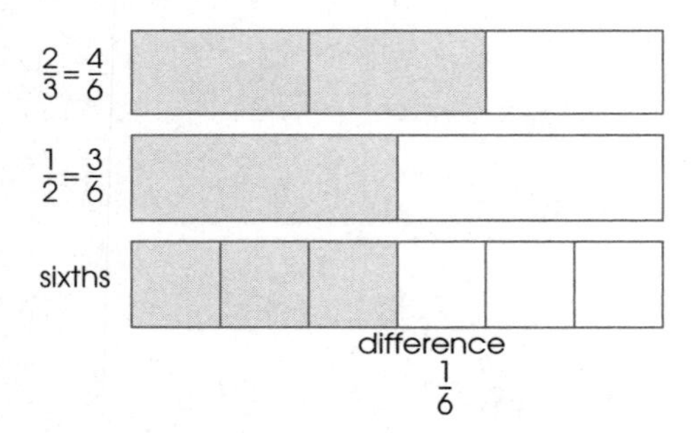

Therefore, $\frac{2}{3} - \frac{1}{2} = \frac{1}{6}$

Example 2

Subtract: $3\frac{1}{2} - 1\frac{3}{4}$

Solution

Step 1

Estimate.

The difference must be slightly less than 2, since $3\frac{1}{2} - 1\frac{1}{2}$ would be exactly 2.

Step 2

Find the LCD.

The least common denominator must be 4, because 2 is a factor of 4.

Step 3

Express $3\frac{1}{2}$ in fourths.

$\frac{1}{2} = \frac{2}{4}$, so $3\frac{1}{2} = 3\frac{2}{4}$

Step 4

Subtract.

Method 1: Regrouping

$3\frac{1}{2} - 1\frac{3}{4} = 3\frac{2}{4} - 1\frac{3}{4}$ You can't subtract $\frac{3}{4}$ from $\frac{2}{4}$.

$= 2\frac{6}{4} - 1\frac{3}{4}$ Regroup 1 whole from the first mixed number to make 4 more fourths.

$= 1\frac{3}{4}$

The result is close to the estimated difference, so the answer seems reasonable.

Method 2: Use Improper Fractions

$3\frac{1}{2} = 3\frac{2}{4} = \frac{12}{4} + \frac{2}{4} = \frac{14}{4}$ Rewrite each fraction as an improper fraction with a common denominator.

$1\frac{3}{4} = \frac{4}{4} + \frac{3}{4} = \frac{7}{4}$

$3\frac{1}{2} - 1\frac{3}{4}$

$= \frac{14}{4} - \frac{7}{4}$ Subtract the numerators.

$= \frac{7}{4}$ $\frac{7}{4}$ is an improper fraction.

$= \frac{4}{4} + \frac{3}{4}$ Separate the whole.

$= 1\frac{3}{4}$ Simplify to a mixed number.

Exercises

1. Compare the steps you use to subtract fractions with unlike denominators with the steps you use to add them. What similarities and differences are there?

2. Draw a diagram. You can use pattern blocks, fraction strips, fraction circles, or another model of your choice.

(a) $\frac{1}{3} - \frac{2}{9}$

(b) $\frac{3}{4} - \frac{2}{3}$

(c) $\frac{3}{5} - \frac{1}{10}$

3. Draw a diagram to illustrate a subtraction problem and its solution. Exchange diagrams with a partner. Write the subtraction sentence that describes the diagram.

4. Estimate and then find the difference.

(a) $\frac{5}{6} - \frac{3}{4}$

(b) $\frac{10}{9} - \frac{2}{3}$

(c) $\frac{7}{4} - \frac{3}{2}$

(d) $\frac{5}{3} - \frac{1}{2}$

5. Use addition to check each answer you found for Problem 4.

6. Subtract by regrouping.

(a) $3\frac{1}{2} - 2\frac{7}{8}$

(b) $4 - 1\frac{1}{5}$

(c) $6\frac{1}{3} - 2\frac{5}{9}$

(d) $1\frac{5}{6} - \frac{5}{4}$

7. Subtract using improper fractions.

(a) $1\frac{1}{4} - \frac{1}{2}$

(b) $3\frac{1}{3} - 2\frac{3}{5}$

(c) $2\frac{1}{3} - 1\frac{5}{6}$

(d) $2\frac{1}{2} - 1\frac{5}{8}$

8. Estimate and then find the difference using a method of your choice.

(a) $3\frac{1}{4} - 2\frac{3}{8}$

(b) $5\frac{7}{8} - 3\frac{1}{10}$

(c) $4\frac{1}{3} - 4\frac{1}{4}$

(d) $1\frac{8}{9} - 1\frac{2}{3}$

(e) $7\frac{4}{5} - 4\frac{1}{10}$

9. When you subtract mixed numbers, do you usually use regrouping or improper fractions? Why?

10. Find any errors and correct this solution.

Subtract:

$$4\frac{3}{5}-\frac{7}{8}=4\frac{24}{40}-\frac{35}{40} \quad \text{LCD for 5 and 8 is 40.}$$
$$=4\frac{64}{40}-\frac{35}{40}$$
$$=4\frac{29}{40}$$

11. Find the missing number.

(a) $1\frac{7}{8}-\square=1\frac{5}{8}$

(b) $\square-2\frac{1}{4}=2\frac{3}{4}$

(c) $2\frac{3}{5}-\square=\frac{3}{10}$

(d) $1\frac{2}{5}-\frac{7}{10}=\square$

12. In each case, write two fractions with unlike denominators whose difference is the fraction shown.

(a) $\frac{1}{2}$

(b) $\frac{2}{3}$

(c) $1\frac{1}{5}$

(d) $2\frac{5}{8}$

13. Write a word problem for each situation that involves subtracting fractions with unlike denominators. Write the solution on another sheet of paper. Then exchange problems with a partner and solve.

(a) time a person spends rehearsing for a play and doing homework

(b) the change in share values on the stock market

(c) the number of laps two people swim in a pool

4.5 Multiplying Fractions

In this lesson, you learned to **multiply fractions and mixed numbers**. Recall:

- One way to interpret fraction multiplication is to think of finding a fraction of another fraction, so $\frac{1}{2} \times \frac{1}{4}$ can mean 1 half of 1 fourth.
- You can model fraction multiplication by using pattern blocks or by drawing rectangle models.
- To multiply two proper or improper fractions, find $\frac{\textit{product of the numerators}}{\textit{product of the denominators}}$ and then write this number in simplest terms.
- To multiply two mixed numbers, first express each number as an improper fraction.

Example 1

Draw a rectangle model to find $\frac{1}{4} \times \frac{2}{5}$.

Solution

First, draw lines one way to show $\frac{1}{4}$.

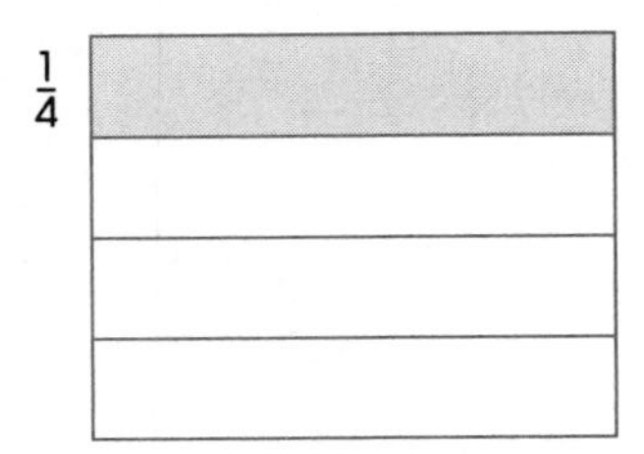

Then draw lines the other way to show $\frac{2}{5}$.

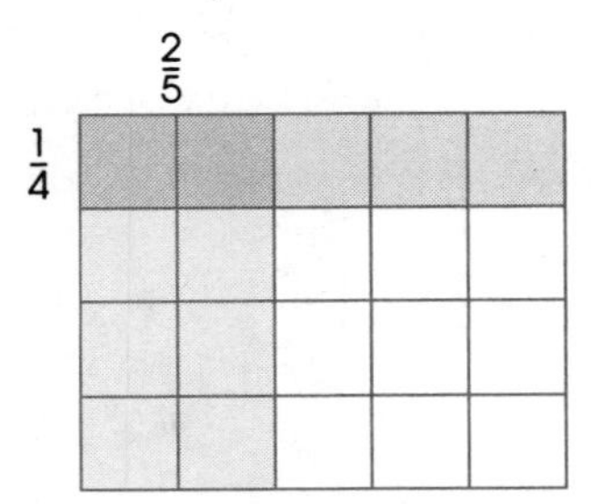

The overlapping sections show that $\frac{1}{4}$ of $\frac{2}{5}$ is $\frac{2}{20}$.

In simplest terms, $\frac{2}{20} = \frac{1}{10}$.

Example 2

Multiply $1\frac{2}{3} \times 1\frac{1}{2}$.

Solution

Step 1

Convert the mixed numbers to fraction form.

$$1\frac{2}{3} = \frac{3}{3} + \frac{2}{3} = \frac{5}{3}$$

$$1\frac{1}{2} = \frac{2}{2} + \frac{1}{2} = \frac{3}{2}$$

Step 2

Find the product of the numerators and denominators.

$$\begin{aligned}\frac{5}{3} \times \frac{3}{2} &= \frac{5 \times 3}{3 \times 2}\\ &= \frac{15}{6}\\ &= \frac{15 \div 3}{6 \div 3} && \text{Simplify by dividing the numerator and denominator by 3.}\\ &= \frac{5}{2}\\ &= \frac{4}{2} + \frac{1}{2} && \text{Simplify by converting the improper fraction to a mixed number.}\\ &= 2\frac{1}{2}\end{aligned}$$

Step 3

Check for reasonableness.

$1\frac{2}{3} \times 1\frac{1}{2}$ is a bit less than $2 \times 1\frac{1}{2}$, so the answer should be a bit less than 3.

The product $2\frac{1}{2}$ seems reasonable.

Exercises

1. When you add or subtract fractions, you must write the numbers with a common denominator.

   (a) Why is this not necessary for multiplication?

   (b) If you write the numbers with a common denominator, will the answer be wrong? Explain.

2. Estimate each product. Explain your thinking.

   (a) $\frac{9}{10}\times\frac{7}{8}$

   (b) $\frac{1}{2}\times\frac{11}{10}$

   (c) $\frac{1}{4}\times 7\frac{7}{8}$

   (d) $2\frac{1}{8}\times 10$

3. Write the steps you use to multiply a fraction by a whole number. Give an example.

4. Write the steps you use to multiply two mixed numbers. Give an example.

5. Complete the diagram. Give the product in simplest terms.

   (a) $\frac{1}{2}\times\frac{1}{5}$

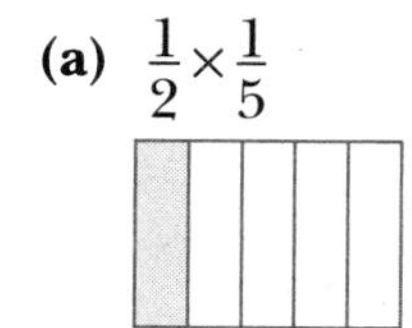

   (b) $\frac{1}{4}$ of $\frac{2}{3}$

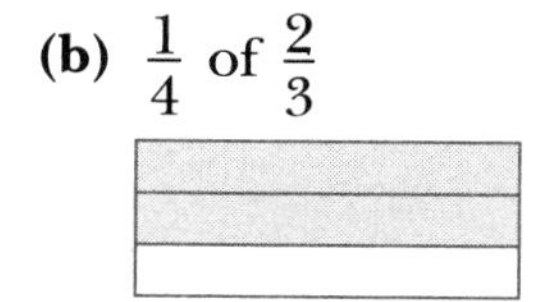

   (c) $\frac{3}{4}$ of $\frac{2}{5}$

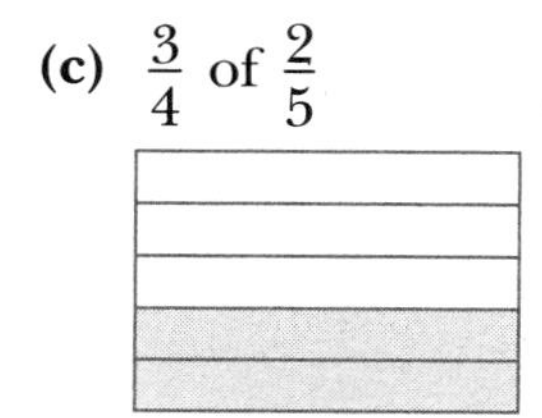

   (d) $\frac{1}{2}$ of $1\frac{1}{4}$

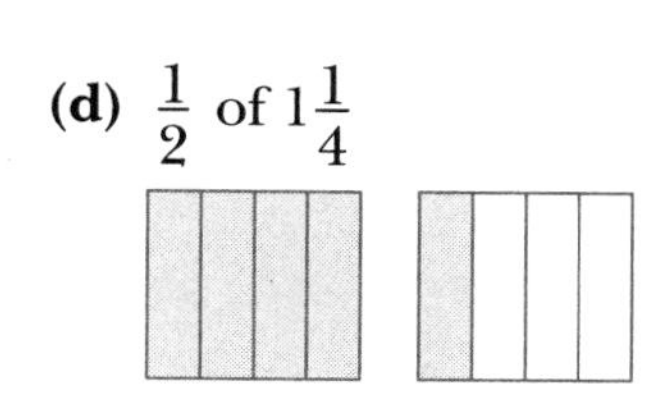

6. Draw a rectangle diagram. Give the product.

(a) $\frac{6}{7} \times \frac{1}{3}$

(b) $1\frac{1}{2} \times \frac{1}{4}$

(c) $2\frac{1}{4} \times \frac{3}{4}$

7. Write the product in simplest terms.

(a) $\frac{5}{6} \times \frac{3}{5}$ **(b)** $\frac{3}{10} \times \frac{1}{2}$ **(c)** $\frac{1}{6} \times \frac{9}{10}$

(d) $\frac{3}{4} \times \frac{3}{4}$ **(e)** $\frac{2}{3} \times \frac{7}{12}$ **(f)** $\frac{7}{3} \times \frac{3}{2}$

(g) $\frac{8}{3} \times \frac{3}{4}$ **(h)** $\frac{12}{9} \times \frac{9}{10}$

8. Write the product in simplest terms.

(a) $\frac{3}{8} \times 3$ **(b)** $\frac{1}{5} \times 7$ **(c)** $\frac{3}{4} \times 8$

(d) $4 \times \frac{2}{3}$ **(e)** $\frac{1}{4} \times 6$ **(f)** $8 \times \frac{4}{5}$

9. Write the product in simplest terms.

(a) $3\frac{1}{2} \times 2\frac{2}{3}$ **(b)** $2\frac{2}{5} \times 4$

(c) $1\frac{3}{5} \times 3\frac{1}{3}$ **(d)** $2\frac{5}{8} \times 1\frac{1}{2}$

(e) $2\frac{3}{4} \times 1\frac{1}{5}$

10. During summer vacation, Noah practices his golf swing for $\frac{1}{4}$ of an hour each day. How many hours does he spend practicing in the month of July?

11. Estimate the prices of ten things you would like to own. Write each price to the nearest dollar.

(a) What would be the total cost of buying one of each item?

(b) What would be the cost of each item at a $\frac{1}{4}$-off sale?

(c) What would be the cost of each item at a $\frac{1}{3}$-off sale?

(d) How much money would you save in total by buying one of each item on sale for $\frac{1}{4}$ off? for $\frac{1}{3}$ off?

12. Compare your way of solving Problem 11(d) with a classmate's. Describe the most efficient way to solve this problem.

13. Find and correct the error in this solution.

$$5\frac{2}{8} \times \frac{10}{5} = \frac{42}{8} \times \frac{10}{5}$$
$$= \frac{420}{40}$$

14. Create five multiplication exercises that involve fractions and mixed numbers less than 5. Write the answers to your problems on another sheet of paper.

Exchange exercises with a partner and find the products. Draw a rectangle diagram to show why each product is correct.

4.6 Dividing Fractions

In this lesson, you learned to **divide fractions and mixed numbers**. Recall:

- One way to interpret a division question such as $\frac{1}{2} \div \frac{1}{4}$ is to say "How many $\frac{1}{4}$s can you make from $\frac{1}{2}$?"
- You can model fraction division by using pattern blocks or by drawing fraction models.
- To divide two proper or improper fractions, multiply the dividend by the reciprocal of the divisor. For example, to find $\frac{1}{2} \div \frac{1}{4}$, multiply $\frac{1}{2} \times \frac{4}{1}$.
- To divide two mixed numbers, first express each number as an improper fraction.

Example 1

Find $\frac{1}{2} \div \frac{1}{4}$.

Solution

Draw diagrams to show $\frac{1}{2}$ and $\frac{1}{4}$.

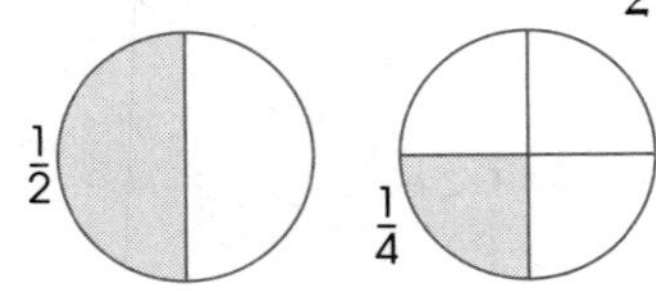

$\frac{1}{2} \div \frac{1}{4}$ means "How many $\frac{1}{4}$s can you make from $\frac{1}{2}$?"
It would take two $\frac{1}{4}$s to fill $\frac{1}{2}$, so $\frac{1}{2} \div \frac{1}{4} = 2$.

Example 2

Find $\frac{1}{10} \div \frac{1}{5}$.

Solution

Draw diagrams to show $\frac{1}{10}$ and $\frac{1}{5}$.

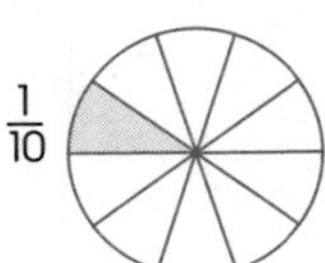

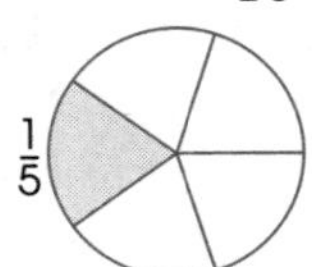

$\frac{1}{10} \div \frac{1}{5}$ means "How many $\frac{1}{5}$s can you make from $\frac{1}{10}$?"

$\frac{1}{10}$ is smaller than $\frac{1}{5}$, so you can fit less than one $\frac{1}{5}$ into $\frac{1}{10}$.

Two tenths are equal to one fifth, so you can fit $\frac{1}{2}$ of a fifth into $\frac{1}{10}$.

$$\frac{1}{10} \div \frac{1}{5} = \frac{1}{2}$$

Example 3

Find $3 \div \frac{1}{4}$ without using a diagram.

Solution

To find out how many $\frac{1}{4}$s are in 3, start by finding the number of $\frac{1}{4}$s in 1. This is easy to do, because 1 divided by any number is equal to the reciprocal of the number.

$1 \div \frac{1}{4} = \frac{4}{1}$ or 4.

There are four $\frac{1}{4}$s in 1, so there must be three times as many in 3.

$3 \times 4 = 12$, so there are 12 sets of $\frac{1}{4}$ in 3.

Check by multiplying.

If $3 \div \frac{1}{4} = 12$, then $12 \times \frac{1}{4}$ must equal 3.

$$12 \times \frac{1}{4} = \frac{12}{4} = 3$$

The quotient is correct.

Example 4

Find $2\frac{1}{2} \div 1\frac{3}{4}$.

Solution

Write each mixed number as an improper fraction.

$$2\frac{1}{2} = \frac{4}{2} + \frac{1}{2} = \frac{5}{2} \qquad 1\frac{3}{4} = \frac{4}{4} + \frac{3}{4} = \frac{7}{4}$$

Now the division becomes $\frac{5}{2} \div \frac{7}{4}$.

There are $\frac{4}{7}$ sets of $\frac{7}{4}$ in 1, so there must be $\frac{5}{2} \times \frac{4}{7}$ sets in $\frac{5}{2}$.

$$\frac{5}{2} \div \frac{7}{4} = \frac{5}{2} \times \frac{4}{7}$$ To multiply two fractions, find $\dfrac{\textit{product of the numerators}}{\textit{product of the denominators}}$.

$$= \frac{20}{14}$$ Simplify.

$$= \frac{10}{7}$$ Write $\frac{10}{7}$ as a mixed number. Check by multiplying.

$$= 1\frac{3}{7}$$

If $2\frac{1}{2} \div 1\frac{3}{4} = 1\frac{3}{7}$, then $1\frac{3}{7} \times 1\frac{3}{4}$ must equal $2\frac{1}{2}$.

$$1\frac{3}{7} \times 1\frac{3}{4} = \frac{10}{7} \times \frac{7}{4}$$
$$= \frac{70}{28}$$
$$= \frac{5}{2} = 2\frac{1}{2}$$

The quotient is correct.

Exercises

1. Define each term.

(a) dividend

(b) divisor

(c) quotient

(d) reciprocal

2. What math skills do you need to divide fractions?

3. Estimate each quotient. Explain your thinking.

(a) $\frac{7}{10} \div \frac{1}{9}$

(b) $\frac{3}{4} \div 5$

(c) $7 \div \frac{2}{3}$

(d) $12\frac{1}{4} \div \frac{3}{8}$

4. Write the steps you use to divide two fractions. Give an example.

5. Write the steps you use to divide two mixed numbers. Give an example.

6. Write the quotient in simplest terms.

(a) $\frac{7}{12} \div \frac{1}{6}$ (b) $\frac{11}{12} \div \frac{1}{3}$ (c) $\frac{5}{6} \div \frac{1}{3}$

7. Draw a diagram. Write the quotient in simplest terms.

(a) $4 \div \frac{1}{5}$ (b) $6 \div \frac{2}{3}$ (c) $\frac{3}{4} \div \frac{1}{2}$

8. Divide without using a diagram.

(a) $\frac{2}{3} \div \frac{6}{7}$ (b) $\frac{1}{5} \div \frac{3}{8}$ (c) $\frac{7}{8} \div \frac{1}{2}$

(d) $\frac{9}{4} \div \frac{2}{3}$ (e) $\frac{5}{6} \div \frac{5}{12}$ (f) $\frac{11}{12} \div \frac{2}{3}$

9. Find and correct the error in this solution.

$$\begin{aligned} 2\tfrac{7}{8} \div 1\tfrac{2}{3} &= 2\tfrac{7}{8} \times 1\tfrac{3}{2} \\ &= \frac{23}{8} \times \frac{5}{2} \\ &= \frac{115}{16} \\ &= 7\frac{3}{16} \end{aligned}$$

10. Divide.

(a) $1\frac{1}{4} \div 3$ (b) $3\frac{1}{5} \div 5$ (c) $8 \div 2\frac{2}{3}$

(d) $2\frac{2}{3} \div 1\frac{3}{5}$ (e) $1\frac{3}{4} \div 1\frac{1}{8}$ (f) $3\frac{1}{2} \div 1\frac{2}{5}$

11. Marcus takes 17 min to type $8\frac{1}{2}$ pages. How many pages does he type per minute?

12. Joanne has a roll of ribbon 6 ft long. She needs pieces that are $\frac{3}{4}$ ft long. If she cuts the whole roll, how many pieces can she make?

13. When you divide 0 by a fraction, the quotient is always the same. What is the quotient? Why?

14. Is this statement true or false? Give reasons and an example to support your answer.

    When you divide any number by a fraction less than 1, the quotient is always less than the dividend.

15. Create two fraction division problems and write a solution for each one. Then rewrite one of the solutions so it contains an error.

    Copy the problems and solutions — one correct and one incorrect — onto another sheet of paper. Exchange with a classmate to find and correct the errors.

4.7 Problem Solving with Fractions

You have seen how to solve fraction problems using the **problem-solving model**. Recall the four steps:

- Think about the problem.
- Make a plan.
- Solve the problem.
- Look back.

Example

It snowed for $3\frac{1}{2}$ hours on Monday, $2\frac{3}{4}$ hours on Tuesday, and $1\frac{4}{5}$ hours on Wednesday. Find the mean snowfall duration over the three-day period.

Solution

Step 1: Think about the problem.

Information given

Day	Snowfall Duration (h)
Monday	$3\frac{1}{2}$
Tuesday	$2\frac{3}{4}$
Wednesday	$1\frac{4}{5}$

Information needed

the mean duration of snowfall over the three days

Step 2: Make a plan.

The mean duration is the length of each snowfall if the times had been the same for all three days.

To find the mean, find the total amount of snowfall time and then divide it by 3.

Step 3: Solve the problem.

Total snowfall time: $3\frac{1}{2}+2\frac{3}{4}+1\frac{4}{5}$

$=\frac{7}{2}+\frac{11}{4}+\frac{9}{5}$ Convert mixed numbers to improper fractions.

$=\frac{70}{20}+\frac{55}{20}+\frac{36}{20}$ Convert to a common denominator.

$=\frac{161}{20}$

The total snowfall time was $\frac{161}{20}$ hours. Keep this number in improper fraction form.

Divide the total snowfall time by 3.

To divide by 3, multiply by the reciprocal $\frac{1}{3}$.

Find $\frac{\textit{product of the numerators}}{\textit{product of the denominators}}$.

$$\begin{aligned} \textit{Mean duration} &= \frac{161}{20} \div 3 \\ &= \frac{161}{20} \times \frac{1}{3} \\ &= \frac{161 \times 1}{20 \times 3} \quad \text{Simplify.} \\ &= \frac{161}{60} = 2\frac{41}{60} \end{aligned}$$

The mean snowfall duration was $2\frac{41}{60}$ hours, or 2 hours and 41 minutes.

Step 4: Look back.

Has the question been answered?

The problem asked for the mean snowfall duration. This is 2 hours and 41 minutes.

Is the answer reasonable?

Snow fell for about 3 h on Monday, about 3 h on Tuesday, and about 2 h on Wednesday, so it fell for about 8 h in all. $8 \div 3 = \frac{8}{3}$ or $2\frac{2}{3}$, so the mean should be about $2\frac{2}{3}$ hours or 2 hours and 40 minutes. The calculated answer, 2 hours and 41 minutes, is very close to this amount, so the answer is reasonable.

Exercises

1. Compare operations with proper fractions to operations with mixed numbers. What similarities and differences can you find?

2. If a problem involves several operations, you must follow the rules for ordering operations. What are these rules?

3. About $\frac{1}{4}$ of the tapes in Mariah's collection are alternative rock and about $\frac{1}{3}$ are rap music. All the rest are classic rock. What fraction of Mariah's tapes are classic rock?

4. The average person spends $\frac{1}{3}$ of his lifetime sleeping. How many

(a) hours will a person sleep in one day?

(b) hours will a person sleep in one week?

(c) minutes will a person sleep in one year?

5. For dessert, Mark's family ate $\frac{1}{3}$ of a blueberry pie and some apple pie. If they ate $1\frac{1}{12}$ pies in all, what fraction of the apple pie did they eat?

6. Check Erin's solution to this problem. Correct any errors.

Sonya bought $2\frac{1}{3}$ bushels of apples and gave $\frac{1}{5}$ of them to her sister. What fraction of a bushel does she have left?

Solution

$$\begin{aligned} \frac{1}{5} \times 2\frac{1}{3} &= \frac{1}{5} \times \frac{7}{3} \\ &= \frac{1 \times 7}{5 \times 3} \\ &= \frac{7}{15} \end{aligned}$$

Sonya has $\frac{7}{15}$ of a bushel left.

7. Marnie has 8 popcorn bowls, each of which holds $\frac{7}{8}$ of a bag of popcorn. How many bags of popcorn will she need to fill each bowl once?

8. Suppose Marnie's bowls were a different size. Calculate the number of bags she would need in each situation.

(a) She has 6 bowls. Each bowl holds $\frac{5}{6}$ of a bag.

(b) She has 9 bowls. Each bowl holds $\frac{8}{9}$ of a bag.

(c) She has 10 bowls. Each bowl holds $\frac{9}{10}$ of a bag.

9. Describe the pattern you see in Problems 7 and 8. Explain why it occurs. Then show how you could use the pattern to predict a different product.

10. The chart shows the rainfall duration for one week.

Day	Rainfall Duration (h)
Monday	$1\frac{1}{2}$
Tuesday	$3\frac{3}{5}$
Wednesday	0
Thursday	$2\frac{1}{3}$
Friday	0
Saturday	0
Sunday	$\frac{3}{4}$

(a) How many hours longer did it rain on Tuesday than on Thursday? How many minutes?

(b) How many times as long as Sunday's rainfall was Monday's rainfall?

(c) How many hours of rain in all fell during the week?

(d) What is the mean rainfall duration for this week?

11. Doctors recommend that people spend 20 minutes per day exercising. Sarah kept track of the time she spent exercising over three days. The chart shows her results.

Type of Exercise	Total Time Spent (min)
Sit-ups	$12\frac{1}{2}$ min
Chin-ups	$10\frac{1}{2}$ min
Running on the spot	$15\frac{1}{3}$ min

(a) Determine the mean number of minutes per day Sarah spent exercising.

(b) By how many minutes must she increase her mean time to meet the recommended requirement?

12. Create a problem of each of these types. Show how you would use the four problem-solving steps to solve each problem.

(a) add, subtract, multiply, or divide two proper fractions

(b) add, subtract, multiply, or divide two mixed numbers

(c) perform more than one operation with fractions and/or mixed numbers

4.8 Adding and Subtracting Rational Numbers

In this lesson, you reviewed addition with fractions and learned how to add and subtract rational numbers in decimal form. Recall:

- A **rational number** is a number that can be written in the ratio form $\frac{a}{b}$, where a and b are integers, and $b \neq 0$. The set of rational numbers includes natural numbers, whole numbers, integers, positive and negative fractions, and terminating and repeating decimals.
- To add a negative rational number, you can subtract the opposite of the number.
- To subtract a negative rational number, you can add the opposite of the number.

Example 1

Add 6.213 + (–8.735).

Solution

Step 1: Estimate.

Adding a negative number is like subtracting its opposite, so this number sentence can be restated as 6.213 – 8.735.

This is about 6 – 8, or about –2.

Step 2: Compute.

Use a calculator to find the sum.

Addition Method: Enter 6.213 + 8.735 [+/-] [=].

Subtraction Method: Enter 6.213 – 8.735 [=].

In each case, the display shows –2.522.

This is close to the estimated result, –2, so it is likely correct.

Step 3: Verify.

If 6.213 + (–8.735) = –2.522, then (–2.522) – (–8.735) = 6.213.

Use a calculator.

Subtraction Method:
Enter 2.522 [+/-] [-] 8.735 [+/-] [=].

Addition Method: Subtracting a negative number is like adding its opposite. Enter 2.522 [+/-] [+] 8.735 [=].

Each result is 6.213, so the calculations were correct.

Example 2

Draw a number line to find 1.4 – (–0.8).

Solution

Both numbers in the problem are in tenths. Create a number line marked in tenths. Subtracting –0.8 is like adding its opposite.

To add +0.8 to 1.4, start at 1.4 and move 0.8 to the right.

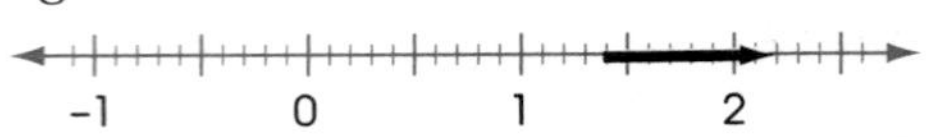

The arrow stops at 2.2. So 1.4 – (–0.8) = 2.2.

Example 3

Find $\left(-\frac{1}{8}\right)-\frac{9}{10}$.

Solution

To find the difference with a calculator, convert the fractions to decimal form.

$-\frac{1}{8} = -(1 \div 8) = -0.125 \qquad \frac{9}{10} = 9 \div 10 = 0.9$

Step 1: Estimate.

Now the problem becomes (–0.125) – 0.9. This is close to 0 – 1 or –1.

Step 2: Compute.

Method 1: Use the [+/-] key. Enter 0.125 [+/-] [-] 0.9 [=].

Method 2: Subtracting 0.9 is like adding –0.9. Now the number sentence becomes (–0.125) + (–0.9). On a number line, you could start at 0, move 0.125 to the left and then move another 0.9. To find the distance traveled from 0, enter 0.125 + 0.9 [=]. Then press [+/-] to make the sum negative.

Each result is –1.025. This is close to the estimated result.

Step 3: Verify.

If (–0.125) – 0.9 = –1.025, then (–1.025) + 0.9 = (–0.125).

Enter 1.025 [+/-] [+] 0.9 [=].

The display shows –0.125, so the calculations were correct.

Exercises

1. Describe a situation where someone might need to add or subtract negative decimals or fractions.

2. Explain two different ways to use a calculator to find $0.275 - (-4.16)$.

3. Explain how you would move right or left on a number line to find each result. Start at 0 each time.

   (a) positive + positive

   (b) positive + negative

   (c) negative + positive

   (d) negative + negative

   (e) positive – positive

   (f) positive – negative

   (g) negative – positive

   (h) negative – negative

4. Is each statement true or false? Give an example to support your answer.

   (a) When you subtract a positive number from a negative one, the result is always negative.

   (b) When you subtract a positive number from a positive one, the result is sometimes negative.

   (c) When you subtract a negative number from a positive one, the result is sometimes negative.

   (d) When you add two negative numbers, the result is always positive.

   (e) When you add one negative number and one positive one, the sign is the same as the sign of the greater number.

5. Write the decimal equivalent. Round to the nearest thousandth if necessary. Do as many as you can from memory and then use a calculator for the rest.

   (a) $\frac{1}{2}$ **(b)** $\frac{1}{5}$ **(c)** $\frac{1}{3}$ **(d)** $\frac{1}{10}$

   (e) $\frac{1}{6}$ **(f)** $\frac{1}{4}$ **(g)** $\frac{1}{8}$ **(h)** $\frac{2}{3}$

   (i) $\frac{4}{5}$ **(j)** $\frac{3}{8}$ **(k)** $\frac{3}{4}$ **(l)** $\frac{5}{6}$

6. Write each decimal to the nearest thousandth and then add or subtract.

(a) $\frac{3}{4}+\left(-\frac{1}{8}\right)$ **(b)** $\left(-\frac{3}{5}\right)-\frac{2}{3}$ **(c)** $\frac{4}{5}-\frac{7}{8}$

(d) $\left(-\frac{11}{12}\right)+\left(-\frac{4}{3}\right)$ **(e)** $4\frac{1}{5}+\left(-3\frac{1}{2}\right)$ **(f)** $\left(-2\frac{3}{8}\right)-\left(-2\frac{9}{24}\right)$

7. Estimate or calculate mentally. Explain your thinking.

(a) 0.45 – (–0.926) **(b)** (–7.2) + (–11.36) **(c)** (–1.3) – (–1.326)

(d) 5.2379 – 10.016 **(e)** (–3.2) – 5.67 **(f)** (–4.333) + 2.715

8. Estimate, calculate, and verify.

(a) 85.3 – 12.19 **(b)** 17.65 + (–35.44) **(c)** (–22.657) + (–48.15)

(d) (–182.34) – 134.667 **(e)** 215.2 – (–72.8) **(f)** (–175.823) + 154.13

(g) 154.321 + 947.25 **(h)** (–38.5) – (–1942.71)

9. Verify. Correct any errors.

(a) 85.3 – (–52.1) = 33.2 **(b)** 25.4132 + (–119.27) = –93.8565

(c) (–47.254) + 47.254 = –282.221 **(d)** 72.35 – 184.11 = –111.76

10. Justin said, "When you add or subtract two decimals, always look to see which number has more decimal places. The sum or difference will have the same number of decimal places as this number." Is Justin correct? Explain why his rule does or does not work.

11. Kyra forgot to bring her calculator to school, so she tried to subtract 6.213 – 8.735 on paper. She ran into a problem when she tried to subtract 8 ones from 6 ones, and there were no more numbers to regroup.

$$\begin{array}{r} 6.213 \\ \underline{-\ 8.735} \\ -\ ?.478 \end{array}$$

(a) Use a calculator to subtract. What do you notice about the other digits Kyra found?

(b) Think of how this subtraction would look on a number line. How could Kyra find the difference by adding or subtracting on paper?

12. The temperature was 4.56°F at noon and –8.67°F at midnight. Show two different ways to find the change in temperature. Express the change as a rational number.

13. In Problem 11, you saw that the traditional pencil-and-paper subtraction method doesn't work if you subtract more than the starting amount.

Try using pencil and paper to add and subtract some other rational numbers. Which combinations can be added and subtracted in the traditional way? Which cannot? What can you do instead?

14. Create a decimal calculation that fits into each category in Problem 3. For each calculation:

(a) Estimate the result.

(b) Calculate the result.

(c) Check by using the inverse operation.

(d) Draw the number line model.

4.9 Multiplying and Dividing Rational Numbers

In this lesson, you learned that the rules for multiplying and dividing integers **also apply to rational numbers**. Recall:

- If you multiply or divide two positive numbers, the result is positive.
- If you multiply or divide two negative numbers, the result is negative.
- If you multiply or divide one positive number and one negative one, the result is negative.

Example 1

Find $0.95 \times (-3.62)$.

Solution

Step 1: Estimate.

0.95 is a bit less than 1, so the product should be close to $1 \times (-3.62)$ or about -3.5.

Step 2: Compute.

Since one factor is positive and the other negative, the product will be negative. Use a calculator to multiply 0.95×3.62 and make the product negative, or use the [+/-] key.

$0.95 \times (-3.62) = -3.439$

Step 3: Verify.

If $0.95 \times (-3.62) = -3.439$, then $(-3.439) \div (-3.62)$ must equal 0.95.

Enter 3.439 [+/-] [÷] 3.62 [+/-] [=].

The display shows 0.95, so the product was calculated correctly. This makes sense, since the product, -3.439, is close to the estimated product, -3.5.

Example 2

Joe recorded the daily low temperatures for one week and found that the average low temperature for the week was –4.1°F. What was Saturday's temperature?

Day	Low Temperature (°F)
Sunday	+1.7
Monday	-7.4
Tuesday	0
Wednesday	-13.6
Thursday	-5.9
Friday	+2.1
Saturday	

Solution

If the average temperature was –4.1°F, the sum of the seven temperatures must have been –4.1°F × 7 or –28.7°F.

The sum of the six known temperatures is:

$(+1.7) + (-7.4) + (-13.6) + (-5.9) + (+2.1)$
$= -23.1$

So $(-23.1) +$ *unknown temperature* $= -28.7$.

Let t be the temperature.

$$(-23.1) + t = (-28.7)$$
$$t = (-28.7) + 23.1 \quad ❶$$
$$t = -5.6 \quad ❷$$

❶ Add 23.1 to both sides.

❷ The temperature on Saturday was –5.6°F.

Exercises

1. Describe a situation where someone might need to multiply or divide with negative numbers.

2. Explain two different ways to use a calculator to find $0.25 \times (-3.6)$.

3. How could you have predicted the product in Problem 2 without using a calculator?

4. Calculate.

   (a) $\frac{7}{8} \times 23$ **(b)** $\frac{3}{4} \div 4$ **(c)** $1\frac{3}{5} \div \frac{2}{3}$

   (d) $1\frac{5}{6} \times 14$ **(e)** $2\frac{1}{2} \times 1\frac{1}{3}$ **(f)** $9\frac{1}{2} \div 4\frac{1}{2}$

5. Estimate. Explain the method you used.

   (a) $1.5 \times (-0.15)$ **(b)** $0.59 \div 0.1$ **(c)** 2.13×0.6

   (d) $4.915 \div (-0.7)$ **(e)** $(-8.2) \div (-0.916)$ **(f)** $(-2.98) \times (-3.714)$

6. Use a calculator to find each product or quotient to the nearest hundredth.

   (a) $4.16 \times (-0.48)$ **(b)** $(-103.2) \div 43.66$ **(c)** $95.17 \times (-3.91)$

   (d) $(-3.85) \times (-2.1)$ **(e)** $8.123 \div 2.45$ **(f)** $(-2.75) \div (-3.33)$

7. Verify. Correct any errors.

(a) $7.23 \times 6.1 = 44.103$ **(b)** $(-4) \times 11.11 = 44.44$ **(c)** $2.2 \div 0.22 = 10.36$

(d) $(-4.2) \div (-0.6) = 0.7$ **(e)** $(-8.3) \times (-5.12) = -42.496$ **(f)** $88.88 \div (-2.22) = 44$

8. Eileen said, "To predict the number of decimal places in a product, all you have to do is count the total number of decimal places in both factors."
Is Eileen correct? Explain why her rule does or does not work.

9. Calculate the amount of each discount.

(a) $\frac{1}{3}$ off \$66.75 **(b)** $\frac{1}{2}$ off \$44.36

(c) $\frac{1}{4}$ off \$59.96 **(d)** $\frac{3}{4}$ off \$380.52

10. Find the average.

(a) $-3.7, -11.6, -0.7, -4.0$ **(b)** $\frac{1}{3}, \frac{7}{12}, 1\frac{1}{2}, 6\frac{1}{3}, 3\frac{3}{4}$

(c) $5.61, -5.88, 6.24, -5.03, -4.12, 2.7$

11. A piece of wood is $4\frac{1}{2}$ ft long. How many pieces can be cut from it if each piece is to be $\frac{3}{8}$ ft long?

12. Doug earns half as much as Marika. Marika earns $\frac{3}{4}$ as much as Greg. If Greg earns $20.00/h, how much does Doug earn?

13. Bill earns $15.85/h. If he earned $570.60 in one week, how many hours did he work?

14. To attract customers, a store offered dresses on sale for less than cost price. Alicia calculated that for each dress sold the store had a (negative) profit of –$8.50. On the day of the sale, she found that the store's total profit on these dresses was –$76.50. How many dresses were sold?

15. Create five multiplication and division exercises that involve rational numbers. Write the solutions and then rewrite several solutions so they contain an error. Exchange exercises and solutions with a partner and find the errors.

5 THE LANGUAGE OF ALGEBRA

5.1 Patterns and Relations

This lesson introduced different methods for **solving pattern problems**. For example:
- Use a **diagram** or a model.
- Make a **table**.
- Write a **mathematical expression.**

Example 1

A parking ticket dispenser accepts only quarters and dimes. Write an expression you can use to find the total value of the coins, in dollars, in the dispenser, no matter how many there are of each type.

Solution

Let q represent the number of quarters and d represent the number of dimes.

Each quarter is worth $0.25. Each dime is worth $0.10.

$$\begin{aligned} \textit{Total value} &= (0.25 \times \textit{number of quarters}) \\ &\quad + (0.10 \times \textit{number of dimes}) \\ &= (0.25 \times q) + (0.10 \times d) \\ &= 0.25q + 0.10d \end{aligned}$$

Example 2

Two hundred coins are arranged in this pattern:

$1, 50¢, 25¢, $1, 50¢, 25¢, …

What is the total value of the coins?

Solution

The repeating pattern has three coins: $1, 50¢, and 25¢.

There are 200 coins in all.

Therefore, there must be 200 ÷ 3 sets of coins.

200 ÷ 3 = 66, with 2 left over

Therefore, in a set of 200 coins, there are 66 full sets of coins, followed by $1 and 50¢.

Each full set of three coins has a value of
$1 + 50¢ + 25¢ or $1.75.

Total value = 66 × $1.75 + $1 + 50¢
= $115.50 + $1 + 50¢
= $117

The total is $117.

Example 3

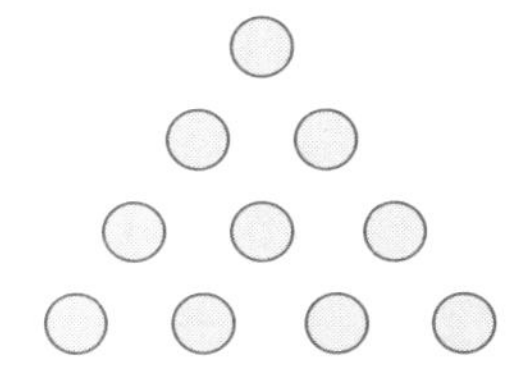

Paul is a window designer. A store manager asked him to create a triangular display of balloons, with 1 on the top, 2 in the next row, 3 in the next row, and so on.

How many balloons does Paul need for a triangle 10 rows high?

Solution

$1 + 2 + 3 + 4 + 5 + 6 + 7 + 8 + 9 + 10 = 55$
Paul needs 55 balloons.

Example 4

The manager from Example 3 also wants a giant balloon triangle, 40 rows high, to be displayed inside the store. How many balloons will Paul need for this display?

Solution

It would take a long time to solve this problem by adding the number of balloons in each row. Paul notices that, in the small triangle, he can match small rows with larger ones to make equal sets of balloons.

$1 + 2 + 3 + 4 + 5 + 6 + 7 + 8 + 9 + 10$

$3 + 8 = 11$

$2 + 9 = 11$

$1 + 10 = 11$

As long as the number of rows is even, Paul can find the number of balloons by adding the smallest row to the largest and then multiplying the sum by the number of row pairs.

In a 40-row display, the smallest row has 1 balloon and the largest row has 40. Paul can match rows to make sets of 41.

Since there are 40 rows, Paul can make 20 row pairs.

$20 \times 41 = 820$

To create a display with 40 rows, Paul will need 820 balloons.

Exercises

1. A pattern can be formed in two ways — by repeating the same pattern stem over and over or by changing the pattern stem in a repetitive way. Give an example of a repeating pattern and a changing pattern.

2. In the expression $n + 3$, n is called a *variable* and 3 is called a *constant.* What do these terms mean?

3. Describe each pattern rule and then use the rule to complete the pattern.

(a) 7, 8, 10, 13, 17, ☐, ☐, ☐

(b) 100, 98, 94, 88, 80, ☐, ☐, ☐

(c) 10, 10, 20, 60, 240, ☐, ☐, ☐

(d) 1024, 512, 256, 128, 64, ☐, ☐, ☐

(e) ZY, AB, XW, CD, VU, ☐, ☐, ☐

(f) $3, 3^2, 3^3, 3^4, 3^5$, ☐, ☐, ☐

(g) ☐, 4^3, ☐, 4^7, ☐, 4^{11}

(h) ☐, $n + 4$, ☐, $n + 2$, ☐, $n + 0$

(i) 121, 1221, ☐, ☐, ☐, 12222221

(j) A, 1, BC, 2, DEF, 3, ☐, ☐, ☐

4. Apple seeds are arranged in this formation.
(a) What relationship can you find between the figure number and the number of seeds?

Figure 1 Figure 2 Figure 3

(b) How many seeds are needed to make the next figure?

(c) What expression could you use to find the number of seeds for any figure?

(d) How many seeds would you need to make Figure 50?

5. Jamilla has some pennies. Pete has five more than triple Jamilla's number. Write an expression you could use to find out how many pennies Peter has if you knew how many Jamilla has.

6. If p is an odd number, is $5p - 3$ odd or even? Explain.

7. A moving company charges $150 for each hour of labor. It pays its workers $100 per person for each job completed. Write an expression you can use to find:
(a) the cost of h hours of labor.

(b) the amount the company will have to pay w workers for one job.

(c) the amount the company can expect to earn on a job.

8. One hundred coins are arranged in this pattern. What is the value of the 100th coin?

10¢ 10¢ 25¢ 25¢ 25¢ 5¢ 10¢ 10¢ 25¢ 25¢ 25¢ 5¢

9. If a month begins on a Tuesday, can the 28th day of the same month be on a Wednesday? Explain.

10. Eight teams are competing in a chess tournament. How many rounds must there be if each team is to play each other team once?

11. Write a mathematical expression.

(a) the product of 4 and n

(b) John's age in 1999, if he was n years old in 1987

(c) the number of hours Marlene spends at school if she arrives at x a.m. and leaves at y p.m.

(d) the quotient of n and 16

(e) the perimeter of a picture frame measuring 20 in. by w in.

12. A coin sorter accepts only quarters, dimes, or nickels. Write an expression you can use to find the total value of the coins in the sorter, no matter how many there are of each type. Your expression should give the value in dollars, not cents.

13. Jamal has six photographs. Each one measures 3 in. × 5 in. How many different-sized rectangular displays can Jamal create by combining all six photographs? Include diagrams with your answer.

14. Create a pattern with numbers or shapes. Write a mathematical expression to match your pattern. Write two other expressions that do not match. Exchange patterns with a classmate to identify the correct expression. Use the expression to predict another part of the pattern.

5.2 Graphing Relations

In this lesson, you learned to interpret a table of values, graph a relation, analyze the graph, and use the graph to predict values. Recall:

- A relation connects two or more sets of numbers or quantities. You can show a relation in a table or on a graph.
- The horizontal axis on a graph is called the x-axis. The vertical axis is called the y-axis.
- Estimating a related value between two known values is called **interpolation.** To interpolate a value, first extend a line to the graph from a known coordinate on one axis. Then read the other coordinate of the point where the line meets the graph.
- Estimating a value beyond the known values is called **extrapolation.** To extrapolate a value, extend the graph, and then use the same procedure you used to interpolate.

Example 1

The graph represents a car trip from Cambridge to Washington.

Find the distance traveled after 2.5 h.

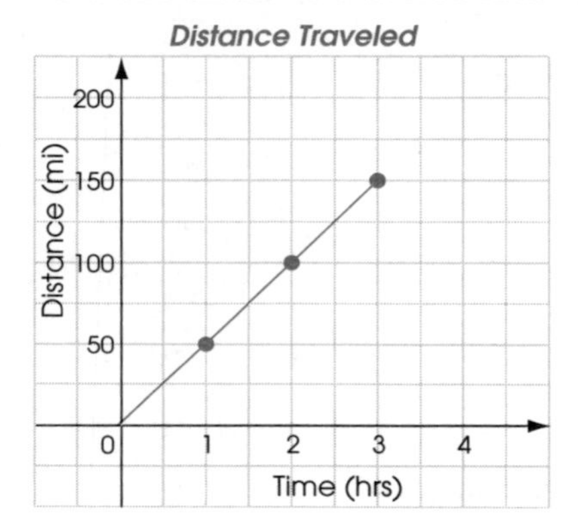

Solution

The ordered pairs are (*time, distance*).

At the point where the 2.5 h line meets the graph, the y-coordinate shows the distance driven.

The dotted lines show that (2.5, 125) is a point on the graph. In 2.5 h, the car has traveled 125 mi.

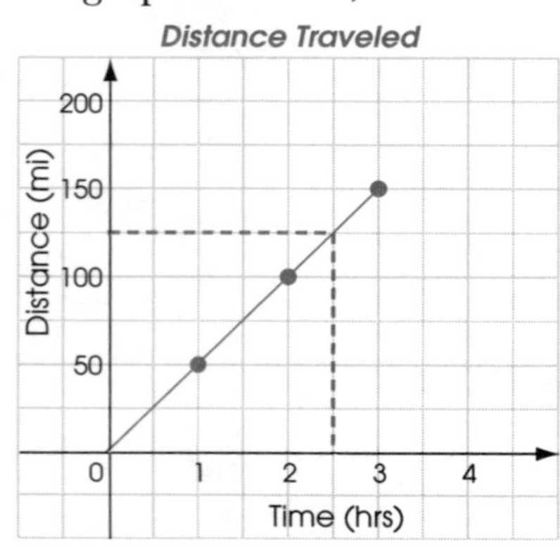

Example 2

How far will the car from Example 1 travel in 5 h?

Solution

The ordered pairs are (*time, distance*).

Extend the graph until it passes 5 h on the x-axis.

At the point where the 5 h line meets the graph, the y-coordinate shows the distance driven.

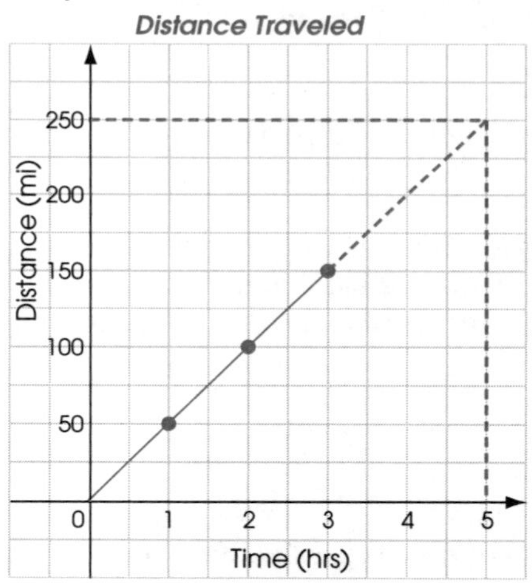

The dotted lines show that (5, 250) is a point on the extended graph. In 5 h, the car will travel 250 mi.

Exercises

1. What are some advantages and disadvantages of showing a relation in each of these ways?

   (a) table of values

   (b) graph

2. The graph shows how fast a hot air balloon rises after take-off.

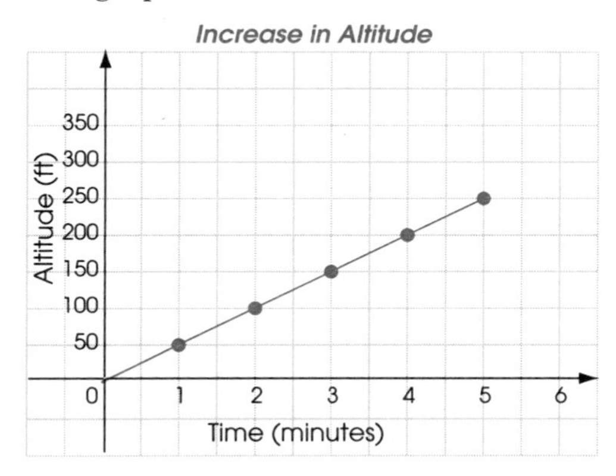

   (a) How high is the balloon after 4.5 min?

   (b) How long does it take the balloon to reach an altitude of 125 ft?

3. The graph shows the amount of water used to water a lawn.

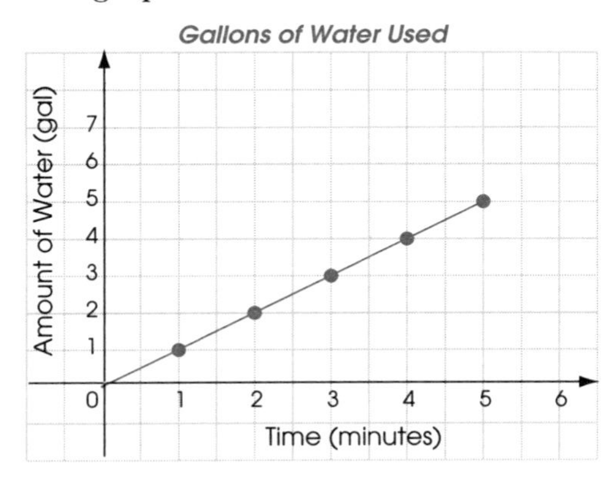

   (a) How much water was used in 1.5 min?

   (b) How long will it take to use up 5.5 gal of water?

4. Amanda has a part-time job at a bagel shop.

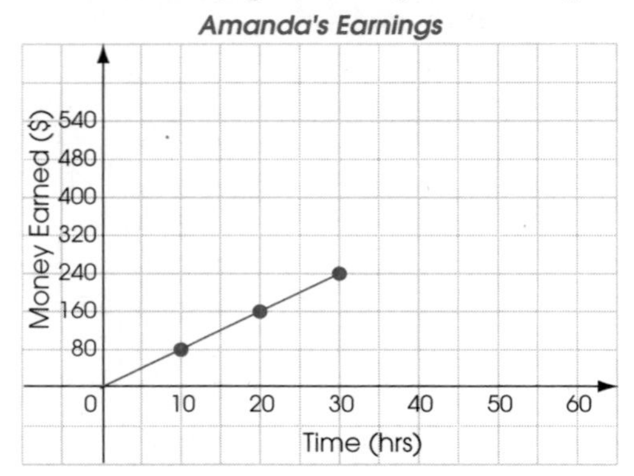

(a) How long would it take Amanda to earn $480?

(b) What is Amanda's hourly wage?

5. This graph represents a car traveling down a road. Describe what the graph tells you about the trip.

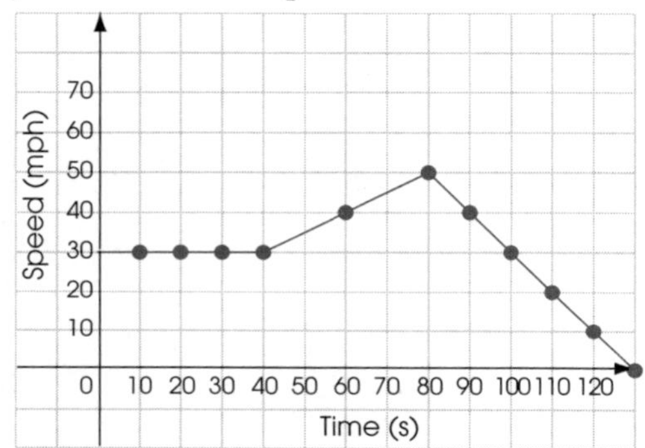

6. The distance a skier lands from the end of the jump depends on the height of the start. Graph the data. Since the length of the jump depends on the starting height, plot starting heights along the x-axis. Be sure to title your graph and label the axes.

Starting Height (ft)	Length of jump (ft)
5	10
15	30
25	50
35	70

7. A bus travels from El Paso to Dallas. The table shows the total time elapsed and the total distance traveled at each stop.

Towns	Total Time (min)	Total Distance (mi)
El Paso	0	0
Pecos	314	209
Odessa	420	285
Midland	460	307
Abilene	642	459
Dallas	878	644

(a) Graph the relationship between time and distance. Plot the time along the x-axis.

(b) Between which two towns does the bus seem to be moving at the fastest speed? Explain your choice.

(c) What else would you need to know to confirm your choice in part (b)? (Hint: If the difference in speed is difficult to see on your graph, try changing the scales.)

5.3 Writing and Evaluating Expressions

In this lesson, you learned how to write algebraic expressions to solve problems, and to simplify expressions by substituting values for variables. Recall:

- Mathematical expressions are made up of **variables** and **constants**.
- **Variables** are letters or symbols that represent numbers. **Constants** are numbers that do not change.
- Algebraic expressions are often used to describe and extend patterns.

Example 1

Complete the table for the given values of p.

p	$3p - 5$
0	-5
1	
2	
3	
4	

Solution

Substitute each given number for p in the expression.

For 1:

$$3p - 5 = 3(1) - 5 = 3 - 5 = -2$$

For 2:

$$3p - 5 = 3(2) - 5 = 6 - 5 = 1$$

You can see a pattern in the results:

Each time the value for p increases by 1, the value for $3p - 5$ increases by 3. Use this pattern to complete the rest of the table.

p	$3p - 5$
0	-5
1	-2
2	1
3	4
4	7

Check the results you predicted by substituting the values for p into the expression.

$$3p - 5 = 3(3) - 5 = 9 - 5 = 4$$

$$3p - 5 = 3(4) - 5 = 12 - 5 = 7$$

Example 2

There are w red triangles in a diagram; write an expression for the number of black triangles in terms of w.

Solution

Make a table of values.

Red Triangles	Black Triangles
1	0
2	1
3	2
4	3

The number of black triangles is always one less than the number of red ones. The expression for the number of black triangles is $(w - 1)$.

Example 3

What mathematical expression shows the relationship between the number on the left and the one on the right?

t	?
1	3
2	5
3	7
4	9
5	11

Solution

Each time t increases by 1, the partner number increases by 2. Part of the operation must be $\times 2$.

If you multiply each value for t by 2, the product is always one less than the partner number. The operation must be $\times 2 + 1$.

The expression that describes this relationship is $2t + 1$.

Exercises

1. What are three different ways to represent a number pattern? Give examples.

2. Describe the steps you use to simplify an expression when you know the value of the variable. Give an example.

3. Complete the table for the given values of d.

d	$4d + 2$
1	6
2	
3	
4	
5	
6	

4. Complete the table for the given values of s.

s	$3s - 9$
−2	−15
−1	
0	
1	
2	
3	
4	

5. Simplify each expression if $a = 3$.

(a) $a + 6$

(b) $10 - 4a$

(c) $6a$

(d) $3a - 10$

(e) $3a + 3$

(f) $(-5) - a$

(g) $10 + (-a)$

(h) $7 - 2a$

6. Simplify each expression if $x = 3$ and $y = -4$.

(a) $x + y$

(b) $4x - 2y$

(c) $2x + y$

(d) $2(x + y)$

(e) $x + 2y$

(f) $5x - y$

(g) $5x + 4y$

(h) $3(y - x)$

7. Find and correct any errors in the solution.

Problem
Simplify $4n - n$ if $n = 3$.

Solution

$$\begin{aligned} 4n - n &= 4 \times 3 - 3 \\ &= 4(0) \\ &= 0 \end{aligned}$$

8. Find an expression to describe this relationship.

Then complete the table.

m	?
5	16
10	
15	
20	76
25	96
30	116

9. **(a)** Make a table of values to show the relationship between the number of squares and the number of triangles.

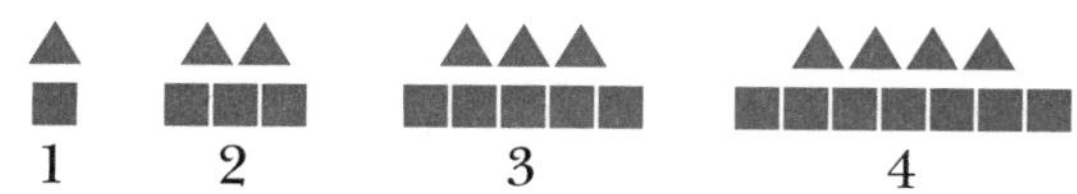

(b) Write an expression for the number of triangles if there are s squares.

10. Make a table of values to show the relationship between the diagram number and the number of triangles.

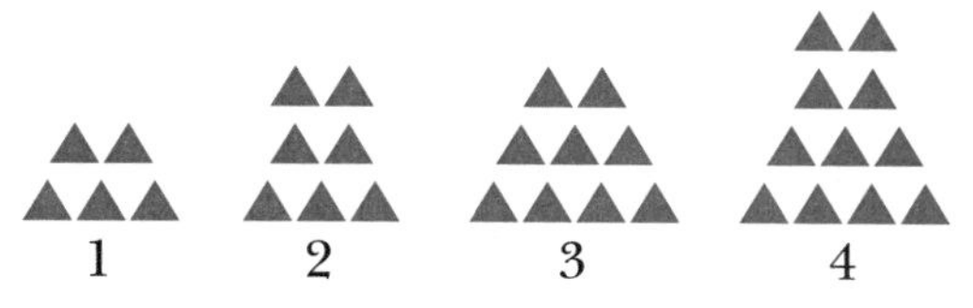

(a) Write an expression for the number of triangles in diagram number n.

(b) Use your expression to find out how many triangles will be in diagram 8.

11. Write an expression with one or two operations and use it to create a table of values. Exchange tables with a classmate to identify each other's expression.

12. Create a pattern with one or two shapes so the number of shapes increases in the same way in each successive diagram. Write an expression that describes your pattern.

Trade with a classmate and try to identify each other's pattern.

5.4 Solving Equations by Addition and Subtraction

In this lesson, you learned to use models and number sentences to **solve addition and subtraction equations**. Recall:

- An equation is like a balanced scale. The value of the right side must always **equal** the value of the left side.
- To keep the equation balanced, you need to perform matching operations on both sides.
- The symbol that represents the unknown in an equation is called a **variable**.
- To solve an equation, add or subtract numbers on both sides in a way that will help you **isolate** the variable on one side. The **solution** will be on the other side.
- To check a solution, replace the variable in the original equation with the solution. If the right side equals the left side, then the solution is correct.

Example 1

Three less than a number is five. Find the number.

Solution

Step 1: Write the equation.

Let n be the unknown number.
Three less than a number is five.
3 less than $n = 5$
$n - 3 = 5$

Make a tile model.

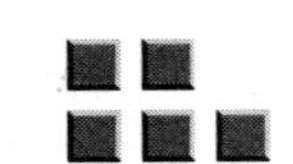

n + 3 negative tiles = 5 positive tiles

Step 2: Isolate the variable on one side.

Add three positive tiles to both sides.

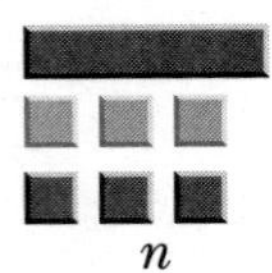

n = 8 positive tiles

The value of n is +8.

Step 3: Check the solution.

The question stated that three less than a number is five.
Three less than eight is five.
Therefore, 8 is the correct solution.

Example 2

Shaina takes the temperature outside. If it were five degrees warmer, the temperature would be negative eight degrees. What is the temperature?

Solution

Step 1: Write the equation.

Let t be the unknown temperature.
If it were five degrees warmer, the temperature would be negative eight degrees.

$$temperature + 5 = -8$$
$$t + 5 = -8$$

Step 2: Isolate the variable on one side.

$$t + 5 = -8$$
$$t + 5 - 5 = -8 - 5$$
$$t = -13$$

The temperature is $-13°$.

To isolate t on the left side, you have to subtract 5. To keep the equation balanced, you must also subtract 5 on the right side. Simplify both sides.

Step 3: Check the solution.

Substitute $t = -13$ into the original equation.

L.S.	R.S.
$t + 5$	-8
$=(-13) + 5 = -8$	

The left side equals the right side, so the solution to this equation is correct.

To make sure the equation was written correctly, compare with the original problem.

The problem says that if it were five degrees warmer, the temperature would be $-8°$. The solution is correct since $-13° + 5° = -8°$.

Exercises

1. Why is an equation like a balanced scale?

2. Explain how you could use a bag of marbles on a scale to solve the equation $x + 3 = 5$.

3. Use or draw tiles to model each equation. Write the solution.

(a) $x + 3 = 5$

(b) $t + (-4) = 6$

(c) $a + 3 = -8$

(d) $x - 5 = -15$

(e) $n + 3 = -10$

4. Use a method of your choice to solve each equation.

(a) $x + (-3) = 5$

(b) $y + 4 = -6$

(c) $p - 3 = 2$

(d) $d - (-5) = -9$

(e) $n - 12 = 0$

(f) $n - 13 = -20$

(g) $t - 3 = -4$

(h) $t + (-12) = -11$

5. Create two different tile models for each solution.

(a) $x = +4$

(b) $m = -2$

6. Write two different equations that have each solution.

(a) $n = +6$ **(b)** $n = -7$

7. Check each solution.

(a) Problem: When you increase a number by 4, the result is 12.
Solution: The number is 8.

(b) Problem: Five less than a number is –7.
Solution: The number is 2.

(c) Problem: Four more than a number is –3. Solution:
The number is –1.

(d) Problem: A number decreased by 5 is –8.
Solution: The number is –3.

8. Check each solution.

(a) The solution to $x - 6 = -3$ is –3.

(b) The solution to $p - 3 = 3$ is 0.

(c) The solution to $b + 3 = -8$ is –5.

(d) The solution to $m + 2 = -12$ is –14.

(e) The solution to $y - 3 = -21$ is –18.

(f) The solution to $s - 8 = 20$ is 12.

9. Five less than Elena's age is twelve. How old is Elena?

10. Find each number.

(a) When you increase a number by six, the result is negative five.

(b) A number decreased by three gives negative twelve.

(c) A number decreased by fifteen gives nine.

11. If the morning temperature increases by five degrees, it will reach two degrees. Find the morning temperature.

12. A plane descends 100 ft to a height of 1500 ft above the ground. What was the height of the plane when it began its descent?

13. After midnight, the temperature decreased 12°F to a low of –28°F. What was the temperature at midnight?

14. Create an equation word problem. Write the solution on another sheet of paper. Exchange problems with a classmate and solve.

15. On a separate sheet of paper, create three different equation word problems where the solution is 0. Check your problems by solving them.

16. Which types of equations do you find most difficult to solve? Why? Compare answers with a classmate. Can you offer any suggestions to each other?

5.5 Solving Equations by Division and Multiplication

In this lesson, you learned to use models and number sentences to **solve multiplication and division equations**. Recall:

- To keep the equation balanced, you need to perform matching operations on both sides.
- To solve an equation, multiply or divide the quantities on both sides in a way that will help you **isolate** the variable.
- To isolate the variable, you need to **undo** the operation in the equation. For a multiplication equation, undo by dividing. For a division equation, undo by multiplying.
- To check a solution, replace the variable in the original equation with the solution. If the right side equals the left side, then the solution is correct.

Example 1

The product of 8 and a number is 56. Find the number.

Solution

Step 1: Write the equation.

Let n represent the number.
The product of 8 and a number is 56.
The product of 8 and n is 56.

$$8 \times n = 56$$
$$8n = 56$$

Step 2: Isolate the variable on one side.

Isolate n on one side.
In the equation, n is multiplied by 8.
To undo this operation, divide by 8.

$$8n = 56$$
$$8 \times n \div 8 = 56 \div 8$$
$$n = 7$$

Divide both sides by 8 to keep the equation balanced.
Simplify both sides.

Step 3: Check the solution.

Substitute $n = 7$ into the original equation.

L.S.	R.S.
$8n$	56
$= 8 \times 7 = 56$	

The left side is equal to the right side, so the solution to this equation is correct.

To make sure the equation was written correctly, compare with the original problem. The problem states that the product of 8 and a number is 56. The solution is correct, since $8 \times 7 = 56$.

Example 2

When you divide a number by 7, the result is 3. Find the number.

Solution

Step 1: Write the equation.

Let n represent the number.
A number divided by 7 = 3

$$n \div 7 = 3$$
$$\frac{n}{7} = 3$$

Step 2: Isolate the variable on one side.

In the equation, n is divided by 7.
To undo this operation, multiply by 7.

$$\frac{n}{7} = 3$$
$$\frac{n}{7} \times 7 = 3 \times 7$$
$$n = 21$$

Multiply both sides by 7 to keep the equation balanced. Simplify both sides.

Step 3: Check the solution.

Substitute $n = 21$ into the original equation.

L.S.	R.S.
$\frac{n}{7}$	3
$= \frac{21}{7} = 3$	

The left side is equal to the right side, so the solution to this equation is correct.

To make sure the equation was written correctly, compare with the original problem. The problem states that when you divide a number by 7, the result is 3. The solution is correct, since $21 \div 7 = 3$.

Exercises

1. Define each term. Give an example.

(a) variable

(b) equation

(c) solution

2. Explain each phrase. Give an example.

(a) isolate the variable

(b) check by substitution

3. Leslie was asked to solve the equation $4n = 16$. In order to isolate n on one side, she multiplied both sides of the equation by 4. Was this the correct operation? Explain.

4. What mathematical operation would you use to isolate each variable? Give the operation sign and the number.

(a) $5n = 15$

(b) $6x = 18$

(c) $\frac{n}{9} = 3$

(d) $6n = -30$

(e) $\frac{n}{-4} = 12$

(f) $\frac{x}{5} = -8$

(g) $-6n = 36$

(h) $\frac{n}{5} = 20$

5. Solve and check.

(a) $6n = 42$

(b) $8x = 32$

(c) $9x = 54$

(d) $4n = 28$

(e) $-6n = 48$

(f) $9x = -36$

6. Solve and check.

(a) $\frac{n}{5} = 6$

(b) $\frac{n}{3} = 5$

(c) $\frac{x}{7} = 3$

(d) $\frac{x}{4} = -9$

(e) $\frac{n}{-5} = 7$

(f) $\frac{x}{8} = 4$

7. Solve and check.

(a) $4n = 28$

(b) $5n = -35$

(c) $-16 = \frac{x}{4}$

(d) $27n = 108$

(e) $\frac{n}{6} = 8$

(f) $15n = 75$

(g) $\frac{x}{5} = -20$

(h) $\frac{n}{40} = 4$

Write an equation to solve Problems 8 to 12. Show your work.

8. When a number is multiplied by 4, the result is 56. Find the number.

9. When a number is divided by 7, the quotient is 6. Find the number.

10. The perimeter of an equilateral triangle is 45 cm. Find the length of each side.

11. The perimeter of a square is 44 in. Find the length of each side.

12. In a skating competition, Mark scored 56 points. This was one-third as many points as were scored by the winning skater. How many points did the winner have?

13. Karla created a computer spreadsheet and programmed it to perform a mystery operation on any numbers that were entered. To test her spreadsheet, she entered these numbers. The results are shown. What was the mystery operation?

Input	Output
–5	-1
–4	–0.8
–3	–0.6
–2	–0.4
–1	–0.2
0	0
1	0.2
2	0.4
3	0.6
4	0.8
5	1

14. When Karla entered another number into her spreadsheet, the result was 10. What number did she enter?

15. Create an equation word problem you can solve using multiplication or division. Write the solution. Then rewrite the solution on another sheet of paper so it contains an error. Exchange problems with a classmate and find the errors.

16. Create three different equation word problems, each involving multiplication or division, where the solution is –5. Check your problems by solving them.

5.6 Substitution and Relations

In this lesson, you learned to evaluate an expression by **substituting a number for the variable**. For example, to evaluate $2x - 3$ for $x = 5$:

$2x - 3$	$= 2(5) - 3$	Substitute 5 for x.
	$= 10 - 3$	Multiply first and then subtract.
	$= 7$	The value of $2x - 3$ is 7 when $x = 5$.

The number you substituted (5) and the result (7) are connected by the **relation** $2x - 3 = 7$. To record a pair of related numbers, you can use:

- a row in a table of values, where the first column represents the number you substituted for the variable.
- an ordered pair where the first number is the one you substituted.

Example 1

To remodel her basement, Jenna needs two different types of drywall nails. Each long nail costs \$0.07 and each short nail costs \$0.05. What will it cost to buy 110 long nails and 85 short ones?

Solution

Step 1

Write an expression to represent the cost of an unknown number of nails.
Let l represent the number of long nails.

Let s represent the number of short nails.

$$0.07l + 0.05s$$

Step 2

Substitute known values for variables.
The problem asks for the cost of 110 long nails and 85 short ones. Substitute 110 for l and 85 for s in the expression $0.07l + 0.05s$.

$$0.07(110) + 0.05\,(85) = 7.70 + 4.25$$
$$= 11.95$$

It will cost \$11.95 to buy 110 long nails and 85 short ones.

Example 2

A car rental agency charges \$35 per day plus 9¢ per mile driven. Make a table of values to show how much it would cost to rent a car for one day to drive 100 mi, 200 mi, 300 mi, 400 mi, and 500 mi. Graph the relationship.

Solution

Step 1: **Write an expression to show the cost of driving an unknown number of miles.**
Let d represent the distance driven. Since you will be adding the flat rate to the cost per mile, you need to express both prices in dollars.

$$35 + 0.09d$$

Step 2: **Substitute known values for variables.**
Substitute the values 100, 200, 300, 400, and 500 for d in the expression $35 + 0.09d$ and calculate each cost. Create a table of values or a list of ordered pairs. In either case, the value you substituted for d should come first.

Number of miles (d)	Total cost (c)	Ordered Pairs (miles, cost)
100	44	(100, 44)
200	53	(200, 53)
300	62	(300, 62)
400	71	(400, 71)
500	80	(500, 80)

Step 3: **Graph the relationship.**
Graph the numbers you substituted for d along the horizontal axis.

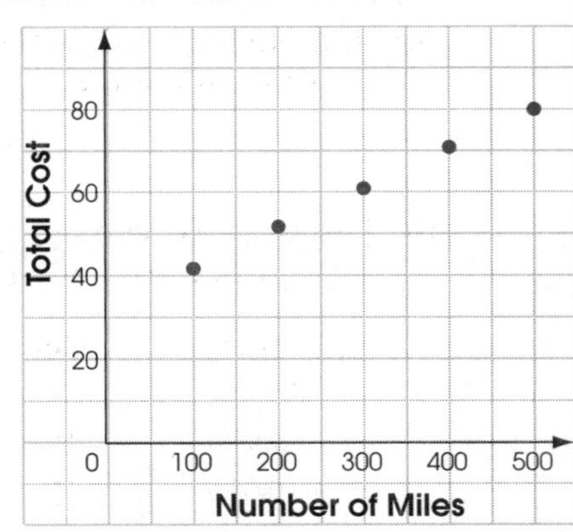

Exercises

1. Define each term.

(a) expression

(b) variable

(c) substitute

(d) evaluate

(e) ordered pair

2. Evaluate for $x = 7$.

(a) $3x$ **(b)** $-5x$ **(c)** $4x + 2$ **(d)** $x^2 + 1$

3. Evaluate $3g + 4h - 2$ for each pair of values.

(a) $g = 1$ and $h = 5$ **(b)** $g = -3$ and $h = 2$

4. Evaluate each expression for $a = 3$ and $b = 5$.

(a) $-4a$ **(b)** $5b - 7$ **(c)** $a^2 + 3$ **(d)** $b^2 + a$

5. A hockey team earns 2 points for a win and 1 point for a tie.

(a) Find an expression for the total number of points a team can earn in any number of games.

(b) Use your expression to calculate the total number of points each team would earn for the games represented in the table.

Team	Wins	Ties
Cougars	8	3
Saints	12	2
Tigers	15	8
Robins	22	4

6. Chris sells onion rings and fries at a snack bar. Onion rings cost $1.40 and fries cost $1.25.

   (a) Chris often fills orders for groups of office workers. Write an expression he can use to calculate the cost of any order if everyone orders either fries or onion rings.

   (b) After a soccer game, Chris sold 25 orders of onion rings and 30 orders of fries. How much did the customers pay in all?

7. Lani and Sam want to book the Elbow Falls golf course for their wedding reception. The dining room charges a $200 flat rate plus $20 per guest. Calculate the cost of the reception if there are:

   (a) 100 guests

   (b) 150 guests

   (c) 200 guests

8. The distance in feet traveled by an object is equal to $s \times t$, where s is the object's speed in feet per second and t is the time in seconds.

   (a) In June of 1992, Martin Brundle drove a Jaguar XJ220 at a speed of 318.06 ft/s, setting a record for the fastest speed ever attained by a standard production car. At this speed, how far could the car travel in 10 s?

   (b) What was the speed of the Jaguar in miles per hour?

9. Rona sells magazine subscriptions. She earns \$100 per week, plus \$3 for each subscription she sells.

(a) Complete the table of values.

Number of Subscriptions (s)	Dollars Earned in One Week (d)
5	115
10	
15	
20	

(b) Make a graph to show the relationship.

10. Fran works at a printer's, putting spiral bindings on books. She binds about 30 books per hour. When she first began working for the company, she was offered her choice of pay plan from this list:

Plan 1: \$8.75 per hour

Plan 2: \$4.75 per hour plus 13¢ for each book bound

Plan 3: 30¢ for each book bound

(a) Write an expression to determine how much money Fran will make in a week with each plan, if she works a different number of hours each week.

(b) Which pay plan do you think Fran chose? Why?

5.7 Translating Written Phrases

In this lesson, you learned how to:
- use numbers and variables to **translate a verbal phrase** into a mathematical expression or equation.
- **rewrite a mathematical expression** as a verbal phrase.
- **write an expression or equation** to reflect a problem situation.

Example 1

Write a mathematical expression that means "three times a number, increased by five."

Solution

Use the variable n to represent the unspecified number.

Words	three times	a number	increased	by five
Expression	$3 \times$	n	$+$	5

The mathematical expression is $3 \times n + 5$ or $3n + 5$.

Example 2

The length of a rectangular yard is four feet less than double the width. Write an equation you could use to find the length of the yard if you knew the width.

Solution

Step 1

Identify the variables.
Let w represent the width of the yard.
Let L represent the length of the yard.

Step 2
Rewrite the problem information in the form of a verbal equation.
Place the information to be calculated on one side of the = sign, and the expression you will use to calculate it on the other.

Length of yard = four feet less than double the width

Step 3
Rewrite the left side of the equation using a variable.
Length of yard = L

Step 4
Rewrite the right side of the equation.
Multiplication is done before addition, so you can change the order of the phrase.

four feet less than double the width = double the width minus four feet

Now use mathematical symbols to rewrite the expression.

Words	double	the width	minus	four feet
Expression	$2 \times$	w	$-$	4

The expression for the right side is $2 \times w - 4$ or $2w - 4$.

Step 5

Write the entire equation.
Left side: L

Right side: $2w - 4$
To find the length of the yard for a given width, you could use the equation $L = 2w - 4$.

Exercises

1. Define each term.

   (a) variable

   (b) expression

   (c) equation

2. List several phrases associated with each operation.

   (a) $+$ (b) $-$ (c) $\div$ (d) $\times$

3. Write the mathematical expression. Use the variable n to represent the unknown number.

   (a) A number is increased by sixteen.

   (b) Five is decreased by a number.

   (c) Two times a number is added to nine.

   (d) The product of six and a number is reduced by twelve.

   (e) The quotient of a number divided by ten is reduced by five.

   (f) Double a number is diminished by eighteen.

   (g) The sum of a number and eleven is divided by three.

   (h) Four is subtracted from the product of seven and a number.

   (i) Thirteen is added to triple a number.

   (j) Fourteen minus twice a number is divided by nine.

4. Describe each expression in words.

(a) $p - 6$

(b) $7 + k$

(c) $3m$

(d) $\frac{c}{14}$

(e) $2n + 8$

(f) $\frac{a}{4} - 7$

(g) $\frac{g+2}{6}$

(h) $5 - 9x$

(i) $11 + 3s$

(j) $\frac{4n+10}{5}$

5. Write the equation. Use the variable n to represent the unknown number.

(a) A number reduced by fourteen is twenty-six.

(b) The difference between nine times a number and five is fifteen.

(c) Nine increased by the product of sixteen and a number results in ten.

(d) The quotient of a number divided by six is equal to the same number increased by twelve.

(e) Triple a number increased by seventeen gives the sum of the same number and twenty.

6. Describe the equation in words.

(a) $r + 5 = 9$

(b) $7x - 4 = 12$

(c) $\frac{x+5}{3} = 2x$

(d) $3n + 5 = \frac{n}{2} + 22$

(e) $12k - 3 = 10k + 11$

7. Write a mathematical expression for each phrase.
   (a) An object's mass, m, is tripled.

   (b) Henry's age in years, y, is doubled and then increased by nine.

   (c) The amount in your bank account, a, will decrease to this after a fifty dollar withdrawal.

   (d) Your age, x, will increase to this in seven years.

   (e) The length of a race, r, is cut in half.

8. Tony has fifty coins, all pennies and nickels, in his pocket. If p represents the number of pennies, write an expression for:
   (a) the number of nickels

   (b) the total value of all the coins

9. Alex is three years older than triple his son's age. If his son's age is s, write an expression for:
   (a) Alex's age

   (b) the sum of their ages

10. A printing company hired to print a yearbook charges a \$200 flat rate, plus \$1.50 per book.
    (a) Write an equation you can use to calculate the cost of printing any number of yearbooks.

    (b) How much would it cost to print 500 yearbooks?

11. Create a mathematical expression. Describe your expression in words in three different ways. Then exchange with a classmate and compare the results.

12. Look for examples of symbols or signs that are used to communicate a concept, idea, or phrase in a non-verbal way. Invite classmates to guess the meaning of each symbol.

6 LINEAR EQUATIONS

6.1 Exploring Equations

In this lesson, you learned to use the *Algebra Tile Explorer* to **model, solve,** and **check equations.**

Example 1

Use the *Algebra Tile Explorer* to solve $-x - 4 = -2$.

Step 1

Enter the equation.

Drag a red x-tile and four red unit tiles to the left side.

Drag two red unit tiles to the right side.

$-x - 4 = -2$

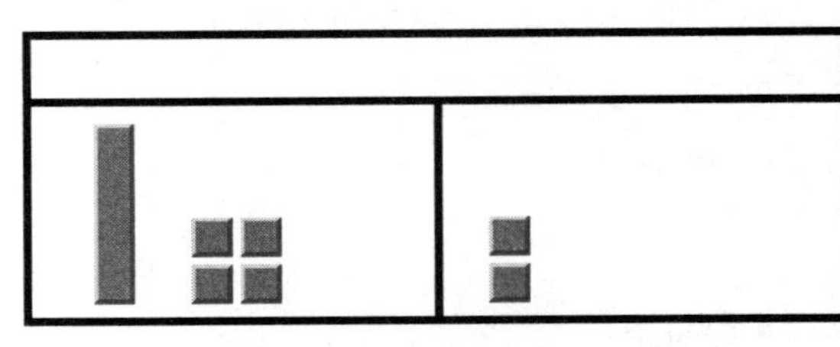

Step 2

Isolate the x-tile.

Recycle the four red unit tiles from the left side.

$-x = +2$

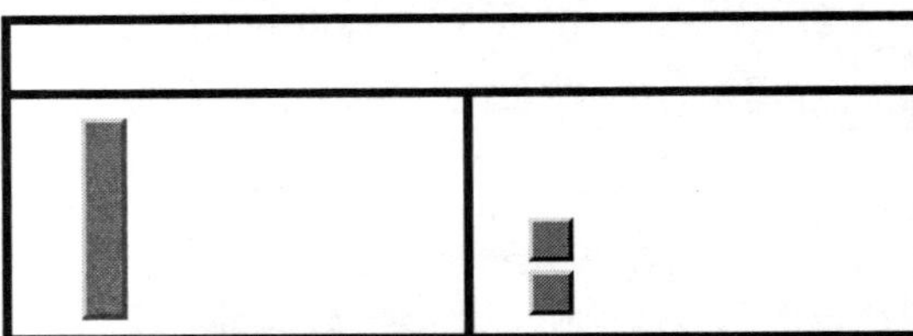

Step 3

Make the x-tile positive.

Multiply both sides by -1.

$x = -2$

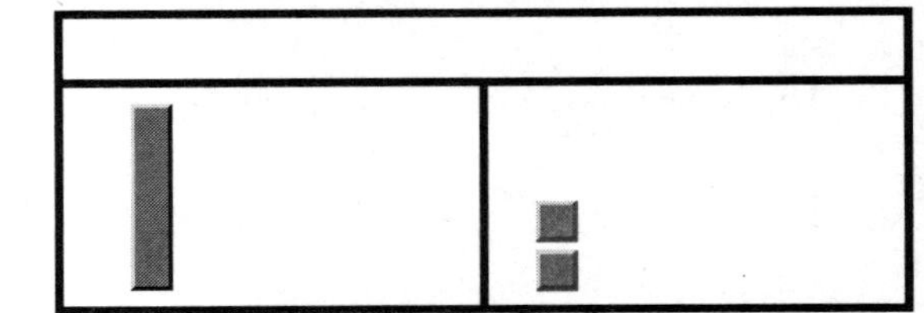

Step 4

Check the solution.

Select CHECK SOLUTION to see if substituting -2 for x will balance the equation.

L.S.	R.S.
$-x - 4$	-2
$= -(-2) - 4$	
$= 2 - 4$	
$= -2$	

Both sides are equal to -2, so the solution $x = -2$ is correct.

Example 2

Use the *Algebra Tile Explorer* to solve $3 = 5 - 2x$.

Step 1

Enter the equation.

Drag three black unit tiles to the left side.
Drag five black unit tiles and two red *x*-tiles to the right side.

$3 = 5 - 2x$

Step 2

Isolate the *x*-tiles.

Recycle the five black unit tiles from the right side.

$-2 = -2x$

Step 3

Make the *x*-tiles positive.

Multiply both sides by -1.

Group the tiles.

$2 = 2x$

Step 4

Simplify.

$1 = x$

Step 5

Check the solution.

Substitute 1 for x in the equation.

L.S.	R.S.
3	$5 - 2x$
	$= 5 - 2(1) = 3$

The solution is correct.

Exercises

1. Whenever you recycle tiles from one side of the *Explorer*, a matching operation happens on the other side.
   (a) Why does this happen?

   (b) When you subtract black tiles, and there are no matching tiles on the other side, why does the *Explorer* add red tiles instead?

2. Explain what is meant by the phrase "isolate the variable."

3. Draw a picture to show how you would use algebra tiles to model the expression $-3x + 4$. Color negative tiles red and positive tiles black.

4. When you use the GROUP function on the *Explorer*, which mathematical operation does it perform? Explain.

$+ \qquad - \qquad \times \qquad \div$

5. Write the equation.

(a)

(b)

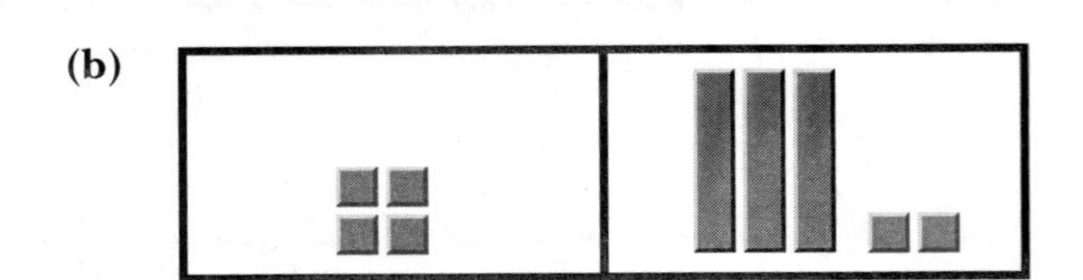

6. Draw a tile model to represent the equation $3x = 2$.

7. Use the *Algebra Tile Explorer* to enter each equation. Draw a picture of the resulting screen.

(a) $7 = 4x$ **(b)** $-3x + 2 = 4$

(c) $5x = -8$ **(d)** $2x - 2 = -5$

(e) $x - 4 = 6$ **(f)** $4 - 6x = 8$

8. Use the *Algebra Tile Explorer* to solve and check each equation.

(a) $2x - 5 = 3$

(b) $6 = -5x - 4$

(c) $8x = 3 - 7$

(d) $-6x = 30$

(e) $-6x = 42 - 6$

(f) $7 = 4 - 7x$

9. Use the *Algebra Tile Explorer* to enter and check each equation. Correct any errors in the solutions.

(a) $24 = 16x - 8$ if $x = 2$

(b) $5 + 6x = 19$ if $x = 3$

(c) $26 - 2x = 14$ if $x = 12$

(d) $3x + 11 = 21$ if $x = \frac{10}{3}$

10. Find and correct any errors in the solution to this equation.

$3x - 6 = 6$	Recycle six red unit tiles on each side.
$3x = 0$	Simplify.
$x = 0$	Therefore, this equation will balance if $x = 0$.

11. Create three equations so that each one involves one or two operations. Predict the solutions. Then use the *Algebra Tile Explorer* to check.

6.2 Linear Equations 1

In this lesson, you learned to solve equations algebraically by **isolating the variable** on one side of the equal sign. Whenever you perform an operation on one side of the equation, you need to perform the same operation on the other side to keep the values balanced.

You have also learned to **verify** your work by substituting the value you found for the variable into the original equation.

Example 1

When you subtract 56 from a number, the result is 124. Find the number.

Solution

$n - 56 = 124$	Write an equation to show the problem situation. Let n represent the unknown number.
$n - 56 + 56 = 124 + 56$	Add 56 to both sides to isolate n on the left side.
$n = 180$	Simplify both sides.

Verify by substituting 180 for n in the original equation.

L.S.	R.S.
$n - 56$	124
$= 180 - 56$	
$= 124$	

The left side is equal to the right side, so 180 is the correct solution for this equation.

Example 2

Two-thirds of a number is 14. Find the number.

Solution

Write an equation to show the problem situation.

$\frac{2}{3}n = 14$	Let n represent the unknown number.
$\frac{2}{3}n \div \frac{2}{3} = 14 \div \frac{2}{3}$	Divide both sides by $\frac{2}{3}$ to isolate n on the left side.
$\frac{2}{3}n \times \frac{3}{2} = 14 \times \frac{3}{2}$	To divide by $\frac{2}{3}$, multiply by the reciprocal, $\frac{3}{2}$.
$n = \frac{42}{2}$	Simplify.
$n = 21$	Express the fraction in simplest form.

Verify by substituting 21 for n in the original equation.

L.S.	R.S.
$\frac{2}{3}n$	14
$= \frac{2}{3}(21)$	
$= 14$	

The left side is equal to the right side, so 21 is the correct solution for this equation.

Exercises

1. Use each word in a sentence that explains its mathematical meaning. Give an example.

   (a) variable

   (b) verify

   (c) isolate

   (d) substitute

2. Explain what it means to solve an equation. Create an example equation with one operation and show how you would solve it.

3. Describe three situations where it would be useful to know how to solve an equation.

4. Explain why it is important to verify your solution to every equation. Write an example equation with one operation. Show your solution and your verification.

5. What operation would you perform on both sides to solve the equation?

   (a) $c + 8 = 37$

   (b) $d - (-5) = -7$

   (c) $p - 5 = 12$

   (d) $v - 8 = 27$

   (e) $t + (-4) = 11$

   (f) $36 + s = 181$

6. Solve and verify.

   (a) $m + 6 = 19$

   (b) $d + (-3) = -8$

   (c) $q - 4 = 12$

   (d) $b - (-5) = -11$

   (e) $x - 15 = -16$

   (f) $-3 + z = 16$

7. By what number would you divide both sides to solve the equation?

(a) $6x = 12$ **(b)** $2x = -8$ **(c)** $7z = -14$

(d) $11r = 22$ **(e)** $6n = -18$ **(f)** $8y = 64$

8. Solve and verify.

(a) $3x = 15$ **(b)** $2y = -12$ **(c)** $-15 = 5p$

(d) $2a = 24$ **(e)** $-25 = 5j$ **(f)** $-7x = -21$

9. By what number would you multiply both sides to solve the equation?

(a) $\frac{n}{3} = 9$ **(b)** $\frac{n}{2} = -4$ **(c)** $\frac{y}{7} = 4$

(d) $\frac{m}{5} = -6$ **(e)** $\frac{t}{4} = -1$ **(f)** $\frac{n}{6} = 0$

10. Solve and verify.

(a) $\frac{x}{2} = 2$ **(b)** $\frac{y}{2} = -4$ **(c)** $\frac{x}{3} = 7$

(d) $-4 = \frac{p}{3}$ **(e)** $\frac{m}{4} = 28$ **(f)** $-3 = \frac{n}{5}$

(g) $\frac{s}{10} = 5$ **(h)** $-2 = \frac{t}{12}$ **(i)** $\frac{u}{11} = 4$

11. Solve and verify.

(a) $\frac{k}{5} = 2.4$ **(b)** $\frac{r}{2} = -1.5$ **(c)** $-4x = -2.8$

(d) $1.6n = 4.8$ **(e)** $4 = \frac{m}{15}$ **(f)** $-8.4 = 0.2x$

12. Solve and verify.

(a) $\frac{2}{3}n = 25.6$ **(b)** $-\frac{4}{5}k = 56$ **(c)** $\frac{x}{5} = -17.5$

(d) $\frac{3}{7}h = -30$ **(e)** $n - 2 = 8.8$ **(f)** $\frac{3}{2}z = 13.5$

(g) $8.8n = 70.4$ **(h)** $-\frac{9}{10}m = -99$

13. One-third of the sum of the side lengths of an equilateral triangle is 4.2 cm. What is the equation you can use to find the sum of the side lengths? Solve and verify.

$3x = 4.2$ $\frac{1}{3}x = 4.2$ $\frac{1}{3}x + 4.2 = 0$ $x = \frac{1}{3}(4.2)$

14. The average yearly snowfall in Amarillo, Texas, is about 15 in. This is about one-sixth of the average yearly snowfall in Buffalo, New York.

Write an equation you can use to find the average snowfall in Buffalo. Solve and verify.

15. At 217 lb, a white-tailed deer is about 1.4 times heavier than an average cougar.

Write an equation you can use to find the weight of an average cougar. Solve and verify.

16. What is the result if you multiply both sides of an equation by 0?

17. Describe the solutions to these equations.

(a) $3x = 0$ **(b)** $0x = 0$ **(c)** $0x = 2$

18. Write an equation you could solve by using each operation. Exchange with a classmate and solve. Make sure each equation has only one operation.

(a) addition **(b)** subtraction

(c) multiplication **(d)** division

6.3 Linear Equations 2

In the previous lesson, you learned to solve equations with one operation. Here, you used the same steps to solve equations with two operations, such as $2n + 4 = 8$.

- First, undo each operation in order to **isolate** the **variable** on one side of the equal sign.
- Remember, whenever you perform an operation on one side of the equation, you must perform the same operation on the other side to keep the values balanced.
- After you have solved the equation, **verify** your work by substituting the value you found into the original equation.

Example 1

Solve the equation $8.2 = 3x - 1.4$.

Solution

$8.2 + 1.4 = 3x - 1.4 + 1.4$	Add 1.4 to both sides to isolate the variable term $3x$.
$9.6 = 3x$	Simplify both sides.
$\frac{9.6}{3} = \frac{3x}{3}$	Divide both sides by 3 to isolate x.
$3.2 = x$	Simplify both sides.

Verify by substituting 3.2 for x in the original equation.

L.S.	R.S.
8.2	$3x - 1.4$ $= 3(3.2) - 1.4$ $= 8.2$

The left side is equal to the right side, so 3.2 is the correct solution for this equation.

Example 2

Solve the equation $\frac{9k}{2} - 6 = 21$

Solution

$\frac{9k}{2} - 6 + 6 = 21 + 6$	Add 6 to both sides to isolate the variable term $9k$.
$\frac{9k}{2} = 27$	Simplify both sides.
$\frac{9}{2}k \div \frac{9}{2} = 27 \div \frac{9}{2}$	Rewrite the left side to separate the variable from the fraction. Divide both sides by $\frac{9}{2}$ to isolate k.
$\frac{9}{2}k \times \frac{2}{9} = 27 \times \frac{2}{9}$	
$k = 6$	To divide by $\frac{9}{2}$, multiply by the reciprocal, $\frac{2}{9}$.

Verify by substituting 6 for k in the original equation.

L.S.	R.S.
$\frac{9k}{2} - 6$ $= \frac{9(6)}{2} - 6$ $= \frac{54}{2} - 6$ $= 21$	21

The left side is equal to the right side, so 6 is the correct solution for this equation.

Exercises

1. How is solving a two-step equation like solving a one-step equation? How is it different?

2. Describe an everyday situation that you could represent with a two-step equation. Write the equation.

3. Match each equation with the correct phrase. Then solve the equation.

Equations	*Phrases*
(a) $\frac{x}{3} + 22 = 4$	**(i)** Four less than three times a number is twenty-two.
(b) $3x - 4 = 22$	**(ii)** Four times a number, diminished by three, is twenty-two.
(c) $4x - 22 = 3$	**(iii)** Twenty-two less than four times a number is three.
(d) $22 = 4x - 3$	**(iv)** Twenty-two less than a number divided by three is four.
(e) $4 = \frac{x}{3} - 22$	**(v)** Four is twenty-two more than one third of a number.

4. Write and solve the equation.
   (a) Six more than one-fifth of a number is twenty-six.

   (b) Negative forty-one is eight less than three times a number.

   (c) One thousand one hundred five is eight times a number, reduced by seventy-two.

   (d) One and five tenths times a number less ten is negative twenty-two.

(e) Two thirds of a number increased by seven is eighty-four.

(f) A number divided by negative three and two tenths is forty-six.

5. Solve and verify.

(a) $2b + 1 = 7$

(b) $23 = 4m + 7$

(c) $4k + 5 = 21$

(d) $3n + 2 = 14$

(e) $22 = 4 + 3m$

(f) $2k + 8 = 12$

6. Solve and verify.

(a) $2b - 3 = 9$

(b) $21b + 12 = 54$

(c) $2j + 7 = 11$

(d) $68 = 2k - 8$

(e) $24 = 5x + 4$

(f) $17 = 3x - 19$

7. Solve and verify.

(a) $\frac{k}{7} - 9 = 40$

(b) $72 = \frac{s}{8} + 8$

(c) $-\frac{m}{6} + 4 = 6$

(d) $18.3 - \frac{c}{2} = 10.3$

(e) $-8-\frac{n}{4}=24$

(f) $-78=\frac{b}{3}-48$

8. Solve and verify each equation. Give a written explanation of each step in your solution, as in Examples 1 and 2.

(a) $3k+1.2=4.2$

(b) $104=-10.3x+1$

(c) $\frac{w}{3}+3=6$

(d) $-\frac{3n}{2.4}-6=42$

9. Write the equation. Then solve and verify.

(a) A taxi ride costs $2.75 plus $1.25 per mile driven. How far can you travel for $25?

(b) Noria paid $18 to rent a canoe for the day. If the rental company charges a flat fee of $6, plus an hourly charge of $2, how long did Noria keep the canoe?

(c) Rick and Jeff both work at the community center. One week, Rick worked for 5 h less than Jeff. The sum of their hours was 28. For how many hours did each person work?

10. Create a problem you can solve using a two-step equation. Write the solution and then rewrite it so it contains an error. Exchange problems with a classmate and find the errors.

6.4 The Coordinate Plane

This lesson introduced the four-quadrant coordinate plane.

Recall:

- You can use an ordered pair of numbers to locate any point on the grid. The first number, called the *x*-**coordinate**, indicates the **horizontal** distance from the origin. The second number, or *y*-**coordinate**, indicates the **vertical** distance from the origin.
- When you are locating or naming a point, always count across the *x*-axis first and then count up or down.
- Points along a horizontal line all have the same *y*-coordinate. If the *y*-coordinate is 0, the line is the *x*-axis.
- Points along a vertical line all have the same *x*-coordinate. If the *x*-coordinate is 0, the line is the *y*-axis.

Example 1

Point *B* is located 5 units to the left of the origin and 2 units below it. In which quadrant is point *B* located?

Solution

To identify the quadrant, start at the origin and then move left and down.

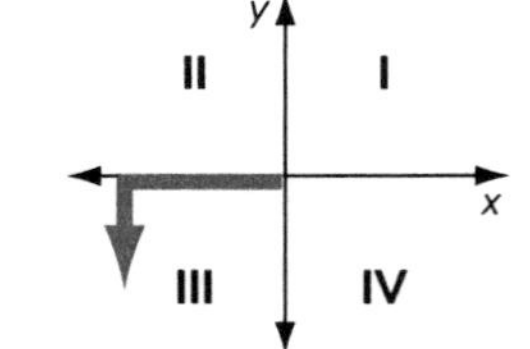

The point is located in quadrant III.

Since the point is in quadrant III, the signs of both coordinates must be negative.

To find the exact coordinates, start at the origin again. Move 5 squares left to –5. This is the *x*-coordinate.

Now move 2 squares down to –2. This is the *y*-coordinate.

Point *B* is located at (–5, –2) in quadrant III.

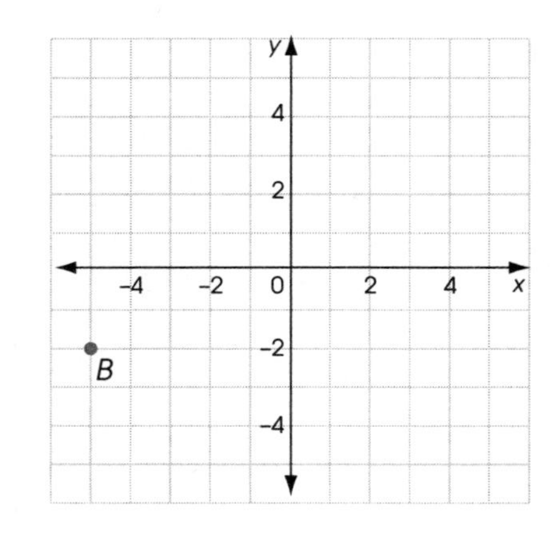

Example 2

A rectangle has vertices *A*(–2, 4), *B*(2, 4), *C*(2, 1) and *D*(–2, 1). Find the area of the rectangle.

Solution

Step 1
Plot the vertices on the coordinate plane.

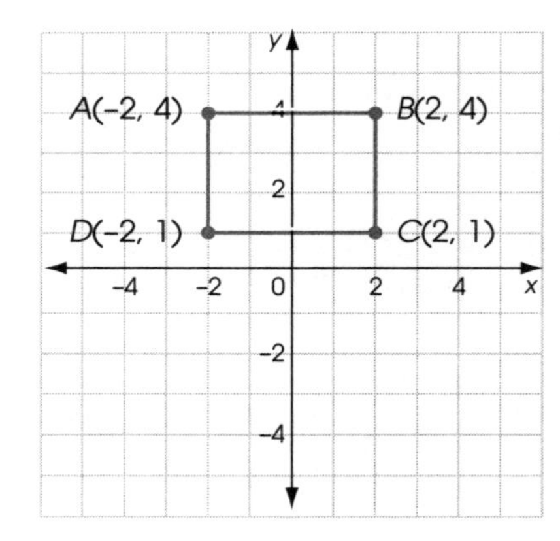

Step 2
Find the area of the rectangle.

The rectangle is 4 units wide and 3 units high.

Area = *Length* × *Width*
= 4 × 3
= 12 square units

The area of the rectangle is 12 square units.

Exercises

1. Choose any term and explain how it relates to another term. Then relate the second term to a third. Continue until you have used all the terms. Relate the last term to the one you started with. Compare your results with a classmate's.

origin	*ordered pair*	*quadrant*
y-coordinate	*coordinate plane*	*x-coordinate*
x-axis	*y-axis*	*point*
horizontal line	*vertical line*	*intersection*

2. State the coordinates of each point.

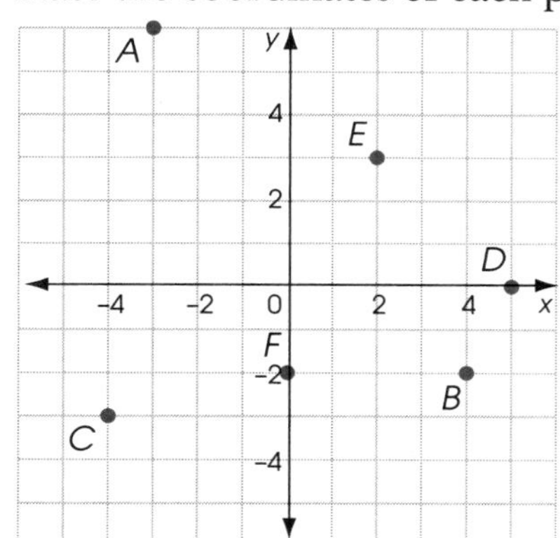

3. Plot each point on a coordinate plane.
$A(-3, 6)$ $B(2, 7)$ $C(1, -4)$
$D(-5, 3)$ $E(2, 0)$ $F(0, -3)$

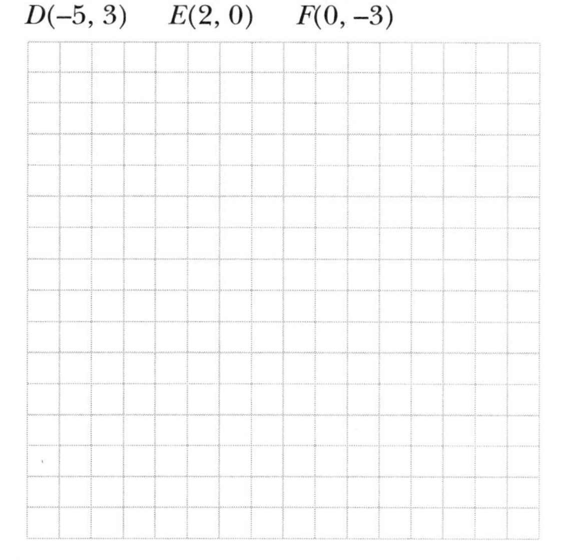

4. In which quadrant is each point located?
$A(2, -4)$ $B(5, 4)$ $C(-7, -9)$ $D(-1, 10)$

5. Select the point that fits each description.

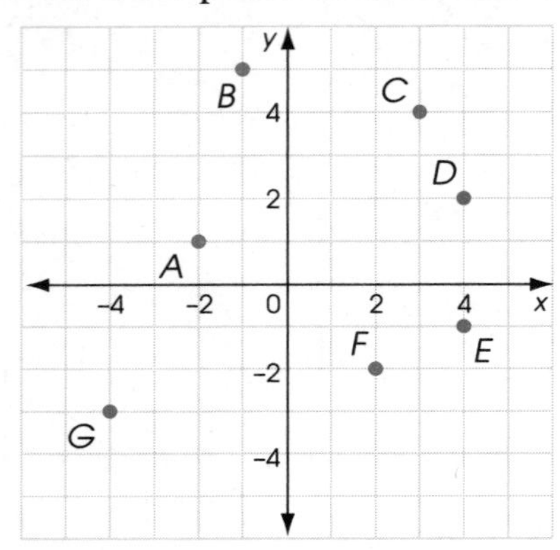

(a) The point is 3 units below the x-axis in quadrant III.
(b) The x-coordinate of the point is two greater than the y-coordinate.

6. Organize these points on the chart:
$A(0, 4)$, $B(-2, 3)$, $C(-3, -2)$, $D(-2, -3)$, $E(-3, 0)$

Description	Points
located on the x-axis	
located on the y-axis	
located along the same horizontal line	
located along the same vertical line	

7. State the coordinates of each point.
Start at the origin each time.
(a) Go left 4 units and up 2 units.

(b) Go right 3 units and down 5 units.

(c) Go left 1 unit to a point on the x-axis.

8. Two vertices of a square are located at $A(0, 0)$ and $B(0, 6)$. What are the possible locations for the other two vertices?

9. A code uses ordered pairs to form letters. Decode the message to find a mystery word.
Join (−7, 1) to (−7, 3) to (−6, 2) to (−5, 3) to (−5, 1).
Join (−4, 1) to (−2, 3) to (0, 1).
Join (−3, 2) to (−1, 2).
Join (1, 3) to (1, 1) to (3, 1).
Join (4, 3) to (4, 1) to (6, 1).

10. Diagonally opposite corners of a rectangle are $A(-4, 3)$ and $C(3, -2)$.

(a) What is the area of the rectangle?

(b) Which quadrant contains the largest portion of the rectangle?

11. Points $A(0, 0)$ and P represent opposite vertices of a rectangle whose area is 32 square units.

(a) Give one set of possible coordinates for P.

(b) Move point P to another location so the area remains 32 square units. Explain why you chose the point you did.

12. Draw a simple design on a coordinate grid so each vertex of the design is at a point where grid lines intersect. Label each vertex with a letter and an ordered pair.

Sit back-to-back with a partner. Describe your design and have your partner plot each point on a coordinate grid. Compare your partner's results with your own. Then repeat with your partner's design.

6.5 Graphing Linear Equations

In this lesson, you combined your knowledge of **substitution** and solving **equations** to graph linear equations. You learned:

- to graph linear equations using **tables of values**.
- to find x- and y-intercepts.
- to graph using x- and y-intercepts.
- to graph equations of the form $x = a$ and $y = b$.
- to identify **dependent** and **independent variables**.

Example 1

Graph $y = 2x + 1$ using a table of values.

Solution

In the equation $y = 2x + 1$, x is the independent variable and y is the dependent variable.
To determine ordered pairs for the graph, substitute some different values for x to find the corresponding values of y.

For $x = 1$
$$\begin{aligned} y &= 2x + 1 \\ &= 2(1) + 1 \\ &= 2 + 1 \\ &= 3 \end{aligned}$$

For $x = -2$
$$\begin{aligned} y &= 2x + 1 \\ &= 2(-2) + 1 \\ &= (-4) + 1 \\ &= -3 \end{aligned}$$

For $x = 0$
$$\begin{aligned} y &= 2x + 1 \\ &= 2(0) + 1 \\ &= 0 + 1 \\ &= 1 \end{aligned}$$

Make a table of values to show the ordered pairs.

x	y	Ordered Pair (x, y)
1	3	(1, 3)
–2	–3	(–2, –3)
0	1	(0, 1)

Plot the ordered pairs on a coordinate plane and join the points with a straight line.

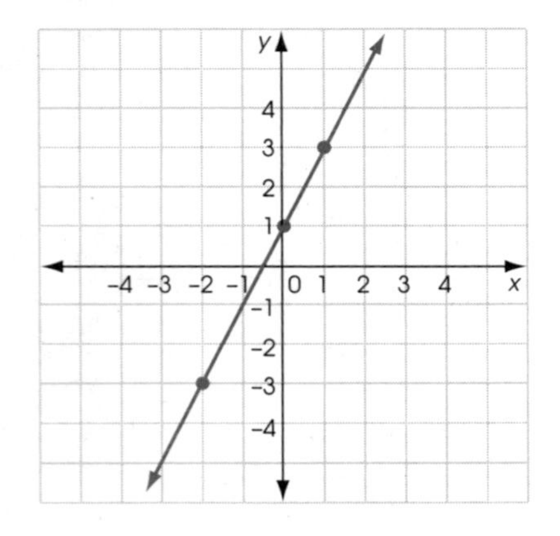

Example 2

Find the intercepts for $4x - 3y = 12$. Then graph the equation.

Solution

The intercept of a graph is the point where the graph crosses the x- or y-axis.

To find the x-intercept, let $y = 0$, and solve for x.
$$\begin{aligned} 4x - 3y &= 12 \\ 4x - 3(0) &= 12 \\ 4x - 0 &= 12 \\ 4x &= 12 \\ x &= 3 \end{aligned}$$
When y is 0, x is 3, so the x-intercept is at (3, 0).

To find the y-intercept, let $x = 0$, and solve for y.
$$\begin{aligned} 4x - 3y &= 12 \\ 4(0) - 3y &= 12 \\ 0 - 3y &= 12 \\ -3y &= 12 \\ y &= -4 \end{aligned}$$
When x is 0, y is –4, so the y-intercept is at (0, –4).

Find a third point as a check.
Let $y = -2$.
$$\begin{aligned} 4x - 3(-2) &= 12 \\ 4x + 6 &= 12 \\ 4x &= 6 \\ x &= 1.5 \end{aligned}$$

When y is –2, x is 1.5, so point (1.5, –2) is on the line.
Plot points (0, –4), (1.5, –2), and (3, 0).

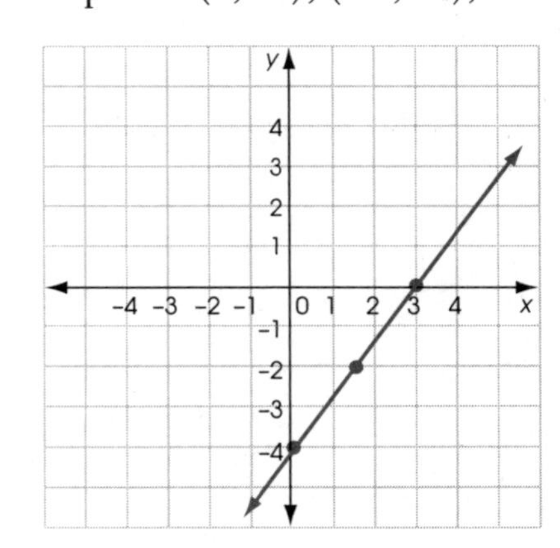

Exercises

1. Complete each statement.

(a) The graph of a linear equation is a ____________________.

(b) The point where the graph of a linear equation crosses the x-axis is called the

____________________.

(c) In the equation $y = 7x + 2$, x is called the ____________________ variable and y is called the ____________________ variable.

(d) The graph of the equation $y = 3$ is a ____________________ line.

(e) The graph of the equation $x = -4$ is a ____________________ line.

2. Use the graph to answer each question.

(a) What is the y-intercept?

(b) What is the x-intercept?

(c) Does the line pass through the point (4, 3)?

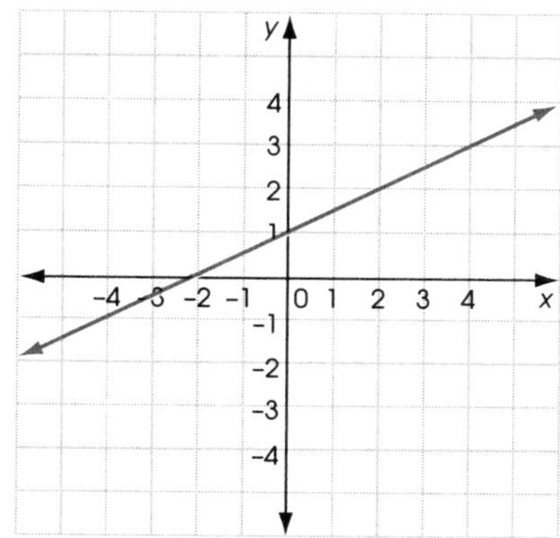

3. Complete the table of values. Graph each equation.

(a) $2x - 5y = 10$

x	y	Ordered Pair (x, y)
–5		
0		
5		

(b) $y = -2x + 3$

x	y	Ordered Pair (x, y)
–2		
3		
4		

4. Find the x- and y-intercepts and the coordinates of one checkpoint. Graph each equation.

(a) $x + y = -6$

(b) $x - y = 3$

(c) $4x + 5y = 20$

(d) $3x - 5y = 15$

5. Make a table of values to show three solutions for each equation. Use your table to graph the equation.

(a) $y = -x - 2$

(b) $3x + 2y = 2$

(c) $y = 3x$

(d) $x = -4$

(e) $y = x - 2$

(f) $y = x$

6. The distance covered by a train traveling at 45 mph is given by the equation $d = 45t$, where d is the distance in miles and t is the time in hours. Graph the equation to show the relationship between distance and time.
(Hint: Plot the independent variable, t, on the horizontal axis.)

7. The coast guard is using radar to track two ships. Their paths are given by the equations $y = -2x + 3$ and $3y = -x - 6$. Graph the two equations to determine whether the two ships will pass through the same point. If so, identify the point.

8. Rita works as a salesperson. Her pay is found using the equation $P = 15 + 8h$, where P is the total amount paid and h is the number of hours worked. Luis works in an office and his pay is found using the equation $P = 13h$.

(a) After how many hours will Rita and Luis be paid the same amount?

(b) How much will they be paid at this point?

9. Create an equation that you think will produce a straight line. Use one or two variables, and do not include any powers.

Test your equation by drawing the graph.

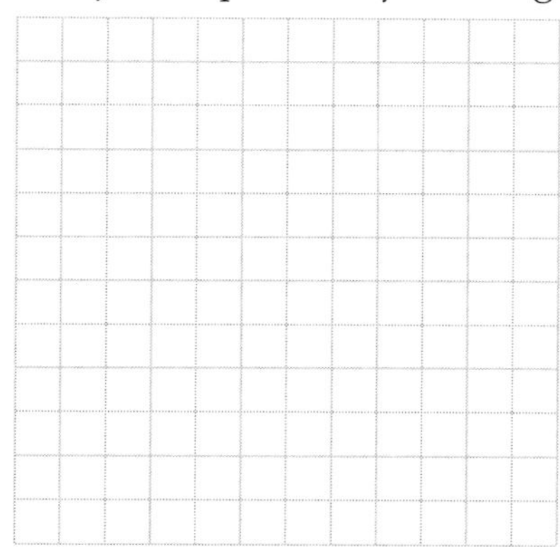

7 POLYNOMIALS

7.1 Terms of Polynomials

You have explored how algebra can be used to **express numbers** and to **describe relationships** between them. You have learned how to:

- identify **parts of a polynomial** (term, coefficient, variable, constant).
- identify **types of polynomials** (monomial, binomial, trinomial).
- **model polynomials** using algebra tiles.
- **write polynomials** based on algebra tile models.

Example

(a) Write the polynomial that these algebra tiles represent.

(b) Complete the table by classifying parts of the polynomial.

Terms	
Coefficients	
Variables	
Constants	

Solution

(a) These tiles are used in the model. Let [tile] represent x and [tile] represent y.

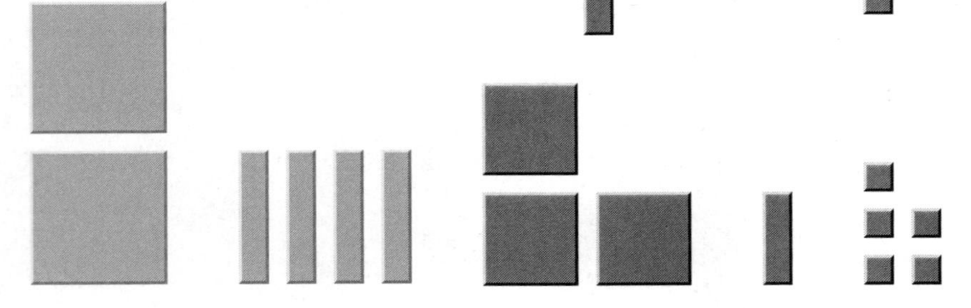

$2 \times (-x^2)$ $4 \times (-x)$ $3 \times (y^2)$ $1 \times y$ 5 unit tiles

The polynomial is $-2x^2 - 4x + 3y^2 + y + 5$.

(b) These are the parts of the polynomial.

Terms	$-2x^2$, $-4x$, $3y^2$, y, 5
Coefficients	−2, −4, 3, 1
Variables	x, y
Constants	5

Exercises

1. **(a)** What is a polynomial?

   (b) Write an example of a monomial, a binomial, and a trinomial.

2. Define each of the following in your own words.

   (a) term

   (b) coefficient

   (c) variable

   (d) constant

3. Write a list of steps that would help someone derive a polynomial from any tile model.

4. For the polynomial $-3 + 4x - 5x^2$, identify:

   (a) terms

   (b) coefficients

   (c) variables

   (d) constants

5. In the polynomial $5a + 3cd + 6$, give the algebraic name for each of these parts.

   (a) 6

   (b) 5 and 3

   (c) $5a$, $3cd$, and 6

6. Classify each polynomial as a monomial, a binomial or a trinomial. Justify your answers.

   (a) $-7a^5b^3$

   (b) $8f - 4g$

   (c) $6a + 4bc - 5$

   (d) $-xy + 3z$

7. What polynomial could each tile model represent?

(a)

(b)

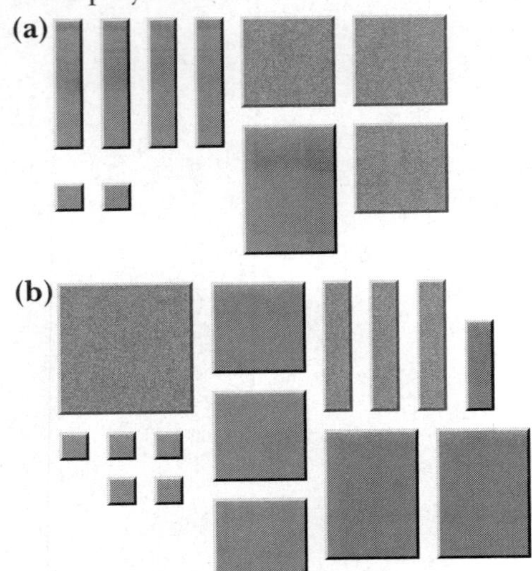

8. Martin is a cab driver. His cab meter starts at \$3 and charges an additional \$0.50 for each mile driven. To represent the total fare, you can use the formula:
$f = 3 + 0.5d$

(a) What do f and d represent in the formula?

(b) For the cab fare formula, list the:

(i) variables

(ii) coefficients

(iii) terms

(iv) constants

9. Classify the polynomial $3 + 0.5d$ as a monomial, binomial, or trinomial. Justify your answer.

10. Use algebra tiles or draw a tile model to represent each polynomial:

(a) $4x - 5$

(b) $-3x - 5y - 2$

(c) $2x^2 + 3x + 1$

(d) $-4y^2 + 4y + 6$

(e) $x^2 - 2xy + 3y^2$

(f) $x^2 + xy - x + y + 2y^2 - 4$

11. Build or draw a tile model of a polynomial that fits each description:

(a) three terms, two variables, coefficients 4 and 5, constant 3

(b) two terms, two variables, coefficients 3 and 2

12. You can estimate the height of a person in centimeters by measuring the length, l, of the humerus bone (shoulder to elbow) and using one of these calculations.

male height $= 2.89l + 70.64$

female height $= 2.75l + 71.48$

(a) Classify the parts of each polynomial as variables, coefficients, terms, and constants.

(b) Classify each polynomial as a monomial, a binomial, or a trinomial. Justify your answer.

(c) Have a partner measure your humerus bone. Use the formula to estimate your height. Check your estimate by measuring your actual height.

13. Create a polynomial involving fractions. One fraction has a variable in the denominator and another is a constant.

14. A word web shows connections among words. For example:

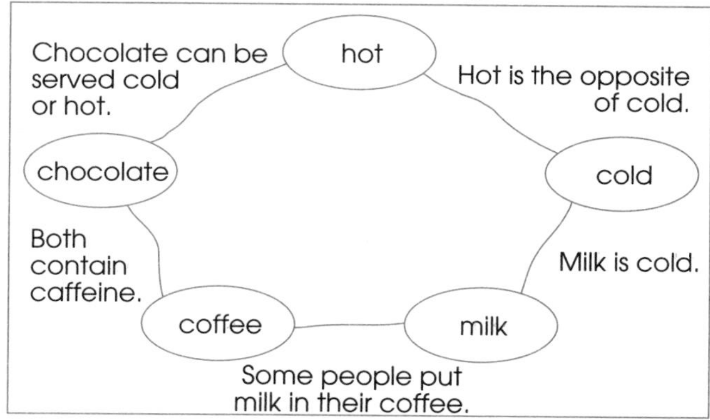

Create a word web based on this word list: polynomial, term, coefficient, variable, constant, algebra, monomial, binomial, trinomial, algebra tiles. Justify your arrangement by explaining your web to a classmate.

7.2 Evaluating Polynomials

You have seen how to **evaluate polynomials** by substituting numbers for variables:
- **Substitute** integers, decimals, and fractions for variables.
- Evaluate polynomials derived from **problem situations.**

Example 1

When you throw an object into the air, you can use the number of seconds of upward travel (t) to estimate the maximum height it reaches in meters (h). This formula shows how to make the estimate:

$$h = -5t^2 + 18.5t + 1.8$$

Use the formula to calculate the maximum height a baseball will reach if it travels upward for 2.5 s.

Solution

Substitute 2.5 s for t in the formula.

$$\begin{aligned} h &= -5t^2 + 18.5t + 1.8 \\ &= -5(2.5)^2 + 18.5(2.5) + 1.8 \end{aligned}$$

Then simplify the expression using the rules for order of operations.

$$\begin{aligned} h &= -5(6.25) + 18.5(2.5) + 1.8 && \text{Exponents} \\ &= -31.25 + 46.25 + 1.8 && \text{Multiplication} \\ &= 16.8 && \text{Addition} \end{aligned}$$

The ball would reach a maximum height of 16.8 m.

Example 2

A vehicle's highway fuel consumption, f, represents the number of gallons of fuel the vehicle needs to travel 100 mi on the highway. When only part of the driving is done on the highway, you can estimate fuel consumption for some cars using an algebraic formula. The variable d represents the fraction of driving that was done on the highway.

$$f = \frac{-36}{5}d + \frac{29}{3}$$

Calculate a car's fuel consumption in gal/100 mi if $\frac{1}{2}$ of the driving was on the highway. Express the result as a decimal.

Solution

Substitute $\frac{1}{2}$ for d in the formula.

$$\begin{aligned} f &= \frac{-36}{5}d + \frac{29}{3} \\ &= -\frac{-36}{5}\left(\frac{1}{2}\right) + \frac{29}{3} \end{aligned}$$

Then simplify the expression using the rules for order of operations.

$$\begin{aligned} f &= -\frac{18}{5} + \frac{29}{3} && \text{Multiplication} \\ &= -3.6 + 9.66 && \text{Division} \\ &= 6.06 && \text{Addition} \end{aligned}$$

The car's highway fuel consumption is about 6.06 gal/100 mi.

Exercises

1. **(a)** Define the terms *variable* and *constant*.

   (b) Why do you think these terms are used?

2. Describe a situation in which you might want to substitute different values into an equation.

3. List the rules for order of operations.

4. Match each expression on the left with the equivalent expression on the right for $x = 2$.

   (a) $4x^3$ **(i)** $3^2 + 3^2 + 3^2 + 3^2$

   (b) $4(3^x)$ **(ii)** $4^3 + 4^3$

   (c) $x(4^3)$ **(iii)** $2^3 + 2^3 + 2^3 + 2^3$

5. Evaluate.

   (a) $-x^4 - 6$, for $x = -3$

   (b) $5 + 3h - h^3$, for $h = -2$

   (c) $ac^2 + b^3$, for $a = -4$, $b = -3$, and $c = 2$

   (d) $5 - 7x + 4x^2$, for $x = 1.5$

   (e) $y^2 - y^3$, for $y = -2.5$

   (f) $3 - mn^1$, for $m = -4$ and $n = 2$

   (g) $30x^2 + 4$, for $x = -\frac{1}{2}$

   (h) $18y^3 - z$, for $y = \frac{1}{3}$ and $z = -\frac{1}{6}$

   (i) $3 + a^1 - b^2$, for $a = 8$ and $b = 4$

6. A paper dart is thrown from a window. The formula describing the path of the dart is $h = t^2 - 3t + 4$, where h is the height of the starting point in yards, and t is the flight time in seconds. If the dart is in flight for 4 s, how high is the starting point?

7. The formula for finding the sum of a series of numbers is $S = \frac{n}{2}(a + b)$, where S is the sum of the numbers, n is the number of terms in the series, a is the first term in the series, and b is the last term in the series.

Calculate the sum.

(a) $1 + 2 + 3 + \cdots + 99 + 100$

(b) $(0.01) + (0.02) + (0.03) + \cdots + (0.99) + (1.00)$

8. Evaluate. Show the steps you used to find each result.

(a) $b - 1 + bc - 5c^2$, for $b = 5$ and $c = -1$

(b) $3xy^1 - x^2y + 4x$, for $x = 2$ and $y = 1$

(c) $8e + 3 + e^2$, for $e = \frac{1}{4}$

(d) $f^3 - 2f^2 + 4f$, for $f = -2$

(e) $8 - ab^1 + b^2$, for $a = -\frac{1}{2}$ and $b = 4$

(f) $-4t^2 + 3x^3$, for $t = 0.25$ and $x = 0.5$

9. James is a television repair technician. His charge for a service call is \$18 plus \$9.50 per hour. To calculate the total charge, he uses the algebraic formula $C = 18 + 9.5t$, where C is the total cost in dollars and t is the time in hours.

Tacia also repairs televisions. To calculate her charges, she uses a different formula: $C = 12 + 12.5t$

(a) Which technician charges less for a 3 h service call? Explain.

(b) Which technician costs less for a 1 h service call? Explain.

10. **(a)** Evaluate the polynomial $2x^2 - 3$ for each value of x. What pattern can you find?

Value of x	Result
1	
2	
3	
4	

(b) Use the pattern to predict the results for $x = 5$ and $x = 7$. Calculate to check.

11. A test driver wants to calculate the distance in feet she will need to stop her car on dry pavement. She uses the formula $d = 0.4s + 0.02\,s^2$, where d is the stopping distance in feet and s is the speed of the car in miles per hour before braking. Calculate the stopping distance for each speed.

Value of s (mph)	Distance Needed
25	
50	
75	
100	
125	

Use the results to make a graph, such as a bar or line graph.

12. **(a)** Create five polynomial expressions. List values for the variables in each expression, using integers, decimals, and fractions.

(b) Write a step-by-step solution for each expression and analyze each step to determine where errors might occur.

(c) On a separate sheet of paper, rewrite the solution for each example so that it contains an error. Trade problems with a classmate and identify the errors.

7.3 Adding and Subtracting Polynomials with Tiles

You can use algebra tiles to model addition and subtraction with polynomials. When you subtract with tiles, you can use the **Take-Away method** or the **Add-the-Opposite method**.

Example 1

Use algebra tiles to add $(3x^2 - 2x + 1) + (-2x^2 + 3x + 4)$.

Solution

Use algebra tiles to model each polynomial.

$(3x^2 - 2x + 1) \quad + \quad (-2x^2 + 3x + 4)$

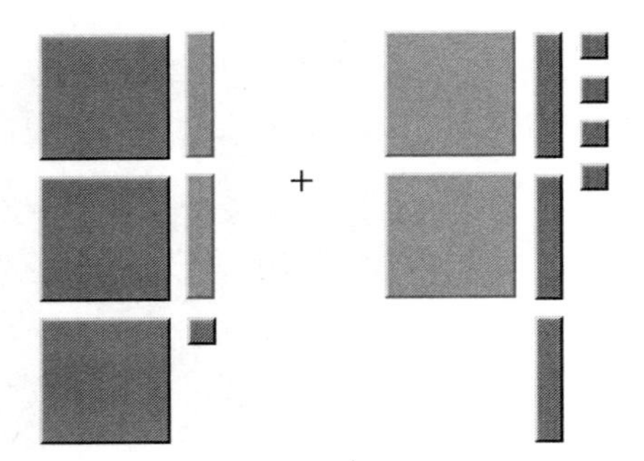

Collect like terms.

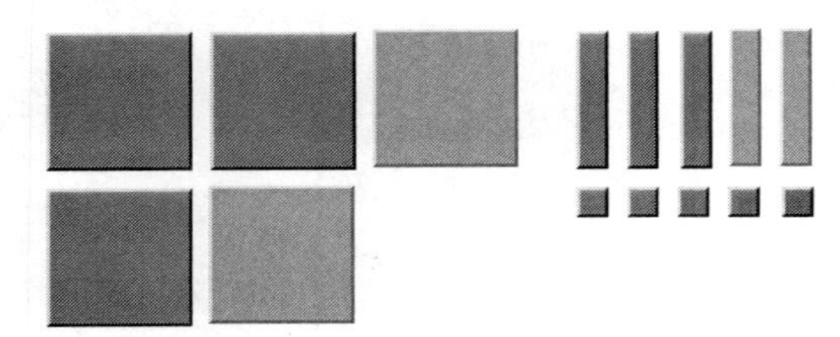

Combine like terms. Use the zero principle if necessary.

Write the simplified expression for the remaining tiles.

$x^2 + x + 5$

Example 2

Evaluate $(-3x^2 + 2x - 4) - (-2x^2 + x - 3)$.

Solution

Take-Away Method

Use algebra tiles to model the starting amount.

Cross out tiles that are to be subtracted.

Write the simplified expression for the remaining tiles.

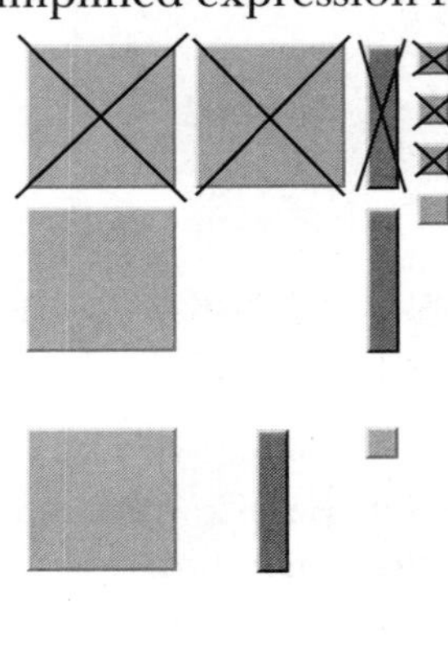

$-x^2 + x - 1$

Add-the-Opposite Method

Use algebra tiles to model each polynomial.

Change the second model so instead of subtracting the given amount, you are adding its opposite.

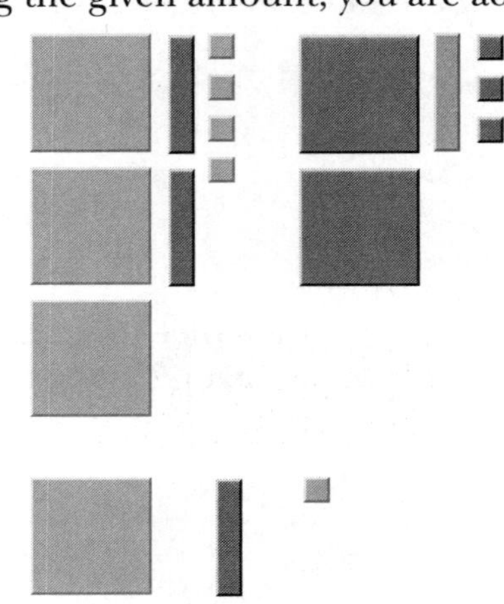

Add, using the steps in Example 1.

Exercises

1. How can you model the opposite of any polynomial?

2. Why do you need to understand opposites before you can add and subtract polynomials?

3. Which subtraction method do you prefer? Why?

4. Add.

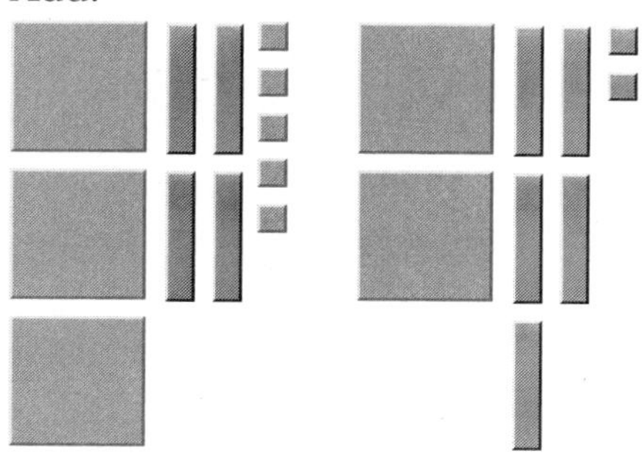

5. Use tile models or pictures to find each sum.

(a) $(6x + 2) + (3x + 4)$

(b) $(8 - 4x) + (-3 - 2x)$

(c) $(3x^2 + 6x - 8) + (-5x^2 - x + 4)$

(d) $(7 - 2x - x^2) + (-5 - x + 4x^2)$

6. Draw the opposite.

(a)

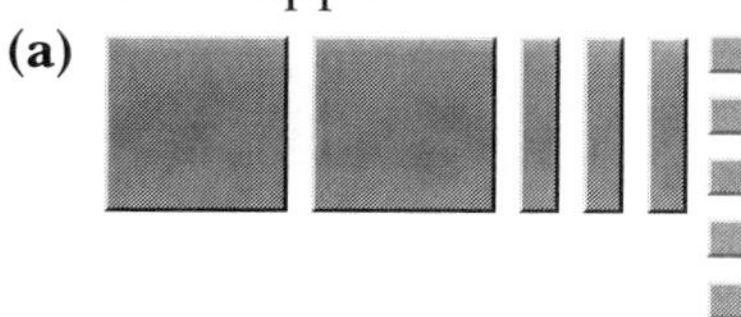

(b) $-x^2 + 4x - 2$

7. Subtract.

8. Use tile models or pictures to find each difference.

(a) $(x + 3) - (7x + 6)$

(b) $(-2x + 2) - (3x + 4)$

(c) $(4x^2 + 2x - 3) - (-6x^2 + 4x + 5)$

(d) $(-6x + 5x^2 + 1) - (4x^2 + 5 - 2x)$

9. Solve using tile models or pictures.

(a) Winnie used tiles to model $(7x - 2)$. What must she add to get the sum $(-x + 4)$?

(b) Henry used tiles to model $(3x - 5)$. He subtracted an amount and got $(-5x + 6)$. What did he subtract?

10. The model represents an addition expression. Write the addition, then find the sum.

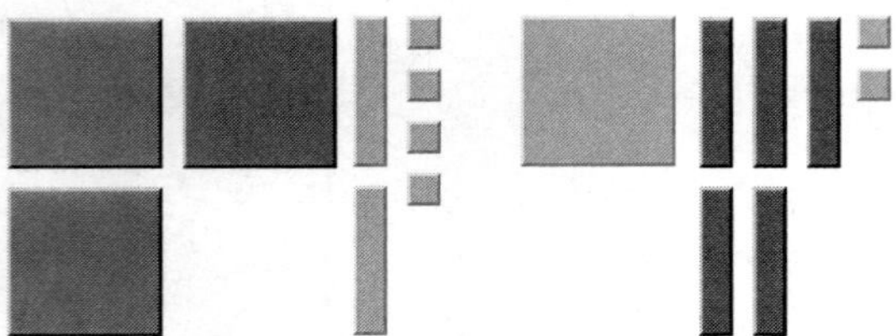

11. The model represents a subtraction expression. Write the subtraction, then find the difference.

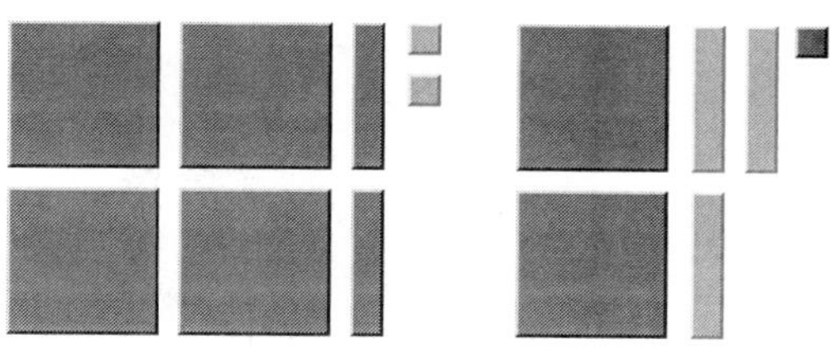

12. Use or draw algebra tiles to find each result.

(a) $(5x^2 - 3x + 1) + (4x - 7 - 3x^2)$

(b) $(-3x^2 + 2x - 4) - (-2x^2 - 4 - 3x)$

13. Create one addition and one subtraction example involving these polynomials. Find each sum or difference.

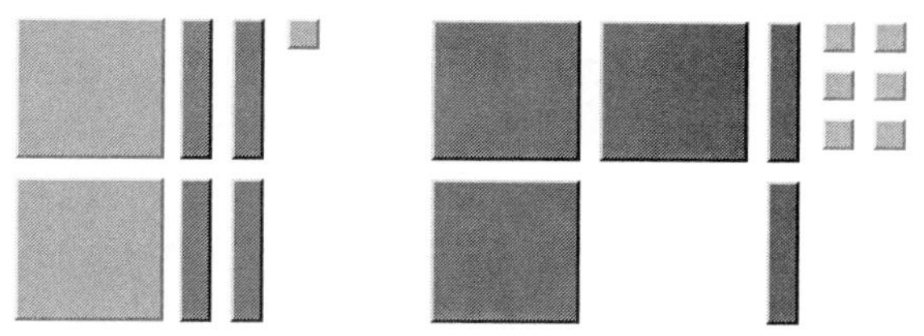

14. Create an addition or subtraction example that gives this result.

7.4 Multiplying Polynomials

You have seen how to use both **algebra tiles** and the **distributive property** to find:
- the product of **two monomials**, such as $(8x)(3x^2)$.
- the product of **monomials** and **binomials**, such as $2x(3x-4)$.
- the product of **two binomials**, such as $(x+4)(x-3)$.
- the **square of a binomial**, such as $(x-5)^2$.

Example 1

Evaluate $3x(x-2)$.

Solution

Using Algebra Tiles

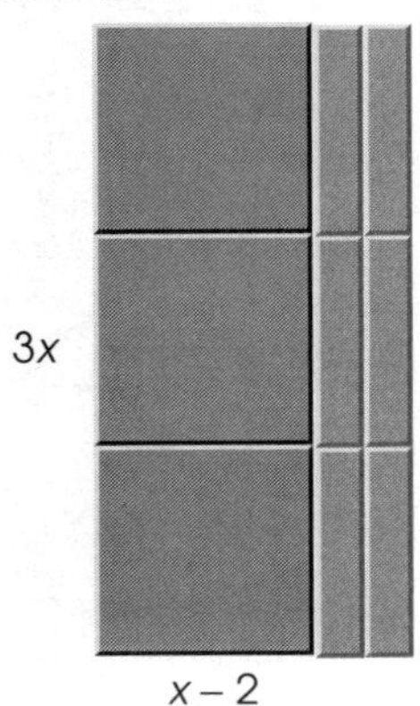

The area of the rectangle is $3x^2 - 6x$.

Using the Distributive Property

$$a(b+c) = ab + ac$$
$$3x(x-2) = 3x^2 - 6x$$

Example 2

What two consecutive odd numbers have squares that differ by 32?

Solution

Consecutive odd numbers, such as 11 and 13, or 23 and 25, always differ by 2.
If you let x represent the first odd number, then the second must be $x+2$.
The squares differ by 32, so $(x+2)^2 - (x)^2 = 32$.
You can find the square of $x+2$ in two ways.

Using Algebra Tiles

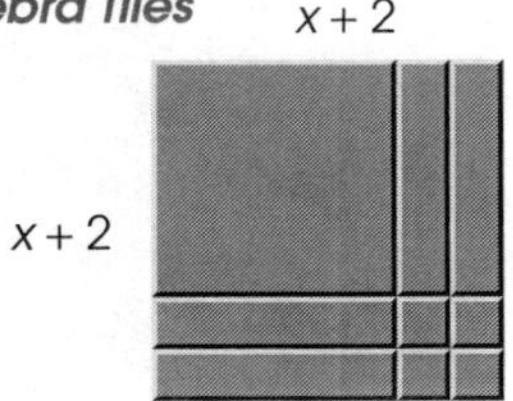

The area of the square is $(x+2)^2$ or $x^2 + 4x + 4$.

Now the equation for the difference between the two odd numbers becomes $x^2 + 4x + 4 - x^2 = 32$.

Solve for x to find the first odd number.

$x^2 + 4x + 4 - x^2 = 32$	
$4x + 4 = 32$	Combine like terms.
$4x = 28$	Isolate the unknown.
$x = 7$	Divide both sides by 4.

The first odd number is 7, so the second is $7 + 2$ or 9.

Using the Distributive Property

$(x+2)(x+2) = x^2 + 2x + 2x + 4$	FOIL
$= x^2 + 4x + 4$	Combine like terms.

Now the equation for the difference between the two numbers becomes $x^2 + 4x + 4 - x^2 = 32$.

Solve the equation as shown for the first method.

Exercises

1. Explain each term in your own words.
   (a) polynomial

   (b) distributive property

   (c) FOIL method

2. **(a)** Why is the x-squared tile called by this name?

   (b) Why is the x-tile called by this name?

3. Explain how you know which type of tiles (positive or negative) to insert to complete the following rectangle.

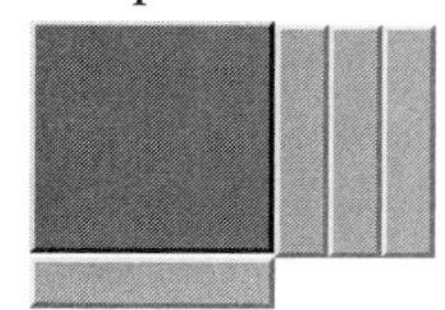

4. Rearrange the tiles to illustrate the product of $x + 2$ and $x - 4$. Draw and label your arrangement, then name the product.

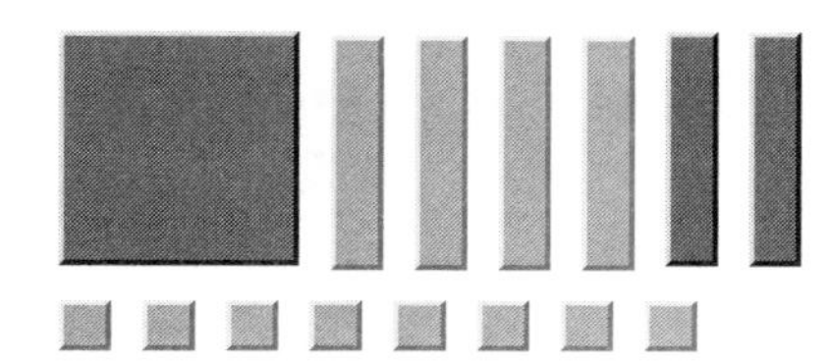

5. Rearrange the tiles to illustrate the product of $x - 1$ and $x - 5$. Draw and label your arrangement, then name the product.

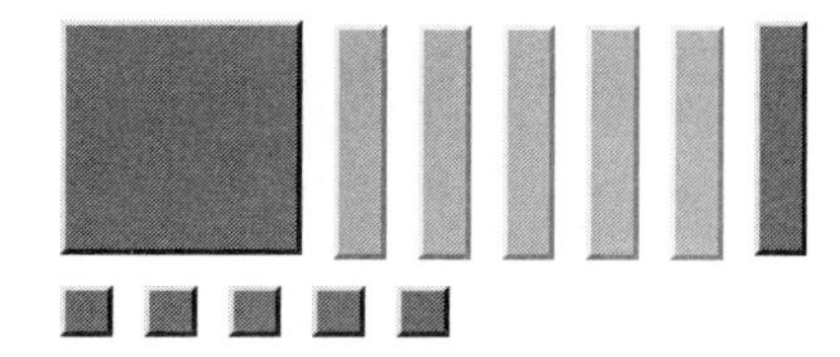

6. Find each product.
   (a) $(4x)(7x)$ **(b)** $(-3u)(u)$

(c) $(0.2x)(1.5x)$

(d) $(0.07a)(0.6a)$

(e) $\left(\frac{1}{4}x\right)\left(\frac{2}{5}x\right)$

(f) $\left(\frac{3}{8}c\right)(4c)$

7. Evaluate.

(a) $2x(3x + 4)$

(b) $3a(1 - 2a)$

(c) $5(7 - x)$

(d) $-2d(3d - 9)$

8. Find each product.

(a) $(x + 4)(x + 6)$

(b) $(2a - 3)(2a - 7)$

(c) $(3h - 5)(h + 8)$

(d) $(1 - 3x)(1 - x)$

(e) $(5x - 2)(5x + 2)$

(f) $(2 + 3x)(1 + x)$

9. Expand.

(a) $(x + 3)^2$

(b) $(m - 6)^2$

(c) $(2x + 7)^2$

(d) $(1 - 3y)^2$

10. The number of tickets sold for the variety concert can be described by the expression $3x$. The price of each ticket can be expressed as $x + 7$. What expression would represent the revenue generated by the variety concert?

11. The MacPhees have decided to increase the size of their rectangular garden by adding 2 ft to the length and 2 ft to the width. The original garden measures $3x$ ft by $(x + 4)$ ft.

(a) Write the expression for the original area.

(b) Write the expression for the enlarged area.

(c) What expression would represent the change in the size of the garden?

12. What mathematical expression could be used to represent the product of three consecutive numbers, where the first number is n?

13. Find each product.

(a) $0.2x(3 - 4x)$

(b) $4h^2(2h + 7)$

(c) $3xy(x^2 + 4y - 1)$

(d) $(2a - 3b)(a + 4b)$

(e) $(x - 3y)^2$

(f) $(x + 1)^3$

14. Create a multiplication problem involving polynomials and this formula:
distance = *speed* × *time*

15. Would you use algebra tiles to help you multiply? Explain.

(a) $(2y + 1)(y - 1)$

(b) $(3x + 7)(2x - 6)$

16. One of the sides of a right triangle is 4 cm shorter than the hypotenuse. The other side length is 8 cm. Find the lengths of the two unknown sides.

7.5 Powers, Bases, and Exponents

This lesson introduced **powers**, which are products of equal factors. You have
- written powers in **expanded form** (e.g., $2 \times 2 \times 2 \times 2$) and **exponential form** (e.g., 2^4).
- identified **bases** (the number to be multiplied) and **exponents** (how many times the number is multiplied by itself).
- worked with bases that are whole numbers, integers, rational numbers, and variables.
- evaluated powers by multiplying the base by itself as indicated by the exponent.
- explored situations where the evaluated power is negative (negative base and odd exponent) or positive (positive or negative base and even exponent).
- used rules for order of operations to evaluate powers.
- found the coefficient of a power (e.g., the coefficient of $-\frac{1}{2}(-3x)^2$ is $-\frac{1}{2}$.

Example 1

Write the simplified form of $1.25 \times g \times h \times (-2) \times h \times h \times g \times g \times g$.
Identify the coefficient, the bases, and the exponents.

Solution

$1.25 \times g \times h \times (-2) \times h \times h \times g \times g \times g$	
$= 1.25 \times (-2) \times g \times g \times g \times g \times h \times h \times h$	Group like terms.
$= 1.25 \times (-2) \times g^4 \times h^3$	Multiply like terms.
$= -2.5 \times g^4 \times h^3$	Find the coefficient by multiplying the other factors.
$= -2.5g^4h^3$	Simplify.

This power has two bases (g and h) and two exponents (4 and 3).

The coefficient is -2.5.

Example 2

Evaluate $(4y)^2$ for $y = 3$. Does $(4y)^2 = 4y^2$? Explain.

Solution

$$(4y)^2 = (4 \times 3)^2 = 12^2 = 144 \qquad 4y^2 = 4(3)^2 = 4 \times 3 \times 3 = 36$$

$(4y)^2$ means $4y \times 4y$ or $4 \times y \times 4 \times y$ or $4 \times 4 \times y \times y$ or $16y^2$.

$4y^2$ means $4 \times y \times y$ or $4y^2$.

The values are not equal.

Exercises

1. What does the base of a power tell you?

2. What does the exponent of a power tell you?

3. What does the coefficient of a power tell you?

4. Use cubes or draw diagrams to show why 5^2 is not equal to 2^5.

5. For the power $12a^3$, identify the:
   (a) base
   (b) exponent
   (c) coefficient
   (d) expanded form
   (e) exponential form

6. Identify the coefficient.
   (a) $\frac{x^3}{4}$
   (b) $-y^3$
   (c) $-\frac{a^2}{6}$

7. State in expanded form.
   (a) $3x^2y^4$
   (b) $6(rs)^2$
   (c) $\left(\frac{3}{7}\right)^5$

8. State in exponential form.
   (a) $4 \times a \times b\ a \times a \times b$
   (b) $5 \times gh \times 2 \times gh \times gh$
   (c) $\frac{2}{11} \times \frac{2}{11} \times \frac{2}{11} \times \frac{2}{11}$

9. Evaluate for $r = 3$ and $s = 4$.
   (a) $3r^2 + s^2$
   (b) $(2r + 1)^2 + 2r^2s$

10. Evaluate.

(a) $3^2 + 4^2$

(b) $(-5)^2 + (-12)^2$

(c) $-2^3 + 10^2$

(d) $\left(\frac{2}{5}\right)^2 \times \left(-\frac{5}{8}\right)^2$

11. Some mutual funds and bonds earn compound interest for the investor. You can use the standard formula $A = P(1 + i)^n$ to calculate the amount of money that will be returned after an investment.

A represents the final amount of money returned; P is the principal or original amount invested; i is the interest rate to be used each time the amount is calculated; and n is the number of times the interest is compounded.

If a customer invests \$10,000 at a rate of 8% or 0.08 per year, how much money would she get back at the end of 10 years if the interest is compounded once per year?

12. Complete the chart.

	Base	Exponent	Exponential Form	Expanded Form	Value
(a)	-10	4			
(b)	5				125
(c)		2			49
(d)	-2			-2 × -2 × -2	

13. **(a)** Does $-7^4 = (-7)^4$?

(b) Does $-7^3 = (-7)^3$?

(c) Which powers from parts (a) and (b) have a negative base? a negative coefficient?

(d) If a power has a negative base, how can you tell if the final answer will be positive or negative?

14. If $a = -1$, $b = 2$, $c = -3$, evaluate the following:

(a) $a^2 - b^2 - c^2$

(b) $a^2 + b^2 + c^2$

15. Evaluate.

(a) $(-4)^2 - (-2)^5$

(b) $(7 + 3)^2 - 5$

(c) $8 \div 4 + 2^4 - 12$

(d) $5 + (6 \times 4 - 15)^2$

16. Design a five-section spinner so that each section contains a power expression. Each power should have one variable in the base and another as an exponent, for example, $-5 + 2a^b$.

Use your spinner to play Spin Ten with a partner or small group. (One way to make the spinner work is to fix a paper clip to the center with a pencil point, then spin the paper clip.)

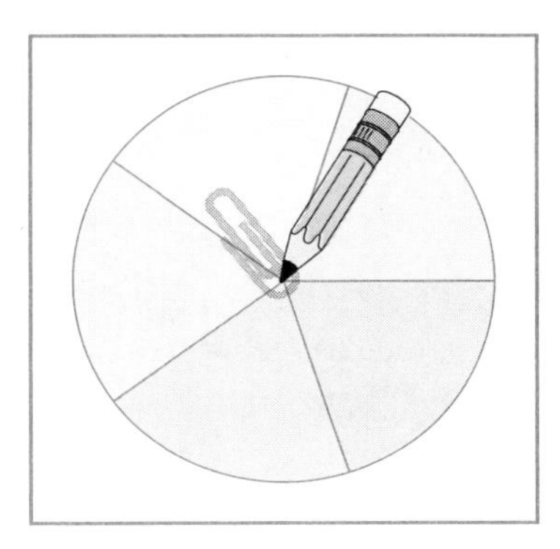

The first player spins the spinner, rolls two dice to determine two values, then substitutes the values into the expression to get the greatest result possible. The second player repeats the same process. The player who gets the greatest result scores a point for the round. The object of the game is to be the first player to score 10 points.

17. Create 20 power expressions with values given for the variables. Evaluate each expression and use the results to create a cross-number puzzle on graph paper. (A cross-number puzzle is a crossword puzzle where the answers are numbers instead of words.) Exchange your puzzle with a classmate.

7.6 Laws of Exponents: Product Laws

In this lesson, you found relationships you can use to simplify multiplication calculations with exponents. The **Product Law**, **Power of a Product Law**, and **Power of a Power Law** can be generalized as follows:

Product Law

To multiply powers with the same base, add the exponents: $x^m \times x^n = x^{m+n}$

Power of a Product Law

To simplify a power of a product, apply the exponent to each term in the product: $(x \times y)^m = x^m \times y^m$

Power of a Power Law

To simplify a power of a power, multiply the exponents: $(x^m)^n = x^{mn}$

Example 1

Write $8^3 \times 8^6$ as a single power.

Solution

Using Expansion

$$\begin{aligned} 8^3 \times 8^6 &= (8 \times 8 \times 8) \times (8 \times 8 \times 8 \times 8 \times 8 \times 8) \\ &= 8 \times 8 \times 8 \times 8 \times 8 \times 8 \times 8 \times 8 \times 8 \\ &= 8^9 \end{aligned}$$

Using the Product Law

$$\begin{aligned} 8^3 \times 8^6 &= 8^{3+6} \\ &= 8^9 \end{aligned}$$

Example 2

Evaluate $(8 \times 5)^2$.

Solution

Using Order of Operations

$$\begin{aligned} (8 \times 5)^2 &= (40)^2 \\ &= 1600 \end{aligned}$$

Using Expansion

$$\begin{aligned} (8 \times 5)^2 &= (8 \times 5)(8 \times 5) \\ &= 8 \times 5 \times 8 \times 5 \\ &= 40 \times 40 \\ &= 1600 \end{aligned}$$

OR

$$\begin{aligned} (8 \times 5)^2 &= (8 \times 5)(8 \times 5) \\ &= 8 \times 8 \times 5 \times 5 \\ &= 64 \times 25 \\ &= 1600 \end{aligned}$$

Using the Power of a Product Law

$$\begin{aligned} (8 \times 5)^2 &= 8^2 \times 5^2 \\ &= 64 \times 25 \\ &= 1600 \end{aligned}$$

Example 3

Evaluate $(16^2)^3$.

Solution

Using Expansion

$$\begin{aligned} (16^2)^3 &= (16^2)(16^2)(16^2) \\ &= (16 \times 16)(16 \times 16)(16 \times 16) \\ &= 16 \times 16 \times 16 \times 16 \times 16 \times 16 \\ &= 16{,}777{,}216 \end{aligned}$$

Using the Power of a Power Law

$$\begin{aligned} (16^2)^3 &= 16^{2 \times 3} \\ &= 16^6 \\ &= 16{,}777{,}216 \end{aligned}$$

Example 4

Simplify $(54w^2)^2$.

Solution

Using Expansion

$$\begin{aligned} (54w^2)^2 &= (54w)(54w) \\ &= 54 \times w \times w \times 54 \times w \times w \\ &= 54 \times 54 \times w \times w \times w \times w \\ &= 2916 \times w \times w \times w \times w \\ &= 2916w^4 \end{aligned}$$

Using the Power of a Power Law

$$\begin{aligned} (54w^2)^2 &= (54)^2 (w^2)^2 \\ &= (54^2)(w^4) \\ &= 2916w^4 \end{aligned}$$

Exercises

1. Explain each law in your own words. Give examples.

(a) Product Law

(b) Power of a Product Law

(c) Power of a Power Law

2. Write as a single power.

(a) $4^1 \times 4^{12}$ **(b)** $10^2 \times 10^7$ **(c)** $8^8 \times 8^{41}$ **(d)** $(-13)^3 \times (-13)^3$

(e) $9^4 \times 9$ **(f)** $2.3^9 \times 2.3^2$ **(g)** $x^4 \times x^3$ **(h)** $w^8 \times w^{41}$

3. Write as the product of two powers.

(a) $(4 \times 3)^5$ **(b)** $(2 \times 8)^0$ **(c)** $(2 \times 5)^8$ **(d)** $(-2 \times 5)^5$

(e) $(4.1 \times 2.3)^2$ **(f)** $(7 \times 7)^4$ **(g)** $(x \times y)^3$ **(h)** $(c \times d)^5$

4. Evaluate.

(a) $4^6 \times 5^6$ **(b)** $10^2 \times 12^2$ **(c)** $2^8 \times 4^8$ **(d)** $3^3 \times (-3)^3$

5. Write as a single power.

(a) $9^4 \times 3^4 \times 4^4$ **(b)** $5.23^8 \times 3.26^8 \times 1.72^8$

6. Write using only one exponent.

(a) $(3^4)^5$ **(b)** $((-5)^3)^4$ **(c)** $(6^2)^{12}$ **(d)** $(4^7)^3$

(e) $((-6)^6)^9$ **(f)** $((-2)^3)^7$ **(g)** $(2.45^3)^{10}$ **(h)** $(8.2^5)^2$

(i) $(2.7^9)^4$ **(j)** $(x^4)^5$ **(k)** $((-y)^3)^4$ **(l)** $(m^2)^{12}$

7. Evaluate.

(a) $3^5 \times 3^7$ **(b)** $(4^2)^6$ **(c)** $(5 \times 3)^6$

(d) $(-2^2)^7$ **(e)** $5^5 \times 5^6$ **(f)** $2^4 \times 7^4$

(g) $9^2 \times 9^4$ **(h)** $(7^3)^2$ **(i)** $(5 \times -3)^6$

8. Examine Zach's evaluation of $(4 \times 5)^3$. Explain why you think his solution is or is not correct.

$$\begin{aligned}(4 \times 5)^3 &= 4 \times 5^3\\ &= 4 \times 125\\ &= 500\end{aligned}$$

9. Use exponent laws and guess and test to find values for n.

(a) $n^4 \times n^2 = 64$ **(b)** $(n \times n)^2 = 1296$ **(c)** $(n^2)^3 = 729$

10. Explain how you could estimate the value of $(3 \times 4)^3$. Use a calculator to check your estimate.

11. Create the greatest power you can using the numbers 8, 3, and 4. Use each number once.

12. Find the value of the exponent that makes each equation true.

(a) $4^3 = 2^{\square}$ **(b)** $6^9 = 216^{\square}$ **(c)** $625^2 = 25^{\square}$ **(d)** $27^4 = 3^{\square}$

13. Create multiplication problems involving exponent laws that have these answers.

(a) $(8)^5$ **(b)** $(-3)^{12}$ **(c)** $(1.5)^3$

14. Earth travels about 10^9 km in its year-long orbit around the sun. About how far does it travel in 100 years?

15. Use exponents to multiply one million by one million.

16. How could you use power rules to help you write 81 as a power of 3?

17. How could you use power rules to help you write 125^4 as a power of 5?

18. Complete.

(a) Does -4^2 equal $(-4)^2$? Explain.

(b) How could you use the power of a product law to rewrite $(-4)^2$ with a base of 4?

19. Explain what is happening at each step in the solution. Identify the exponent laws used.

Simplify $((-4)^5)^3 \times -4^7 \times (-4)^2 \times (-4)^{18}$.

Step 1: $(-4)^{15} \times -4^7 \times (-4)^2 \times (-4)^{18}$

Step 2: $(-4)^{15} \times -4^7 \times (-4)^{20}$

Step 3: $(-4)^{35} \times -4^7$

Step 4: $(-1 \times 4)^{35} \times -4^7$

Step 5: $(-1)^{35} \times 4^{35} \times -1 \times 4^7$

Step 6: $-1 \times 4^{35} \times -1 \times 4^7$

Step 7: 4^{42}

20. Without calculating, use power notation to determine which is greater, 2^{666} or 5^{333}.

8 RATIO AND PERCENT

8.1 Percent

In this lesson, you learned how to **write a fraction in percent form**, and to **estimate and calculate a percent of a given number**. Recall:

- To model percents, you can use a 10×10 grid, a percent strip, or a circle divided in 100 parts.
- To write a fraction in percent form, rewrite the fraction so the denominator is 100. If the denominator of the fraction is a factor of 100, use equivalent fractions: $\frac{3}{25} = \frac{12}{100}$

  If the denominator is not a factor of 100, divide the numerator by the denominator to express the fraction in decimal form, then round to the nearest hundredth:

  $$\begin{aligned}\frac{2}{3} &= 2 \div 3 \\ &= 0.66666666\ldots \quad \text{Round to the nearest hundredth.} \\ &\doteq 0.67 \\ &= 67 \text{ hundredths} \\ &= 67\%\end{aligned}$$

- To calculate a percent of a number, you can use 10% and 1% relationships. For example, 5% of \$29.95 must be equal to $5 \times 1\%$ or about $5 \times 30¢$.

Example 1

There were 20 vehicles in a parking lot, including 8 minivans. What percent of the vehicles were minivans?

Solution

The fraction that represents the minivans is $\frac{8}{20}$.

$$\begin{aligned}\frac{8}{20} \times \frac{5}{5} &= \frac{40}{100} \quad \text{Since 20 is a factor of 100, use equivalent fractions.} \\ &= 40\%\end{aligned}$$

Exactly 40% of the vehicles were minivans.

To check this result with a calculator, enter

[8] [÷] [2] [0] [%]

or

[8] [÷] [2] [0] [×] [1] [0] [0] [=]

Example 2

Calculate 2% of 1500.

Solution

To find 2% of 1500, first calculate 1% of 1500, then multiply by 2.

1% of $1500 = 15$

To find 1% of a number, divide the number by 100 by moving the decimal point 2 places to the left.
2% of 1500 must be twice as much as 1%.

$$2\% \text{ of } 1500 = 15 \times 2$$
$$= 30$$

To check this result with a calculator, enter

[1] [5] [0] [0] [×] [2] [%]

or

[1] [5] [0] [0] [×] [2] [÷] [1] [0] [0] [=]

Exercises

1. The *cent* part of the word *percent* means *hundred.* Brainstorm other words that have *cent* in them. How is each word related to one hundred?

2. Write five statements about your classroom that contain percent estimates. For example, "About 25% of the wall is covered with posters."

3. Estimate each percent.
   (a) percent of a weekday you spend at school

   (b) percent of a weekday you watch television

   (c) percent of a weekend day you watch television

   (d) percent of a day you brush your teeth

   (e) percent of an average day you exercise

   (f) percent of an average day you read

4. Jonathan received the following scores on his science quizzes:

Quiz 1: $\frac{15}{20}$ Quiz 2: $\frac{8}{10}$

Quiz 3: $\frac{28}{35}$ Quiz 4: $\frac{66}{75}$

Estimate which score was Jonathan's best and which was his worst. Then use percents to check your estimates.

5. Why is it useful to convert test scores to percent form?

6. Write the percent. You may need to round to the nearest hundredth.

(a) $\frac{1}{4}$ **(b)** $\frac{3}{5}$ **(c)** $\frac{13}{50}$

(d) $\frac{9}{10}$ **(e)** $\frac{8}{25}$ **(f)** $\frac{11}{12}$

(g) $\frac{9}{13}$ **(h)** $\frac{47}{60}$

7. Write each fraction in simplest form, then calculate the percent. You may need to round to the nearest hundredth.

(a) $\frac{12}{15}$ **(b)** $\frac{15}{30}$ **(c)** $\frac{6}{24}$

(d) $\frac{20}{200}$ **(e)** $\frac{15}{18}$ **(f)** $\frac{25}{35}$

(g) $\frac{21}{33}$ **(h)** $\frac{32}{36}$

8. Estimate or calculate mentally.

(a) 1% of 52 **(b)** 10% of 17

(c) 15% of $19.95 **(d)** 7% of $45.99

(e) 90% of 947

9. Most restaurant diners leave a tip of about 15% of the cost of the meal. Estimate the tip for a meal that costs $29.95.

10. Lien's track team won 15 out of 25 events. What percent of the events did they *not* win?

11. In a bag of 200 jelly beans, 60 are pink or purple. What percent are pink or purple?

12. A lion sleeps 80% of the time. For how many hours per day would you expect a lion to be awake?

13. Which is greater: 27% of 72 or 72% of 27? Explain.

14. In a bad year, a company was forced to cut the workers' salaries by 5%. The next year was better, so the company gave everyone a 5% raise. At the end of the second year, were the workers better off, worse off, or earning the same amount as before the cut? Why?

15. An increase from 5 to 20 is a 300% increase. Is a decrease from 20 to 5 a 300% decrease? Explain.

16. A pair of jeans and a sweater sell for the same price. This week, jeans are on sale for 15% off, and sweaters for 20% off. Next week, jean prices will drop by another 20%, and sweater prices by another 15%. At the end of the two-week sale, will it be cheaper to buy the jeans or the sweater? Why?

17. The sides of a square are quadrupled in length. What percent of the larger square's area would the original square represent?

18. Check your newspaper for advertised sales.
   - **(a)** Estimate the costs of items that are advertised on sale with percent reductions.
   - **(b)** Design an advertisement for a sale with percent reductions.

8.2 Rate and Ratio

In this lesson, you learned to distinguish between **ratios** and **rates**, and to express these relationships in lowest terms. Recall:

- A ratio compares quantities measured with the same unit, so units are not specified.
- A rate compares quantities measured with different units, so units must be included.
- A ratio can be written in three different forms: $\frac{3}{5}$, 3 : 5, or 3 to 5.
- To simplify a rate or ratio, divide both terms by the **greatest common factor**.

Example 1

Tickets to a concert are selling fast in the singer's home town. Each hour, one hundred $55 tickets are sold for seats on the floor. Every two hours, one hundred fifty $40 tickets are sold for seats in the stands.

Write a ratio to compare the cost of a floor seat to the cost of a seat in the stands. Then express the ratio in lowest terms.

Solution

A floor seat costs $55.
A seat in the stands costs $40.

Write the ratio in words.

cost on floor : *cost in stands*

Substitute known values. Since both terms are in dollars, dollar signs are not necessary.

55 : 40

The greatest common factor of 55 and 40 is 5. Divide both terms by 5.

55 ÷ 5 : 40 ÷ 5
11 : 8

The cost ratio is $11 on the floor : $8 in the stands.

Example 2

Use information from Example 1 to compare the rate of floor-seat sales with the rate of sales for seats in the stands.

Solution

Tickets for floor seats are selling at a rate of 100 tickets/h.

Tickets for seats in the stands are selling at a rate of 150 tickets/2 h.

To compare the rates, rewrite the second ratio to see how many tickets are sold in 1 h.

150 tickets : 2 h

Divide both sides by 2 to get 1 h on the right.

150 ÷ 2 tickets : 2 h ÷ 2
75 tickets : 1 h

Now you know that floor seats are selling at 100 tickets/h, and seats in the stands are selling at 75 tickets/h. Each hour, 25 more tickets are sold for floor seats than for seats in the stands.

Example 3

Suppose there are only 250 floor-seat tickets left for the concert described in Example 1.

How long will it take for these tickets to sell out?

Solution

One hundred tickets are sold each hour, so 50 tickets would be sold each half hour.

There are 5 sets of 50 tickets in 250, so it will take 5 half hours, or $2\frac{1}{2}$ hours, to sell out.

Exercises

1. Write three sentences that describe ratios and three sentences that describe rates.

2. Is each situation a rate or a ratio?

(a) It snowed 4 in. on Monday and 6 in. on Tuesday.

(b) Three candy bars were on sale for $1.50.

(c) A take-out lunch for 4 people cost $32.

(d) A week's allowance for two teenagers was $40.

(e) A bobsled traveled at a speed of 60 mph.

(f) A worker was paid $30 for 5 h of work.

(g) The relationship between the numbers of boys and girls in a class was 14 : 12.

3. Express each relationship from Problem 2 in lowest terms.

4. Identify the error(s) in each rate or ratio.

(a) 15/min **(b)** $3 : $10 **(c)** 21 to 39

(d) 6 : 12 = 2 : 1 **(e)** 45¢/3 **(f)** 3 sides to 12 squares

5. A *prime number* is a number whose only factors are 1 and itself. An easy way to find the greatest common factor of two numbers is to write each number as a product of its prime factors. To do this, keep dividing factors until there are no more ways to divide. For example:

$24 = 2 \times 12$, $12 = 2 \times 6$, $6 = (2) \times (3)$

$30 = 5 \times 6$, $6 = (2) \times (3)$

The prime factors 2 and 3 are common to 24 and 30.

$24 = 2 \times 2 \times (2 \times 3)$ and $30 = 5 \times (2 \times 3)$

$2 \times 3 = 6$, so the greatest common factor of 24 and 30 is 6.

Use this method to find the greatest common factor for each pair of numbers.

(a) 12 and 18 **(b)** 21 and 39

(c) 15 and 45

(d) 45 and 81

6. Find the greatest common factor, then write in lowest terms.

(a) 12 : 9

(b) 36 : 45

(c) 99 : 300

(d) 72 : 36

(e) 18 : 48

(f) 87 : 27

7. What is the greatest factor common to all the ratios in Problem 6?

8. Which ratio does not belong in each set? Why?

(a) 3 : 4, 12 : 16, 24 : 30, 36 : 48

(b) 15 mph, 12 ft/min, 10 gal/min, 80 in./s

(c) 4 for $20, 1 for $5, 2 for $10, 10 for $40

(d) 12 to 6, 40 to 20, 8 to 4, 3 to 6

9. Ten people share 2 dozen doughnuts. How many will share 1 dozen?

10. In a class with 16 girls and 12 boys, what is the ratio of the number of girls to the total number of students?

11. A bouquet of 25 flowers costs $2. What is the cost per flower?

12. Corey is now 74 in. tall. His height has increased 2 in. in each of the last 3 years.

(a) How tall was Corey 3 years ago? 2 years ago? 1 year ago?

(b) For each of the three years, write a ratio to compare the increase in Corey's growth to his starting height.

(c) Are the ratios you wrote for part (b) equivalent? Explain.

(d) What rate describes Corey's growth over the past 3 years?

13. In 1996 a survey of American households found that there were about 194,600,000 pets in the United States. Of these, 59,000,000 were cats, 52,900,000 were dogs, and 55,600,000 were fish.

(a) What is the ratio of dogs to the total number of pets? Write the ratio in lowest terms.

(b) What is the ratio of cats to the total number of pets? Write the ratio in lowest terms.

(c) If you add the number of dogs, cats, and fish, the sum is less than the total number of pets. Why?

14. Survey students in your class to find out how many households own pets, and what types of pets they have. Use your data to create and solve some problems about ratios and rates.

8.3 Proportions

When two ratios or rates are **equivalent**, they form a **proportion**. In this lesson, you explored these methods for solving proportions:

- Draw and extend a diagram.
- Use equivalent ratios.
- Look for a multiplication relationship between a numerator and denominator.
- Convert one ratio to decimal form.

Example 1

Express the ratio $\frac{21}{30}$ as a percent.

Solution

A percent is a ratio with a denominator of 100.

To find $\frac{21}{30}$ as a percent, use this proportion:

$$\frac{21}{30} = \frac{n}{100}.$$

The first ratio in this proportion, $\frac{21}{30}$, is not in lowest terms.

To write the ratio in lowest terms, divide the numerator and the denominator by the greatest common factor, 3.

$$\frac{21 \div 3}{30 \div 3} = \frac{7}{10}$$

Now the proportion becomes:

$$\frac{7}{10} = \frac{n}{100}$$
$$\frac{7 \times 10}{10 \times 10} = \frac{n}{100}$$
$$7 \times 10 = n$$
$$70 = n$$

Therefore, $\frac{21}{30} = \frac{7}{10} = \frac{70}{100} = 70\%$.

Dividing by $\frac{3}{3}$ is the same as dividing by 1, so the value does not change.

In the denominator, 10 was multiplied by 10 to get 100.

In the numerator, 7×10 must equal n.

Example 2

A coach ordered 8 pizzas for the 15 players on her soccer team and found that this was exactly the right amount. Two weeks later, she decides to order pizza again. This time, there are only 10 players. How many pizzas should she order?

Solution

You know that 15 players will eat 8 pizzas. You want to find out how many pizzas to order for 10 players. To do this you can use a proportion. Let p represent the number of pizzas.

$$\frac{15 \text{ players}}{8 \text{ pizzas}} = \frac{10 \text{ players}}{p \text{ pizzas}} \quad \text{or} \quad \frac{15}{8} = \frac{10}{p}$$

There are no obvious factor/product relationships, so you can't use equivalent ratios.

Instead, you can use a calculator to convert $\frac{15}{8}$ to decimal form.

Since $\frac{15}{8} = 1.875$, the proportion becomes:

$$1.875 = \frac{10}{p}$$

If $10 \div p = 1.875$, then $10 \div 1.875$ must equal p.

$$10 \div 1.875 = p$$
$$5.\overline{3} = p$$

It isn't possible to order $5.\overline{3}$ pizzas. Even though $5.\overline{3}$ rounds to 5, the coach will need more than 5 pizzas. She should order 6 pizzas.

Exercises

1. Define the term *proportion*. Give an example.

2. In Ravi's aquarium, there are 3 green fish for every 2 yellow fish.
   (a) Write the ratio of green fish to yellow fish.

   (b) If Ravi had 10 yellow fish, how many green fish would there be?

   (c) If Ravi had 9 green fish, how many yellow fish would there be?

3. Find the unknown term.
   (a) $\frac{2}{3} = \frac{x}{15}$

   (b) $\frac{15}{30} = \frac{3}{x}$

   (c) $\frac{4}{x} = \frac{14}{21}$

   (d) 5 is to 3 as 20 is to x

   (e) $x : 5 = 60 : 50$

   (f) $\frac{4}{12} = \frac{x}{9}$

4. Use a proportion to convert each fraction to a percent.
   (a) $\frac{1}{5}$

   (b) $\frac{3}{4}$

   (c) $\frac{1}{8}$

   (d) $\frac{3}{12}$

5. Janessa has 5 pennies, 3 nickels, 4 dimes, and 2 quarters.
   (a) Write the ratio of nickels to pennies.

   (b) Janessa's mother has the same ratio of nickels to pennies in her pocket, except she has 15 pennies instead of 5. How many nickels does she have?

(c) Write the ratio of quarters to dimes.

(d) Suppose Janessa had the same ratio of quarters to dimes, but with 6 quarters instead of 2. How many dimes would she have?

6. Six pounds of apples cost $4.20. How much will it cost to buy 2 lb of apples?

7. Twelve large cans of juice cost $20. How much will it cost to buy 8 large cans? Round the answer to the nearest cent.

8. In a coffee shop, four mugs are sold for $22. How much will it cost to buy 10 mugs?

9. Five concert tickets are sold for $35. How much will it cost to buy 3 tickets?

10. Which goalkeeper had a better season? Why?
    - Goalkeeper A faced 1200 shots. On average, opposing teams scored 5 goals for every 40 shots they took.
    - Goalkeeper B also faced 1200 shots, but let in 6 goals for every 50 shots.

11. The quarterback of a football team has completed 10 out of 15 passes. If this continues, how many passes can he expect to complete if 40 passes are thrown? Round the answer to the nearest whole number.

12. Survey your class to find the number of people who usually:

(a) walk to school.

(b) take a bus.

(c) are driven to school.

(d) ride a bike.

Assume that your class represents the whole school. Use a proportion to estimate the number of students at your school who walk, take the bus, are driven, or ride bikes.

13. Use the data from Problem 12 to answer each question.

(a) What is the most popular means of transportation in your school?

(b) Why might it be useful for the instructors at your school to have this information?

(c) Do you think that the results for your school would be similar to the results for most schools in your area? Why or why not?

14. To conduct a fair survey about students at your school, it is a good idea to choose people who represent the school population. Ask an instructor to help you find the answers to these questions.

(a) How many boys are at your school? How many girls?

(b) How many different grades are there? How many students are in each grade?

15. Suppose you wanted to survey 20 students from your school. Use the information from Problem 14 to help you decide how many boys and how many girls you would ask. How many people from each grade would you include in your group of 20? Remember, the numbers you choose should be in proportion to the number of people in your school.

16. Create a survey about a topic that interests you. Survey 20 students who fit the characteristics you found in Problem 15.

Create a poster to show the data you collected and the conclusions you made about people at your school.

8.4 Mental Mathematics

In this lesson you learned to solve problems mentally by **converting numbers to different forms**. Recall:

- A percent is equal to a fraction in hundredths. For example, $25\% = \frac{25}{100}$.
- To write a fraction in simplest form, divide the numerator and denominator by the greatest common factor. For example, $\frac{25}{100} = \frac{25 \div 25}{100 \div 25} = \frac{1}{4}$.
- A percent can also be expressed as a decimal. For example, 25% = 25 hundredths = 0.25.
- To solve a problem mentally, first identify the calculation that must be made, then round the numbers to make them easier to calculate with.

Example

Mathias scored 35 out of 49 on a test. What is his mark in percent form?

Solution

Think about the problem.

Given: Mathias got 35 marks out of a possible 49, or $\frac{35}{49}$

Find: Mathias' mark in percent form

Make a plan.

Round $\frac{35}{49}$ to a fraction that is easy to convert to hundredths.

Solve the problem.

$\frac{35}{49}$ is about $\frac{35}{50}$.

You would have to double 50 to get 100 in the denominator, so you can double 35 in the numerator to find the percent.

$\frac{35}{50} = \frac{70}{100} = 70\%$

Mathias got about 70% on the test.

Look back.

The question asked for Mathias' mark in percent form.
This mark is approximately equal to 70%.

The answer 70% seems reasonable because $\frac{35}{50}$ is more than $\frac{30}{50}$ (60%) and less than $\frac{40}{50}$ (80%).

If you use a calculator to check, you find that $35 \div 49$ is about 0.71 or 71 hundredths or 71%.

Exercises

Complete these exercises without using a calculator.

1. Write the four steps you used to solve problems in this lesson. List some questions it is important to ask as you complete each step.

2. Write the decimal and percent equivalent for each fraction.

   (a) $\frac{1}{4}$ **(b)** $\frac{1}{5}$ **(c)** $\frac{1}{3}$

   (d) $\frac{1}{2}$ **(e)** $\frac{3}{4}$ **(f)** $\frac{3}{10}$

   (g) $\frac{1}{8}$ **(h)** $\frac{2}{3}$

3. Write the fractional form in simplest terms.

   (a) 22% **(b)** 6% **(c)** 5%

   (d) 99% **(e)** 18% **(f)** 60%

4. Write the decimal form.

   (a) 72% **(b)** 24% **(c)** 40%

   (d) 77.7% **(e)** 49% **(f)** 10%

   (g) 100% **(h)** 82% **(i)** 3%

   (j) 94% **(k)** 33.3% **(l)** 12.5%

5. Choose >, <, or = to make each statement true.

   (a) 0.3 ☐ 30% **(b)** 0.45 ☐ $\frac{7}{20}$ **(c)** $\frac{3}{5}$ ☐ 40%

   (d) 20% ☐ $\frac{4}{25}$ **(e)** 42% ☐ 4.2 **(f)** $\frac{2}{8}$ ☐ 25%

6. Estimate. (Hint: How can you use your answer from part (a) to help you find each other answer?)

   (a) 10% of 143 **(b)** 5% of 143 **(c)** 15% of 143

   (d) 20% of 143 **(e)** 50% of 143 **(f)** 35% of 143

7. Estimate.

(a) 1% of 150 **(b)** 2% of 2500 **(c)** 23% of $4.20

(d) 11% of $149 **(e)** 12% of 340 **(f)** 6% of 200

8. A customer left a 15% tip on a restaurant bill of $39.45. Estimate the amount of the tip.

9. A bike that usually costs $259.99 is on sale for 20% off.

(a) By about how much is the price reduced?

(b) Estimate the sale cost of the bike.

10. Estimate to decide which store offers the best price on a computer.

(a) Store A offers 30% off the regular price of $2499.

(b) Store B offers 40% off the regular price of $3199.

11. An arena holds 4500 people. If 1550 tickets were sold for a hockey game, about what fraction of the seats were empty?

12. Hyndman Peak, in Idaho, is 12,008 ft high. The world's highest mountain, Mt. Everest, has a height of 29,028 ft.

(a) What is the ratio of the height of Hyndman Peak to the height of Mt. Everest?

(b) Round the ratio to make it easier to work with, then write the simplified ratio in decimal form.

(c) About what percent of the height of Mt. Everest is represented by Hyndman Peak?

13. A pair of jeans costs $49.95 in a state where the tax rate is 7%. Estimate the after-tax cost of the jeans.

14. At 5 t, an African bush elephant has a mass equal to about 4% of a blue whale's mass. Estimate the mass of a blue whale.

15. When you are at rest, your lungs still contain about 2500 mL of air after you exhale. With a new breath you take in about 500 mL of fresh air. After a breath, about what percent of the air in your lungs is fresh?

16. Create a ten-question quiz about fractions, decimals, and percents. Answer each question to make sure the solution can be found mentally, then write an answer key on another sheet of paper. Exchange quizzes with a classmate, complete them, and compare results.

Include problems of each of these types:

(a) Express a fraction in decimal and percent form.
(b) Find a given percent of a number.
(c) Compare two numbers by converting them to the same form.
(d) Solve a word problem that involves percent.

8.5 Applications of Percent

In this lesson, you learned about **fractional percents** and **percents greater than 100 percent**. You also practised writing percents in different forms. Recall:

- A percent is a compact way of writing "out of one hundred," so a percent is really a fraction whose denominator is 100.
- Percents greater than 100 represent amounts greater than 1 whole.
- To change a percent to a decimal, move the decimal point 2 places to the left. For example, $128\frac{1}{4}\%$ is 128.25% or 128.25 hundredths or 1.2825.
- To write a decimal in percent form, find the number of hundredths. For example, 1.473 is equal to 147.3 hundredths or 147.3%. This is like moving the decimal point two places to the right.
- To change a percent to a fraction, write the percent with the denominator of 100 and then reduce to simplest terms. For example, 68% becomes $\frac{68}{100}$ or $\frac{17}{25}$.
- To change a fraction to a percent, write the decimal form and then write the number of hundredths as a percent. For example, $\frac{1}{8} = 0.125 = 12.5$ hundredths

  = 12.5%. If you have a calculator, you can enter [1] [÷] [8] [%]. (Some calculators use a different order.)

Example

In a 6 × 6 grid, what percent of the whole grid is represented by the squares around the perimeter?

Write the percent to the nearest tenth.

Solution

Draw a 6 × 6 grid to find out how many squares fit around the perimeter.

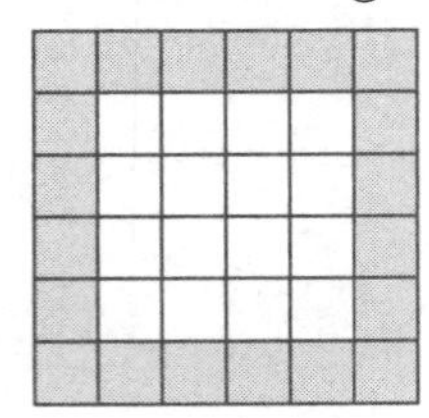

There are 20 squares around the perimeter, and 36 squares in all.

The perimeter squares are equal to $\frac{20}{36}$ or $\frac{5}{9}$ of the whole grid.

$\frac{5}{9} = 5 \div 9$	Use a calculator to write $\frac{5}{9}$ in decimal form.
$= 0.555\ldots$	The decimal form ends in 5, so round up.
$\doteq 55.6$ hundredths	
$= 55.6\%$	

The perimeter squares form 55.6% of the grid.

Exercises

1. Explain how you know that 24.6% is less than one whole and 154.2% is more than one whole.

2. Use an example to show why each strategy works.
(Hint: Remember that the percent represents the total number of hundredths.)

(a) To change a percent to a decimal, move the decimal point two places to the left.

(b) To change a decimal to a percent, move the decimal point two places to the right.

3. Complete the table.

Percent	Decimal	Fraction
	0.555	
$72\frac{1}{2}\%$		
		$\frac{1}{3}$
145%		

4. Write the decimal to the nearest thousandth.

(a) 42% **(b)** $12\frac{1}{2}\%$ **(c)** $133\frac{1}{3}\%$

(d) $68\frac{1}{8}\%$ **(e)** 174% **(f)** $199\frac{9}{10}\%$

5. Write the fraction in simplest terms so the numerator and denominator are both whole numbers.

(a) 36% **(b)** $58\frac{1}{3}\%$ **(c)** $115\frac{1}{2}\%$

(d) $72\frac{3}{4}\%$ **(e)** $125\frac{1}{5}\%$ **(f)** $136\frac{1}{4}\%$

6. Write the percent.

(a) $\frac{5}{6}$ **(b)** $\frac{1}{15}$ **(c)** $2\frac{1}{8}$

(d) $6\frac{1}{6}$ **(e)** $3\frac{3}{8}$ **(f)** $6\frac{1}{5}$

7. If 3 blocks represent $37\frac{1}{2}\%$ of a whole, how many blocks represent the whole? (Hint: How else can you write $37\frac{1}{2}\%$?)

8. Aleta made a large 10×10 grid. She folded it in half, shaded half the squares, and wrote the shaded part of the grid as a percent: $\frac{50}{100} = 50\%$. Then she folded the unshaded part in half and shaded the new half a different color. She counted the shaded parts again and wrote $\frac{25}{100} = 25\%$.

(a) If Aleta repeated this procedure three more times, what percent of the grid did she shade in the final step?

(b) Use a 10×10 grid to recreate Aleta's work. Which shaded parts of the grid represent 150% of another part?

9. Use >, <, or =.

(a) $1.65 \ \square \ 165\%$

(b) $1\frac{2}{7} \ \square \ 127\%$

(c) $3\frac{1}{2} \ \square \ 3.5\%$

(d) $257\frac{1}{3}\% \ \square \ 2.57\overline{3}$

(e) $115\frac{1}{3}\% \ \square \ \frac{11}{9}$

(f) $142.9\% \ \square \ \frac{17}{12}$

For Problems 10 to 12, write each percent as a mixed number in simplest terms.

10. At birth, Americans have a life expectancy of about 77 years. A beaver has a life expectancy of about 5 years. What percent of an American's life expectancy is a beaver's life expectancy?

11. At Neel's Diner, a turkey dinner with vegetables, salad, drink, and dessert costs \$12.50 per serving. The same meal can be prepared at home for about \$3.95 per serving. What percent of the home cost is represented by the restaurant cost?

12. During a real estate boom, the price of a house rose from \$200,000 to \$585,000 in ten years. What percent of the original price is represented by the increase in value?

13. Create one problem of each type on the list. Choose problems that involve fractional percents or percents greater than 100. Write the solutions and then exchange problems with a classmate.

(a) converting percents to fractions and decimals

(b) converting fractions and decimals to percents

(c) using <, >, or = to compare fractions, decimals, and percents

(d) solving word problems

14. (a) Create a word problem about a percent situation.

(b) Write a step-by-step solution and analyze each step to determine where errors might occur.

(c) On a clean sheet of paper, rewrite the solution so that it contains an error. Exchange with a classmate and identify the errors.

9 MEASUREMENT

9.1 Units of Measurement

You have learned about the various units used in the American and metric systems of measurement.

- To convert from one unit to another, you need to know the **conversion** factor. For example, to convert feet to inches, you multiply the number of feet by the conversion factor "12 inches per foot" or $\frac{12 \text{ in.}}{1 \text{ ft}}$.
- The metric system uses the base units **meter** (for length), **gram** (for mass), and **liter** (for capacity).
- The metric system uses a set of standard prefixes to identify units smaller or larger than the base units. You can use a table like the one shown in Example 1 to convert from one metric unit to another.
- To convert between temperatures in degrees **Fahrenheit** and degrees **Celsius**, use these formulas: $F = \frac{9}{5}C + 32$ and $C = \frac{5(F - 32)}{9}$.

Example 1

Convert 3.2 kilometers to meters.

Solution

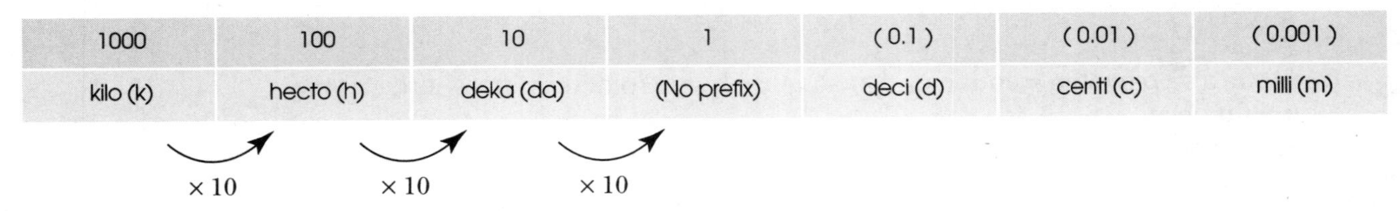

1000	100	10	1	(0.1)	(0.01)	(0.001)
kilo (k)	hecto (h)	deka (da)	(No prefix)	deci (d)	centi (c)	milli (m)

To change kilometers to meters, move three places to the right along the table.

Each time you move one place, you are multiplying the unit size by 10, so you are making 10 times as many units. To convert 3.2 km to meters, multiply 3.2 by 10 × 10 × 10.

3.2 × 10 × 10 × 10 = 3200
Therefore, 3.2 km is equal to 3200 m.

Example 2

Convert 3.5 tons to ounces.

Solution

There are 2000 pounds in each ton, and 16 ounces in each pound.
There must be 2000×16 or 32,000 ounces in 1 ton.
The conversion factor is $\frac{32{,}000 \text{ oz}}{1 \text{ t}}$.

$$3.5 \text{ t} = 3.5 \text{ t} \times \frac{32{,}000 \text{ oz}}{1 \text{ t}}$$

$$= 3.5 \not{t} \times \frac{32{,}000 \text{ oz}}{1 \not{t}}$$

$$= 112{,}000 \text{ oz}$$

Therefore, 3.5 t is equal to 112,000 oz.

Exercises

1. Summarize what you have learned about units of measurement and conversions.

2. Which type of conversion do you find easiest to make? Which do you find most difficult? Explain your answers.

(a) from one American unit to another

(b) from one metric unit to another

(c) from American units to metric units

(d) from metric units to American units

3. Describe some advantages and disadvantages of using each system of measurement.

(a) the American system

(b) the metric system

4. Complete each statement. (Hint: Review the **Tutorial** if necessary.)

(a) 1 ft = ________ in. **(b)** ________ oz = 1 lb

(c) 1 c = ________ fl oz

(d) ________ pt = 1 qt

(e) ________ ft = 1 yd

(f) 1 yd = ________ in.

(g) ________ lb = 1 t

(h) 1 pt = ________ c

(i) ________ qt = 1 gal

5. Use the information you found in Problem 4 to make the conversions.

(a) 4 ft to inches

(b) 24 in. to feet

(c) 8 yd to inches

(d) 5 yd to feet

(e) 1320 ft to miles

(f) 80 oz to pounds

(g) 12.4 t to pounds

(h) 3 qt to pints

(i) 20 qt to gallons

(j) 3 gal to fluid ounces

6. Convert.

(a) 5.7 m = ________ cm

(b) 7.36 km = ________ dam

(c) 0.593 cm = ________ mm

(d) 326 mg = ________ g

(e) 5.6 L = ________ mL

7. Measure each object in inches. Then convert each measurement to centimeters. (Hint: 1 in. = 2.54 cm)

(a) width of a dollar bill

(b) diameter of a quarter

(c) width of a sheet of typing paper

(d) distance between the Q and P keys on a computer keyboard, measured center to center

8. Convert.

(a) 14 ft = ________ m   1 ft = 30.48 cm

(b) 6 m = ________ yd   1 yd = 1.09 m

(c) 10 lb = ________ kg   1 lb = 0.45 kg

(d) 0.75 fl oz = ________ mL   1 fl oz = 29.56 mL

(e) 9 L = ________ gal   1 L = 0.27 gal

9. Complete the solution.

Problem

Convert 50°F to degrees Celsius.

Solution

$$C = \frac{5(F-32)}{9}$$

$$= \frac{5(50-32)}{9}$$

10. Convert.

(a) 36.2°C = ________ °F   **(b)** 70°F = ________ °C

(c) –10°C = ________ °F   **(d)** –20°F = ________ °C

11. With a partner, choose a measurement you can make in your classroom. Make the measurement in American units and have your partner do the same to check.

Review the lesson or use an almanac to find a conversion factor you can use to convert the measurement to an appropriate metric unit. Make the conversion and then compare your result with your partner's. As an additional check, re-measure using the metric units.

9.2 Classifying Angles

In this lesson, you measured angles with a protractor and found their complements and supplements.

- Two angles are **complementary** if they have a sum of 90°.
- Two angles are **supplementary** if they have a sum of 180°.

Example 1

Measure this angle to the nearest 5° and find its complement and supplement.

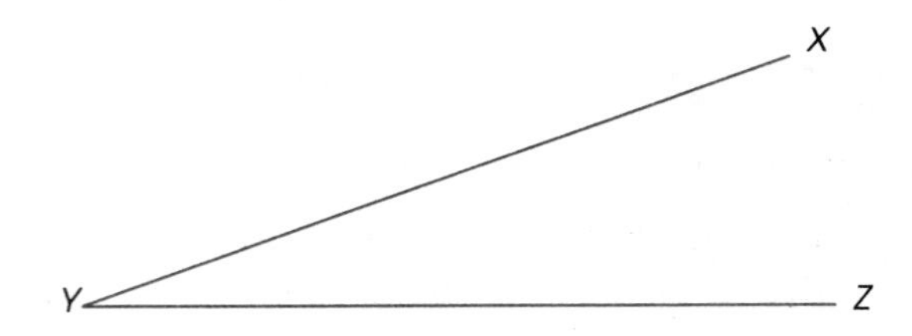

Solution

$\angle XYZ$ is less than 90°, so the measurement must be 20°, not 160°.

You can find the angle's complement by subtracting its measure from 90°.
$90° - 20° = 70°$

The complement of $\angle XYZ$ is 70°.

You can find the angle's supplement by subtracting its measure from 180°.
$180° - 20° = 160°$

The supplement of $\angle XYZ$ is 160°.

Notice that 160° is the measurement that matches 20° on the opposite protractor scale.

Example 2

Construct an angle that is supplementary to $\angle JKL$ and use a protractor to prove that the two angles are supplementary.

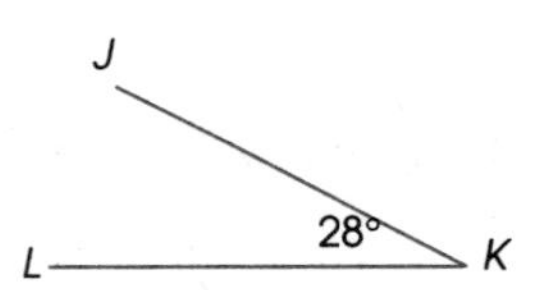

Solution

Extend the base of $\angle JKL$ to form a straight angle, $\angle LKM$.

$\angle JKL$ combines with $\angle JKM$ to fill 180°, so the measure of $\angle JKM$ must be $180° - 28°$ or 152°.

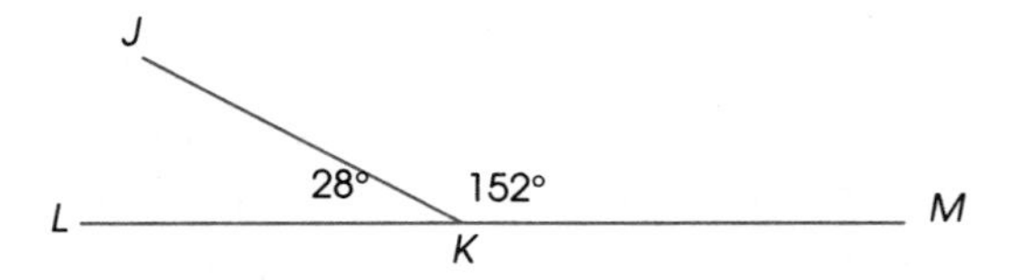

Measuring with a protractor confirms that the supplementary angle is 152°.

Note that two angles can be supplementary even if they don't adjoin to form a straight angle, as long as they *could* adjoin this way.

Exercises

1. Find all the possible pairs of complementary and supplementary angles in the diagram.
   (Hint: Look for a pattern to help you make sure you've found all the supplementary angles.)

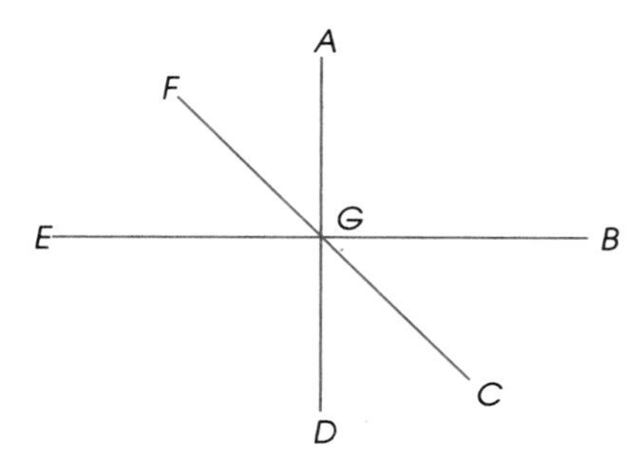

2. Is each statement true or false? Explain how you know.
   (a) A right angle and an acute angle can be supplementary.

   (b) Two acute angles can be complementary.

   (c) Two obtuse angles can be supplementary.

   (d) Two acute angles can be supplementary.

   (e) Two right angles are always supplementary.

3. Measure this angle to the nearest 5° and find its complement and supplement.

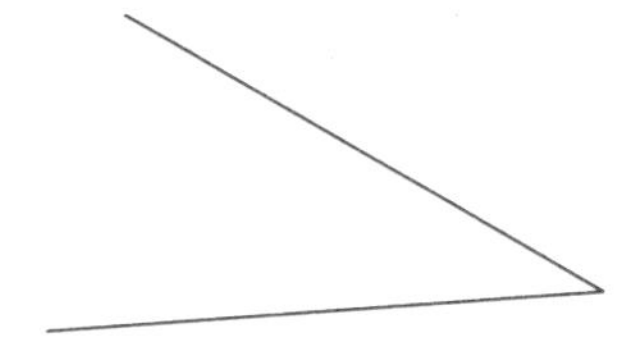

4. An angle measures 82°. Calculate its complement and supplement.

5. What is the value of x? Explain how you know.

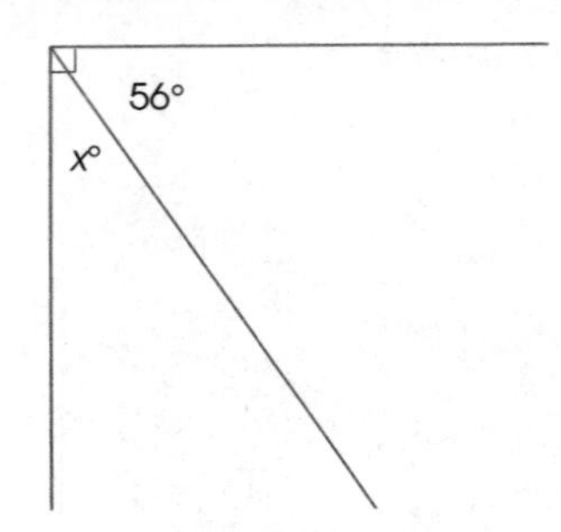

6. Is each pair of angles complementary, supplementary, or neither?

(a) 53°, 37°

(b) 101°, 81°

(c) 56°, 124°

(d) 17°, 173°

(e) 35°, 145°

(f) 29°, 61°

7. Find the supplementary angle.

(a) 34°

(b) 114°

(c) 90°

(d) 79°

(e) 177°

(f) 129°

8. Find the complementary angle.

(a) 34°

(b) 1°

(c) 90°

(d) 77°

(e) 0°

(f) 29°

9. Complete each statement.

(a) 82° and 8° are ____________________ angles because their sum equals _____°.

(b) 151° and 29° are ____________________ angles because their sum equals _____°.

10. Construct this angle. Then construct an angle that is:

(a) complementary.

(b) supplementary.

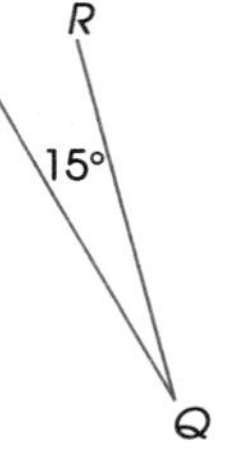

11. In a quadrilateral where the four vertices all touch the circumference of a circle, angles across from each other are supplementary. Find values for x and y.

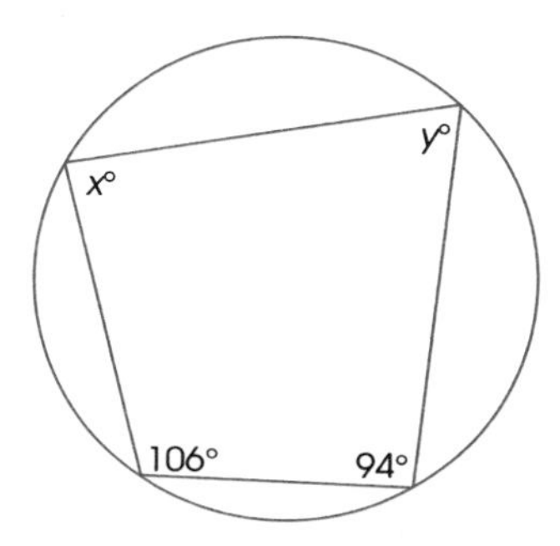

12. Create three problems of your own that are similar to problems on this page. Write each solution and test your problems by exchanging with a classmate.

9.3 Angles and Parallel Lines 1

In this lesson, you examined the angles that form where a **transversal** intersects **two parallel lines**.

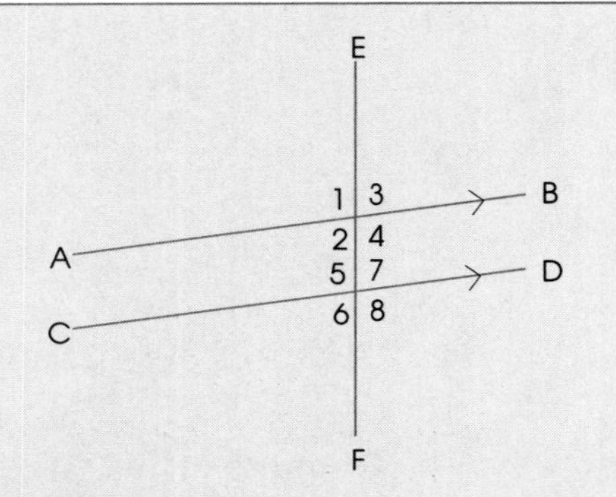

Recall:

- **Corresponding angles** are pairs of equal angles formed on the same side of the transversal, such as ∠3 and ∠7 in the diagram. They occur because the transversal approaches the two parallel lines in the same way.
- **Supplementary angles** are pairs of angles that form a straight angle, such as ∠1 and ∠3 in the diagram. These angles have a sum of 180°.
- **Vertically opposite angles** are equal angles formed by two intersecting lines, such as ∠1 and ∠4 in the diagram.

You can use these relationships to find the measures of angles where some measurements are given.

Example 1

Identify each pair of angles from the diagram above as corresponding, supplementary, or vertically opposite. Explain your reasons.

∠1 and ∠5 ∠2 and ∠3 ∠5 and ∠7
∠6 and ∠7 ∠4 and ∠8 ∠3 and ∠4

Solution

Organize the information in a chart.

Angles	Relationship
∠1 and ∠5	corresponding, since they are equal angles on the same side of a transversal
∠2 and ∠3	vertically opposite, since they are equal angles formed by two intersecting lines
∠5 and ∠7	supplementary, since they form a straight angle
∠6 and ∠7	vertically opposite, since they are equal angles formed by two intersecting lines
∠4 and ∠8	corresponding, since they are equal angles on the same side of a transversal
∠3 and ∠4	supplementary, since they form a straight angle

Example 2

Find the measures of angles *a*, *b*, *c*, *d*, and *e*. Give reasons for your answers.

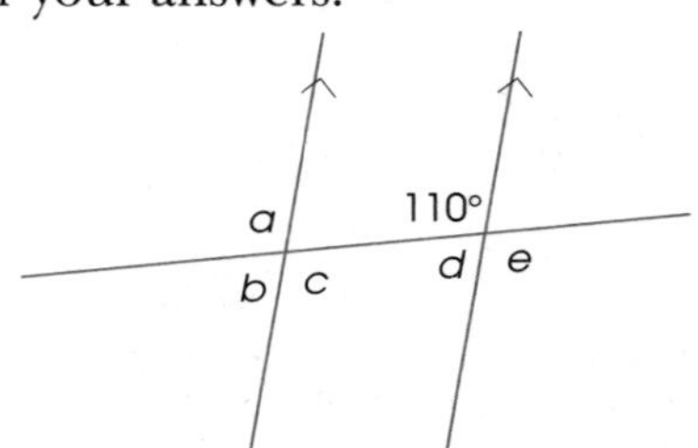

Solution

Organize the information in a chart.

Angle Size	Reason
∠*a* = 110°	∠*a* corresponds to the angle labeled 110°, so ∠*a* = 110°.
∠*b* = 70°	∠*a* and ∠*b* combine to form a straight angle. So ∠*b* = 180° - ∠*a* or 180° -110°.
∠*c* = 110°	∠*c* is vertically opposite ∠*a*, so ∠*c* = ∠*a*.
∠*d* = 70°	∠*d* and the angle labeled 110° form a straight line. So ∠*d* = 180° - 110°.
∠*e* = 110°	∠*e* is vertically opposite the angle labeled 110°, so ∠*e* must also be equal to 110°.

Exercises

1. Define each term.

(a) parallel **(b)** intersect **(c)** transversal

2. Trace along both sides of a ruler to create two horizontal parallel lines. Label the top line *AB* and the bottom one *CD.*

Draw a transversal so it meets *AB* at an angle of 75°. Label the transversal *EF.* Then label the angles in your drawing as shown.

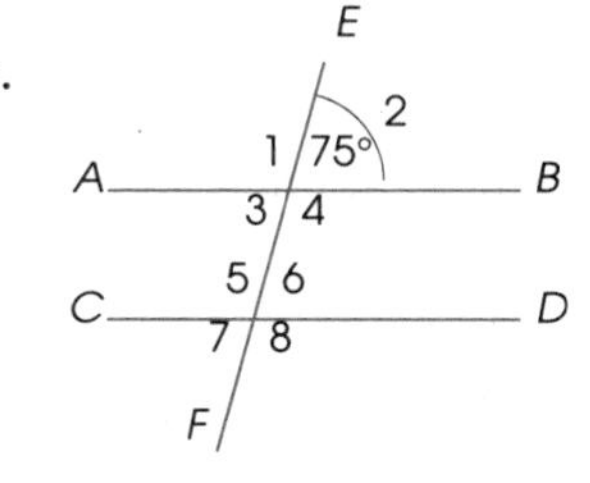

3. Use the diagram from Problem 2.

(a) State three angles that are equal to $\angle 1$. Explain your reasoning.

(b) State three angles that are equal to $\angle 2$. Explain your reasoning.

4. Use the diagram from Problem 2.

(a) The pairs of corresponding angles in the diagram are:

$\angle 1$ and $\angle$ ☐ $\angle 2$ and $\angle$ ☐

$\angle 3$ and $\angle$ ☐ $\angle 4$ and $\angle$ ☐

(b) The pairs of vertically opposite angles are:

$\angle 1$ and $\angle$ ☐ $\angle 2$ and $\angle$ ☐

$\angle 5$ and $\angle$ ☐ $\angle 6$ and $\angle$ ☐

5. Explain how each pair of angles is related.

(a)

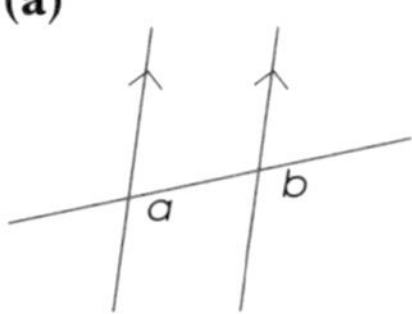

(b)

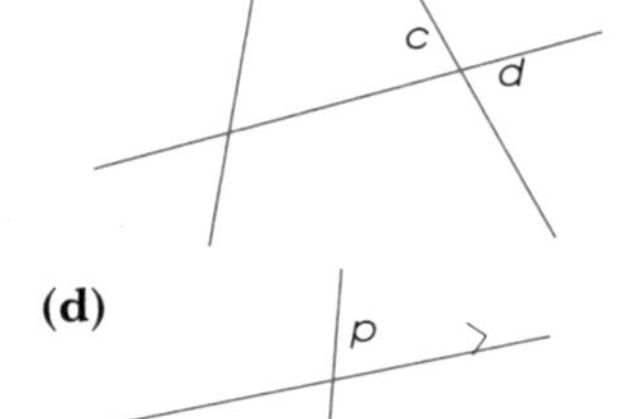

(c)

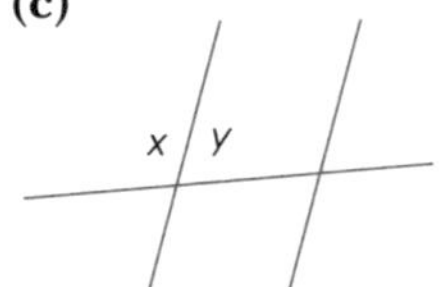

(d)

p

q

6. Which angles in this diagram, besides the labeled angle, measure 65°? Explain your reasoning.

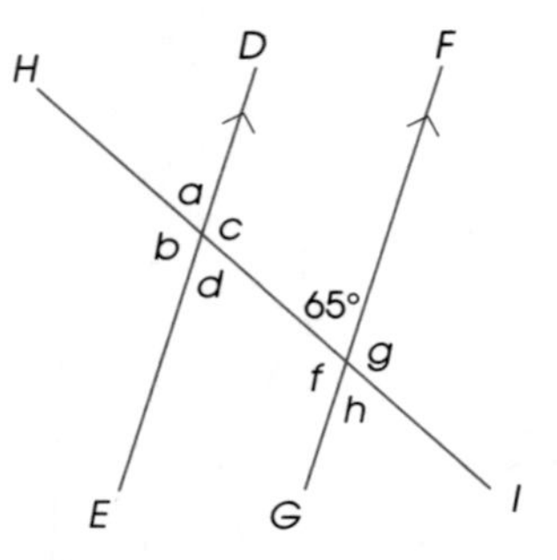

7. Use the diagram to complete each statement.

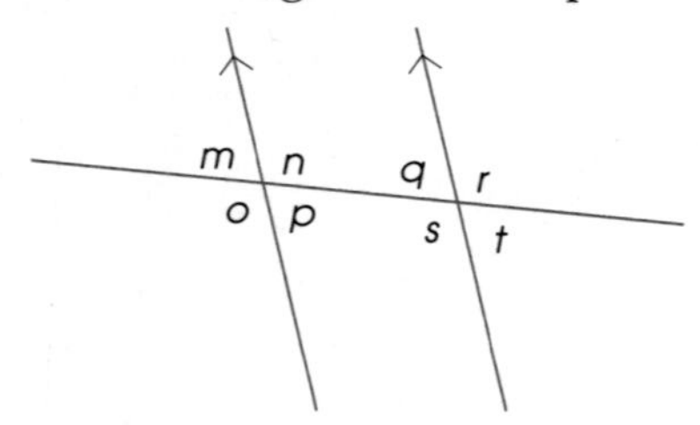

(a) If $\angle m = 35°$, then $\angle p = \square°$ because

(b) If $\angle r = 130°$, then $\angle n = \square°$ because

(c) If $\angle s = 139°$, then $\angle t = \square°$ because

(d) If $\angle q = 56°$, then $\angle m = \square°$ because

(e) If $\angle o = 125°$, then $\angle \square = 125°$ because

(f) If $\angle q = 45°$, then $\angle \square = 45°$ because

8. Why is it impossible for all the statements in Problem 7 to be true at the same time?

9. Complete each statement.

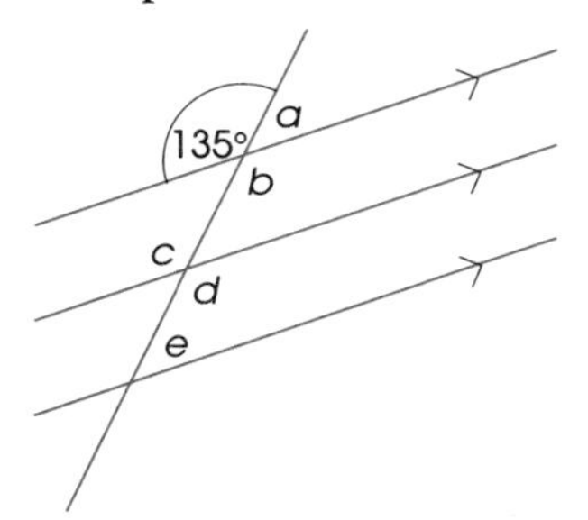

(a) $\angle a =$ ☐° because

(b) $\angle b =$ ☐° because

(c) $\angle c =$ ☐° because

(d) $\angle d =$ ☐° because

(e) $\angle e =$ ☐° because

10. Use the diagram.

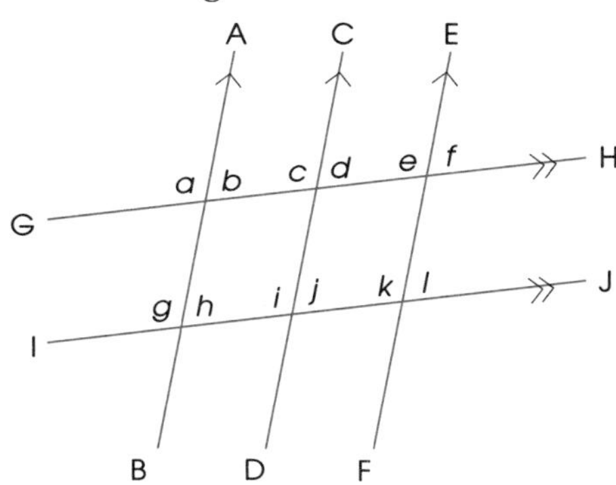

(a) Name four pairs of parallel lines.

(b) Name a transversal for each pair of parallel lines.

(c) Identify all the labeled angles that correspond to $\angle d$.

9.4 Angles and Parallel Lines 2

In this lesson, you examined more angles that form where a transversal intersects two parallel lines.

- **Same-side interior angles**, such as $\angle 2$ and $\angle 5$, are supplementary angles formed inside the parallel lines.
- **Same-side exterior angles**, such as $\angle 1$ and $\angle 6$, are supplementary angles formed outside the parallel lines.
- **Alternate interior angles**, such as $\angle 2$ and $\angle 7$, are equal angles formed inside the parallel lines, on opposite sides of the transversal.
- **Alternate exterior angles**, such as $\angle 1$ and $\angle 8$, are equal angles formed outside the parallel lines, on opposite sides of the transversal.

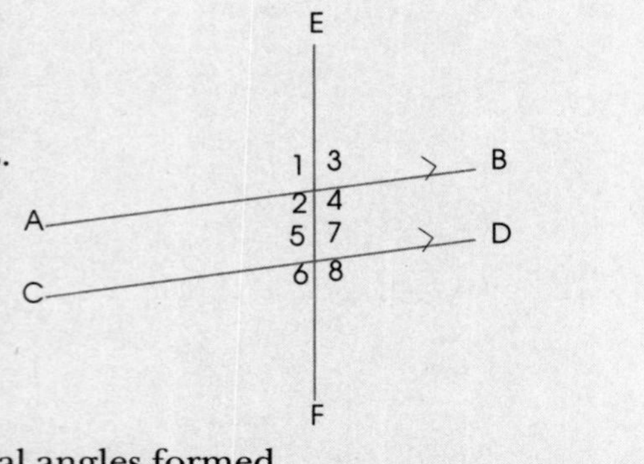

Example 1

Identify each pair of angles from the diagram above as equal or supplementary. Use interior or exterior angle relationships to explain your choices.

$\angle 4$ and $\angle 7$ $\quad$ $\angle 3$ and $\angle 8$
$\angle 4$ and $\angle 5$ $\quad$ $\angle 3$ and $\angle 6$

Solution

Organize the information in a chart.

Angles	Relationship
$\angle 4$ and $\angle 7$	supplementary, since they are same-side interior angles
$\angle 3$ and $\angle 8$	supplementary, since they are same-side exterior angles
$\angle 4$ and $\angle 5$	equal, since they are alternate interior angles
$\angle 3$ and $\angle 6$	equal, since they are alternate exterior angles

Example 2

Find the measures of angles *f*, *g*, *h*, and *i*. Use interior or exterior angle relationships to explain your choices.

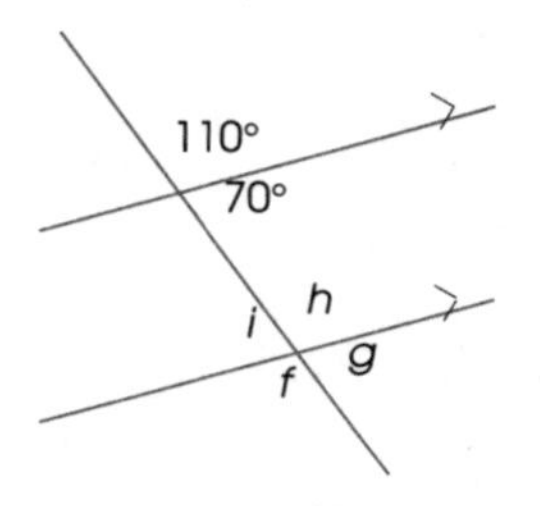

Solution

Organize the information in a chart.

Angle Size	Reason
$\angle f = 110°$	$\angle f$ is equal to the angle marked 110° because these two angles are alternate exterior angles.
$\angle g = 70°$	$\angle g$ and the angle marked 110° are same-side exterior angles, so $\angle g = 180° - 110°$ or 70°.
$\angle h = 110°$	$\angle h$ and the angle marked 70° are same-side interior angles, so $\angle h$ is equal to $180° - 70°$ or 110°.
$\angle i = 70°$	$\angle i$ is equal to the angle marked 70° because these two angles are alternate interior angles.

Exercises

1. Define each term.
   (a) interior angles

   (b) exterior angles

   (c) alternate angles

2. List some everyday situations that involve parallel lines and transversals.

3. Explain three different ways to prove that $\angle 2 = \angle 7$.

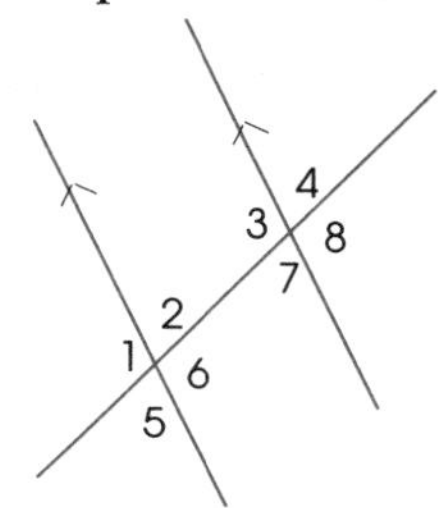

4. Find these angles in the diagram.
   (a) two pairs of same-side interior angles

   (b) two pairs of same-side exterior angles

   (c) two pairs of alternate interior angles

   (d) two pairs of alternate exterior angles

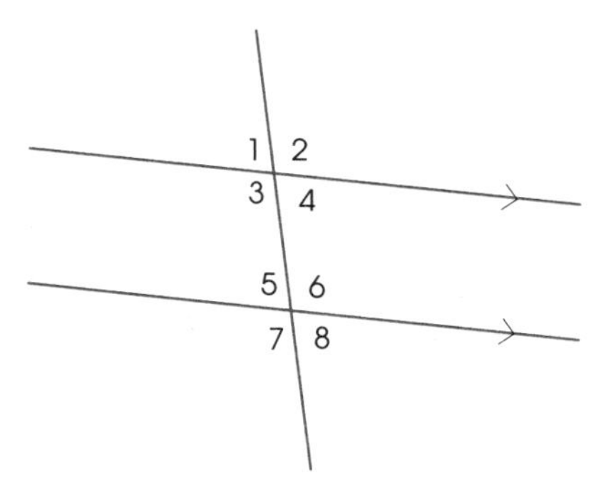

5. Use the diagram to complete each statement.

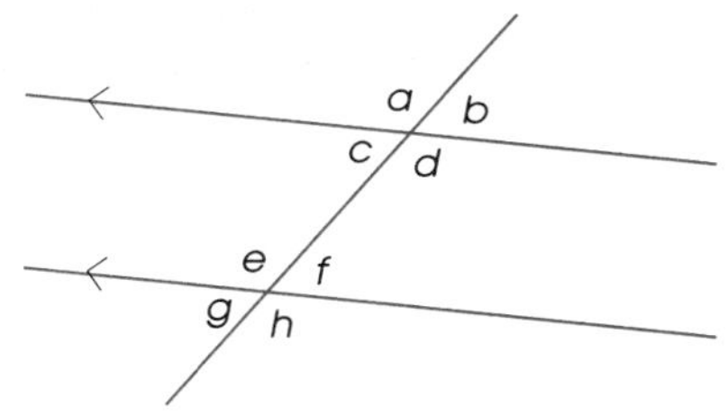

(a) If $\angle b = 32°$, then $\angle g = \square°$, because

(b) If $\angle c = 57°$, then $\angle e = \square°$, because

(c) If $\angle d = 136°$, then $\angle e = \square°$, because

(d) If $\angle a = 117°$, then $\angle g = \square°$, because

6. What size is each labeled angle? Explain how you know.

(a)

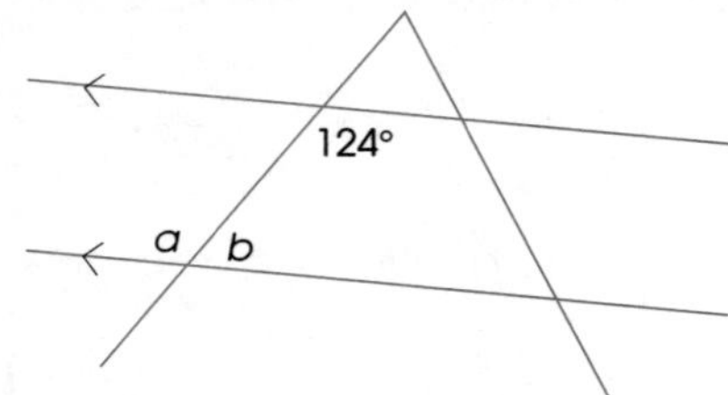

(b)

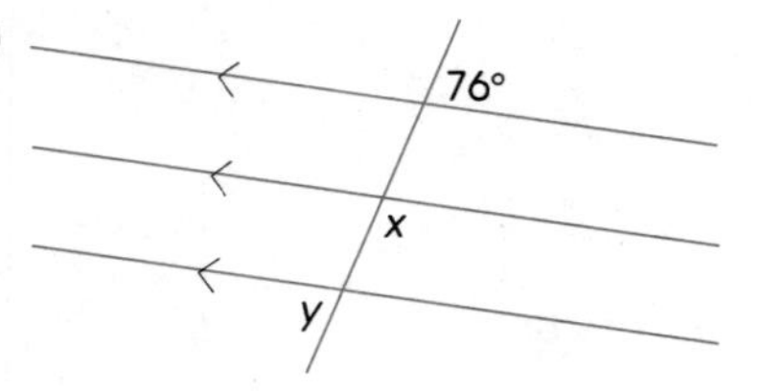

(c)

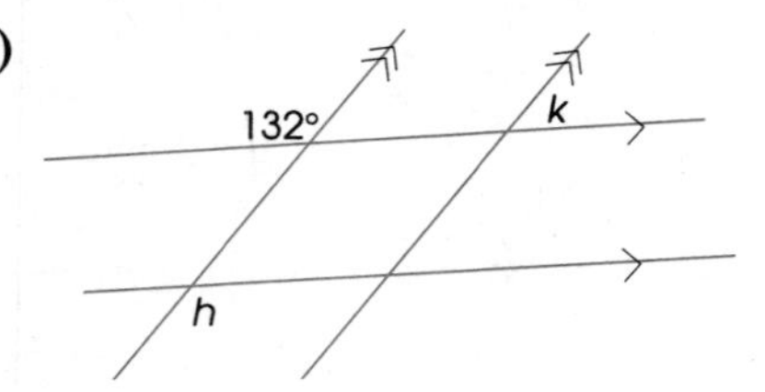

7. Are the lines parallel? Explain how you know.

(a)

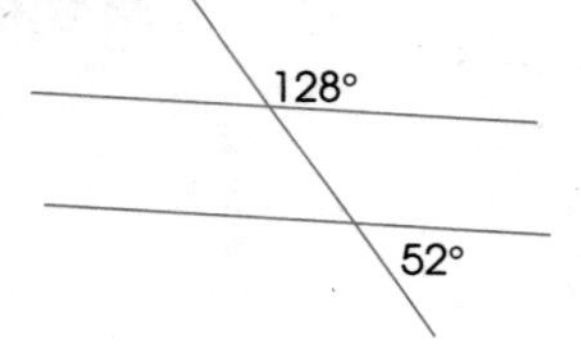

(b)

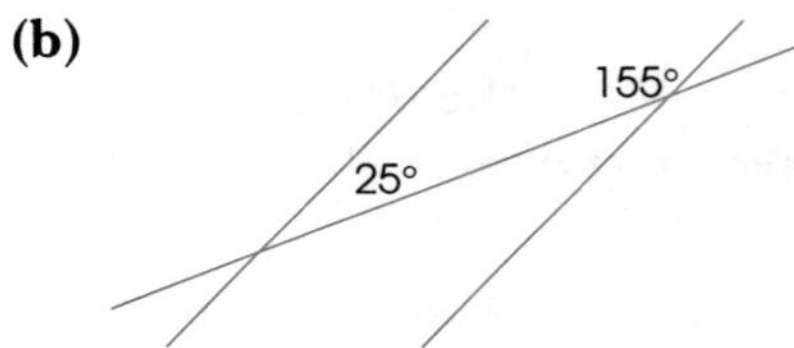

(c)

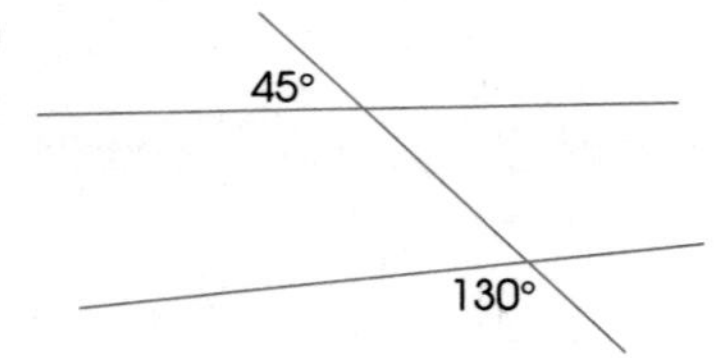

(d)

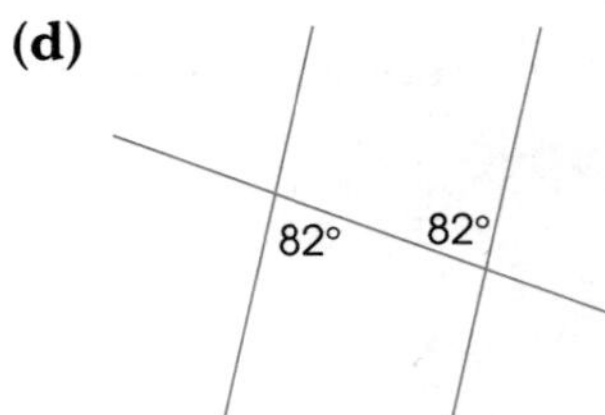

8. A commercial lift is used as a platform.

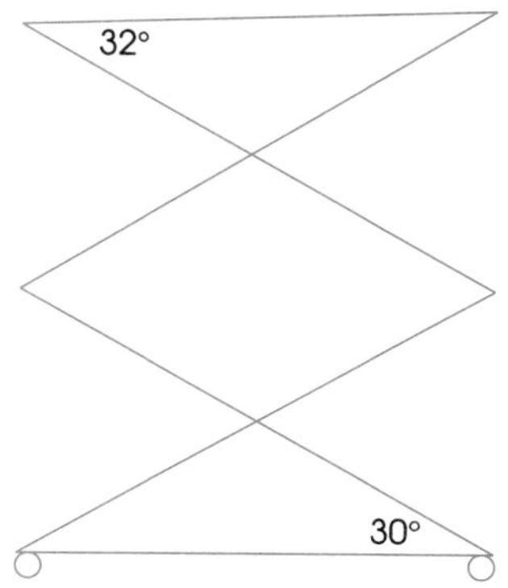

(a) Is the top of the platform level with the base? Explain how you know.

(b) How could you make sure the platform is flat?

(c) If the platform is raised, what will happen to the given angles?

9. George plans to pour concrete into a square form to make a garage floor. How could George use angles and transversals to make sure the form is exactly square? Include a diagram with your answer.

10. For Problem 2, you listed some everyday situations that involve parallel lines and transversals. Choose one situation and create a problem about an unknown angle size. Write a solution to your problem on another sheet of paper and then exchange problems with a partner and solve.

9.5 The Pythagorean Relationship

In this lesson, you developed the Pythagorean relationship for **right triangles** and used it to solve problems. Recall:

- The Pythagorean relationship states that the square of the hypotenuse of a right triangle is equal to the sum of the squares of the other two sides. The algebraic form of this rule is $a^2 + b^2 = c^2$.
- The Pythagorean relationship can only be used with right triangles.

Example 1

What is the length of the hypotenuse, rounded to one decimal place?

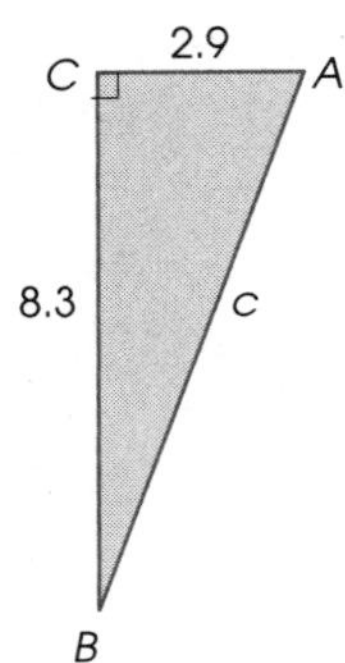

Solution

ABC is a right triangle. AB is the longest side, so it is the hypotenuse.

The Pythagorean relationship states that the length of AB^2 is equal to $BC^2 + AC^2$.

$a^2 + b^2 = c^2$	Write the Pythagorean relationship.
$8.3^2 + 2.9^2 = c^2$	Substitute the known side lengths for a and b.
$68.89 + 8.41 = c^2$	Calculate the square numbers.
$77.3 = c^2$	Add the square numbers.
$\sqrt{77.3} = \sqrt{c^2}$ $8.8 \doteq c$	If two numbers are equal, their square roots must be the same. Find the square root of 77.3. Round to one decimal place.

The hypotenuse is approximately 8.8 units long.

Example 2

Is this a right triangle?

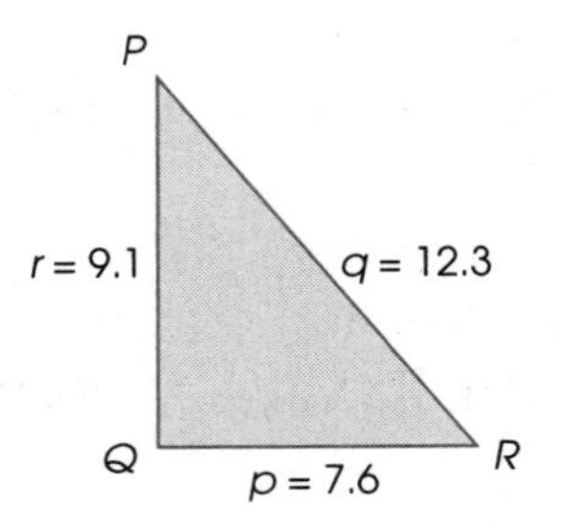

Solution

If this is a right triangle, then the square of the hypotenuse is equal to the sum of the squares on the other two sides. The longest side is 12.3, so q must be the hypotenuse.

L.S.	R. S.
$p^2 + r^2$	q^2
$= 7.6^2 + 9.1^2$	$= 12.3^2$
$= 57.76 + 82.81$	$= 151.29$
$= 140.57$	$= 151.29$

The two sides are not equal, so this is not a right triangle.

Exercises

1. Write the Pythagorean relationship in words and in algebraic form. Draw the squares on the sides of a triangle and compare the areas to illustrate the relationship.

2. Define each term.

(a) right triangle

(b) hypotenuse

(c) side

(d) perimeter

3. Explain how you can use the Pythagorean relationship to find out whether a triangle contains a right angle.

4. Calculate the unknown side length to two decimal places.

(a)

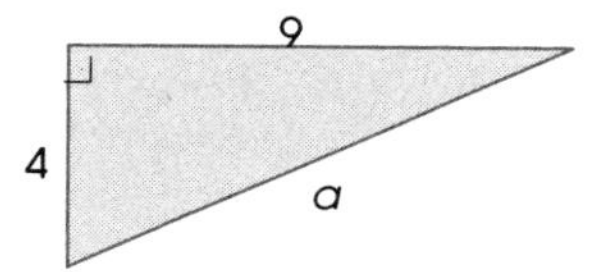

(b)

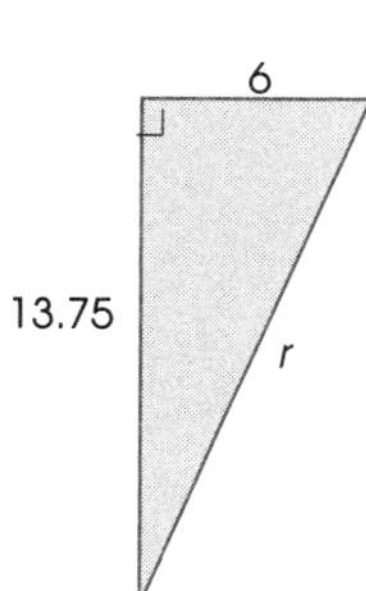

(c)

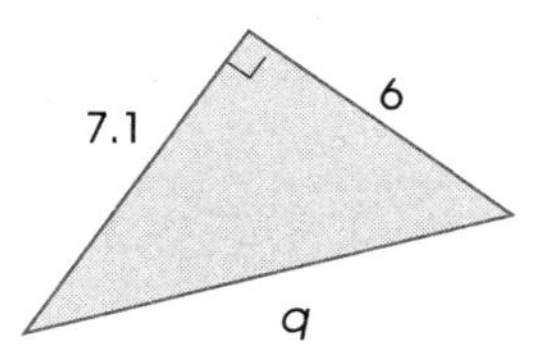

(d)

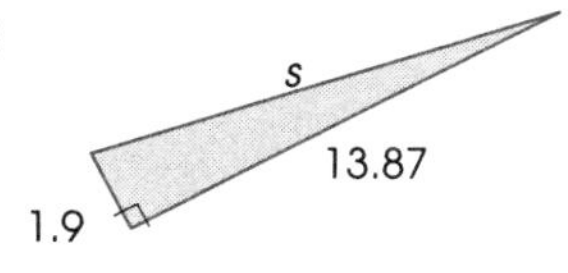

(e)

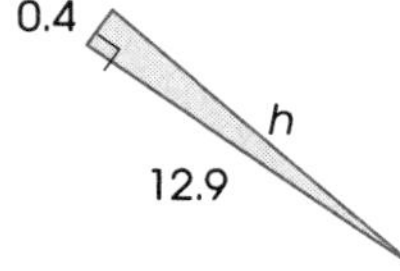

(f)

11.7

8.5

t

5. Solve.

(a) $2^2 + 5^2 = n^2$

(b) $3.1^2 + p^2 = 6.1^2$

(c) $m^2 + 3.9^2 = 4.7^2$

(d) $12.1^2 + b^2 = 17.2^2$

(e) $g^2 + 7.0^2 = 19.5^2$

(f) $16.6^2 + f^2 = 22.0^2$

6. Which triangles contain right angles? Explain how you can tell this without drawing and measuring.

(a)

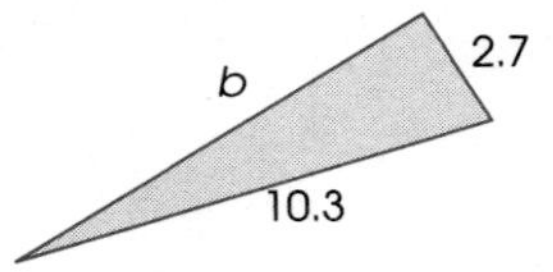

(b)

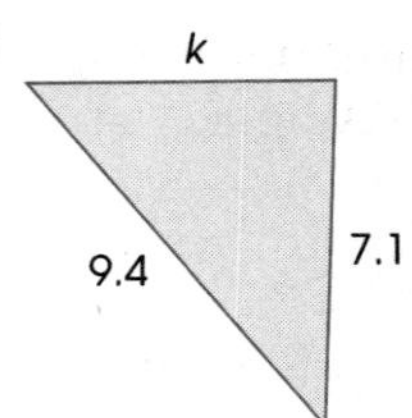

(c)

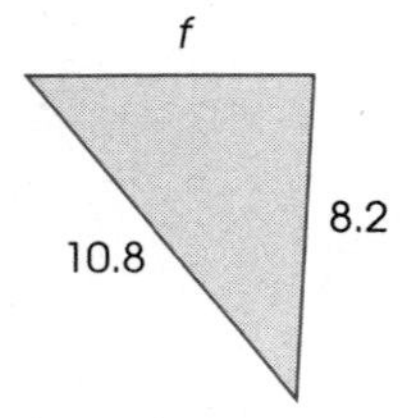

(d)

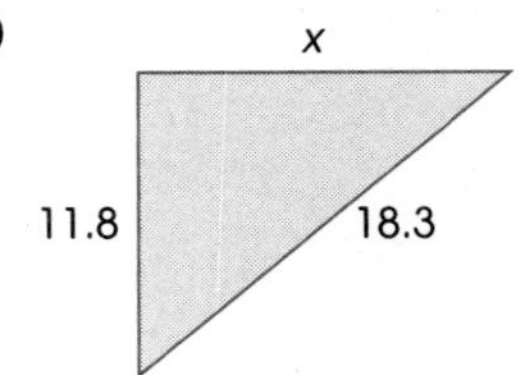

(e)

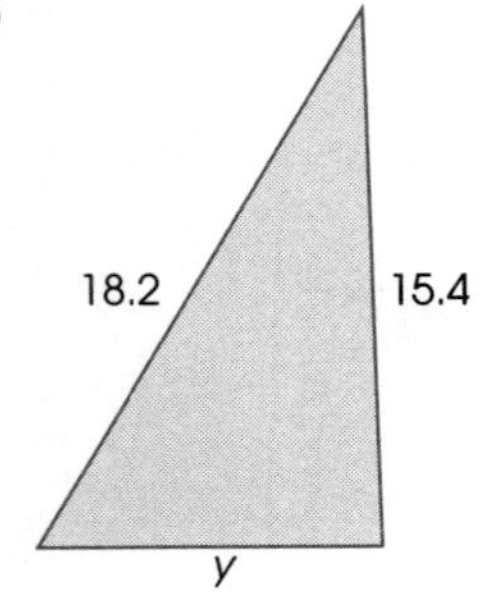

(f)

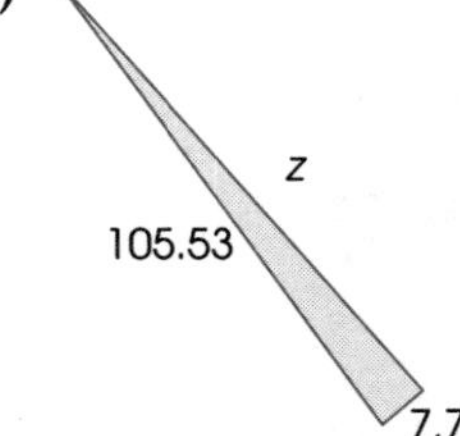

7. Is this a possible right triangle? Explain.

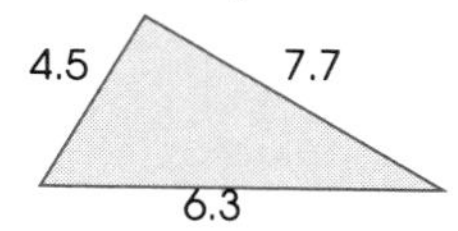

8. Sketch the triangle represented by this equation. Then find and correct the error in the solution.

$$\begin{aligned} 4.3^2 + x^2 &= 8.5^2 \\ 18.49 + x^2 &= 72.25 \\ x^2 &= 72.25 - 18.49 \\ x &= 53.76 \end{aligned}$$

9. Use substitution to verify your corrected solution from Problem 8. Why is the left side of the equation not exactly equal to the right side?

10. Create three problems you could solve using the Pythagorean relationship, so each problem involves a different type of situation.

Write the solution to each problem. Then exchange problems with a classmate, solve, and compare solutions.

10 AREA, PERIMETER AND VOLUME

10.1 Area and Perimeter 1

You have learned how to determine the areas and perimeters of quadrilaterals and circles. Recall:

- The **perimeter** of a figure is the distance around the outside. The **area** is the number of square units inside.
- The perimeter of any figure is equal to the sum of the side lengths. If some sides have the same length, you can shorten your calculations by combining them.
- The perimeter of a circle is called the **circumference.** It is equal to $\pi \times$ *diameter*, or about $3.14 \times$ *diameter.*
- For a square or rectangle, the area is equal to *length* $\times$ *width.*
- For a parallelogram or rhombus, the area is equal to *base* $\times$ *height.*
- For a trapezoid, the area is equal to $\frac{1}{2} \times (base + top) \times height$.
- For a circle, the area is equal to $\pi \times radius^2$, or about $3.14 \times radius^2$.

Example 1

Find the perimeter and area of this parallelogram.

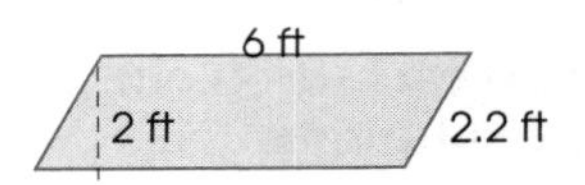

Solution

Step 1: **Find the perimeter.**
The perimeter is the distance around the outside.

For a parallelogram, the perimeter is equal to

$2 \times$ *length* $+ 2 \times$ *width.*

$$\begin{aligned} Perimeter &= 2L + 2w \\ &= (2 \times 2.2 \text{ ft}) + (2 \times 6 \text{ ft}) \\ &= 4.4 \text{ ft} + 12 \text{ ft} \\ &= 16.4 \text{ ft} \end{aligned}$$

The perimeter of the parallelogram is 16.4 ft.

Step 2: **Find the area.**
The area is the number of square units inside the figure.

To find the area of a parallelogram, multiply the base length by the height.

$$\begin{aligned} Area &= b \times h \\ &= 6 \text{ ft} \times 2 \text{ ft} \\ &= 12 \text{ ft}^2 \end{aligned}$$

The area of the parallelogram is 12 ft^2.

Example 2

Find the circumference and area of this circle to two decimal places.

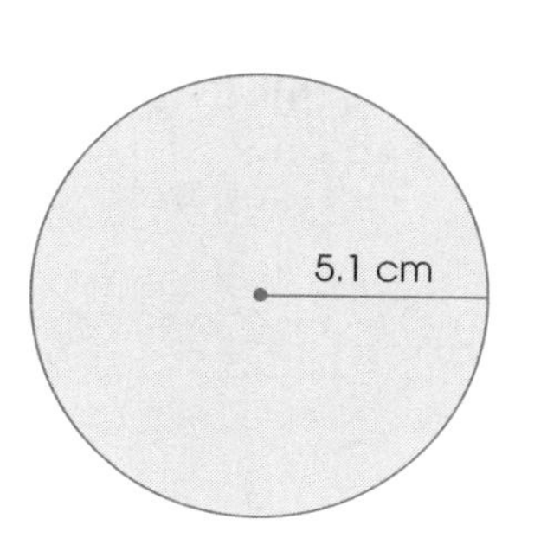

Solution

Step 1: **Find the circumference.**
The circumference of a circle is equal to $\pi \times$ the diameter of the circle. The diameter of this circle is 2×5.1 cm or 10.2 cm.

$$\begin{aligned} \textit{Circumference} &= \pi \times \textit{diameter} \\ &= 3.14 \times 10.2 \text{ cm} \\ &= 32.028 \text{ cm} \\ &\doteq 32.03 \text{ cm} \end{aligned}$$

Round to two decimal places.

The circumference of the circle is about 32.03 cm.

Step 2: **Find the area.**
The area of a circle is equal to $\pi \times \text{radius}^2$.

$$\begin{aligned} \textit{Area} &= \pi \times r^2 \\ &= 3.14 \times 5.1 \text{ cm} \times 5.1 \text{ cm} \\ &= 3.14 \times 26.01 \text{ cm}^2 \\ &\doteq 81.6714 \text{ cm}^2 \end{aligned}$$

Round to two decimal places.

The area of the circle is 81.67 cm^2.

Exercises

1. Write a definition of each type of quadrilateral. Include a diagram.

(a) square

(b) rectangle

(c) rhombus

(d) parallelogram

(e) trapezoid

2. What common feature is shared by all the quadrilaterals in Problem 1?

3. Explain how this diagram demonstrates the formula for finding the area of a parallelogram.

4. On grid paper, draw as many rectangles as you can with a perimeter of 12 cm so each side is a whole number of centimeters.

(a) What is the area of each rectangle?

(b) Which rectangle has the greatest area? the least?

5. Find the perimeter and area of each quadrilateral.

(a)

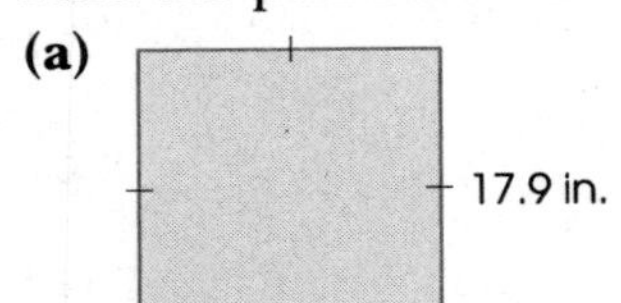

(b)

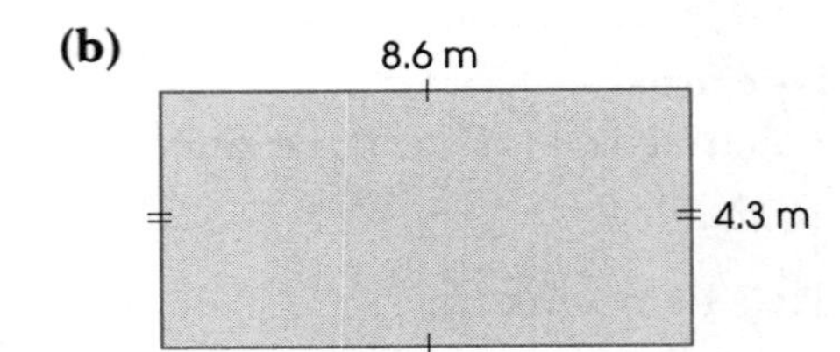

(c)

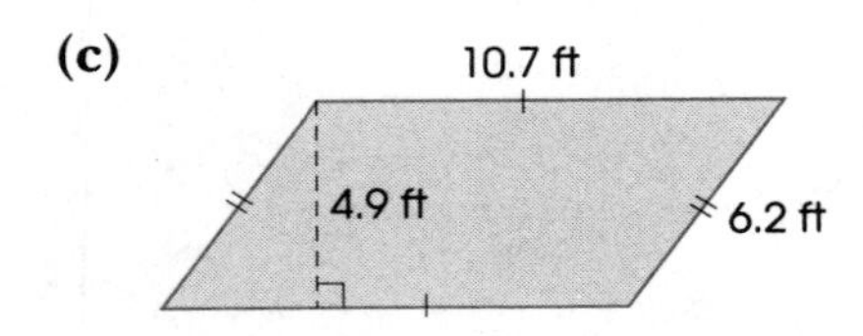

(d)

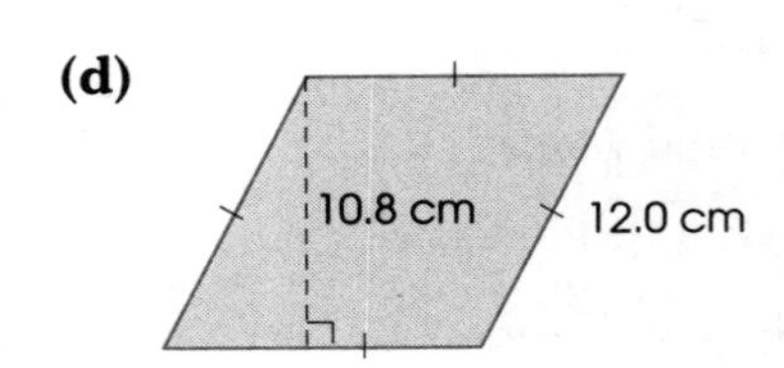

(e)

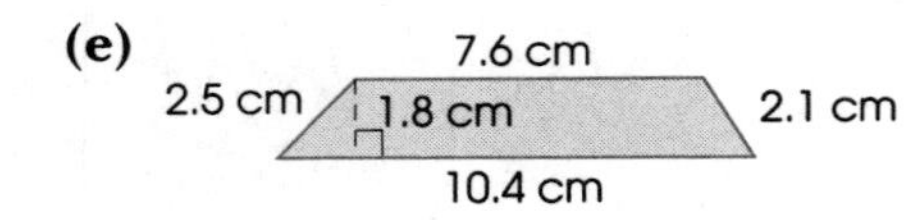

(f)

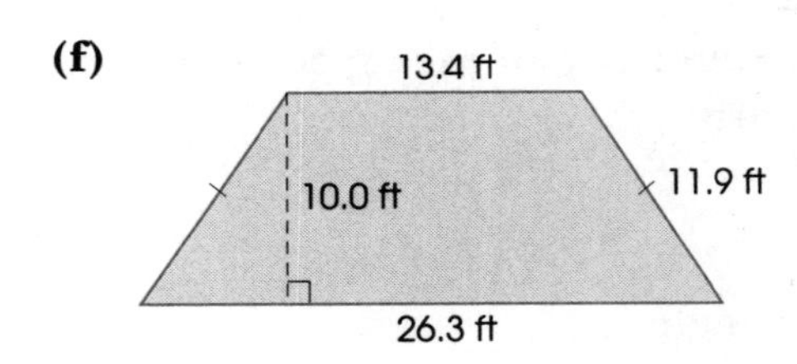

6. Find the circumference and area of each circle.

(a)

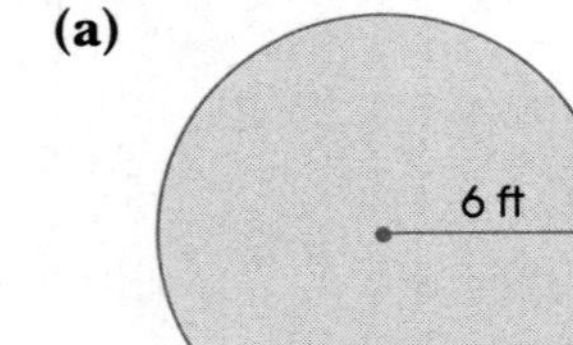

(b)

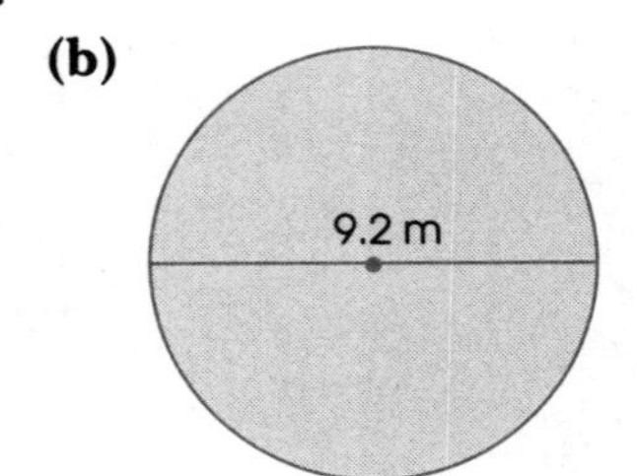

(c)

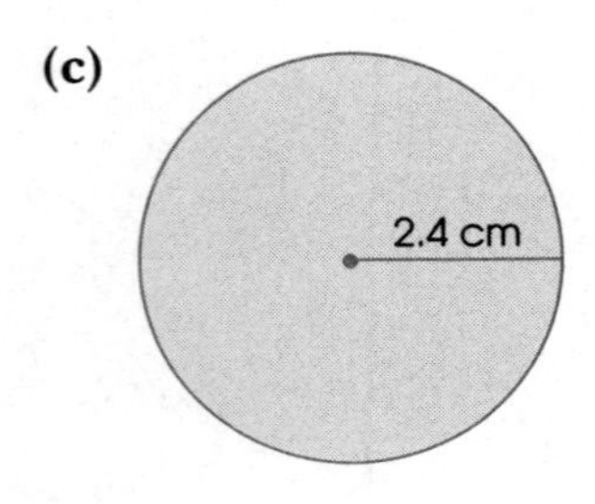

(d)

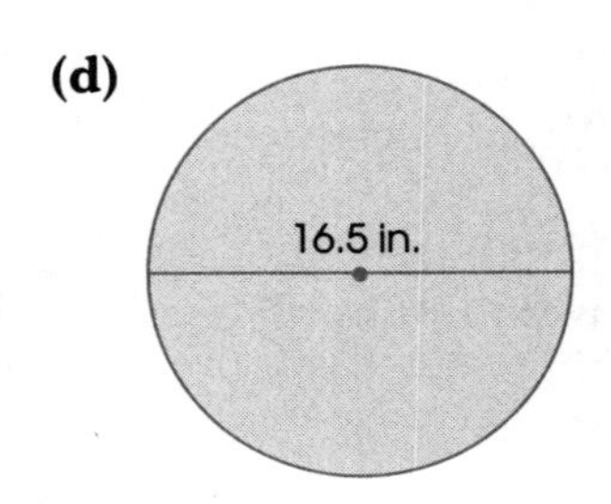

Draw a diagram to solve each problem.

7. A rectangular field is 98.4 yd wide and 109.6 yd long. Find the area of the field.

8. A small circular island has a diameter of 800 ft. If you drew an imaginary circle around the island 50 ft from the shoreline, what area would this larger circle cover?

9. Georgia wants to panel one wall in her den. If one panel covers an area of 24.3 ft^2, and the wall measures 16.2 ft by 7.5 ft, how many pieces of paneling will Georgia need?

10. A square and a circle each have a perimeter of 24 cm. What is the difference in area between these two figures?

11. The world's smallest bicycle has wheels that measure about 0.74 in. in diameter. How far will a point on the wheel travel in 20 revolutions?

12. Create three problems that are similar to problems in this lesson. Write the solutions to your problems on another sheet of paper. Then exchange problems with a classmate, solve, and compare solutions.

10.2 Area and Perimeter 2

In this lesson, you investigated the **relationship between perimeter and area.**

- If the **perimeters** of a non-square rectangle, a square, and a circle are the same, the circle has the greatest area and the rectangle the least.
- If the **areas** of a non-square rectangle, a square, and a circle are the same, the circle has the least perimeter and the rectangle the greatest.

You also examined the effect on perimeter and area when the dimensions of a figure change.

- When the **dimensions are increased** by a factor, the perimeter increases by the same factor (doubled dimensions → doubled perimeter) and the area increases by the square of the factor (doubled dimensions → quadrupled area).

Example

Nathan has 12 square tiles, each 1 cm by 1 cm. How can he arrange the tiles to form the rectangle with the least perimeter?

Solution

Think about the problem.

Since there are 12 centimeter tiles, the area of the rectangle must be 12 cm^2. You need to find the rectangle with the least perimeter.

Make a plan.

You can model rectangles with 12 centimeter tiles and determine the perimeter of each.

Solve the problem.

The factors of 12 are 1 and 12, 2 and 6 and 3 and 4.

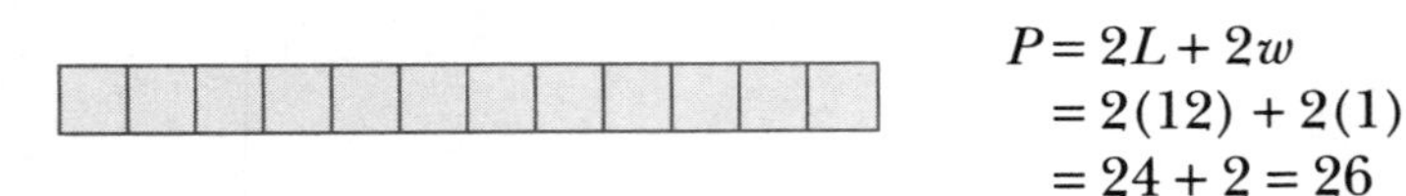

$$\begin{aligned} P &= 2L + 2w \\ &= 2(12) + 2(1) \\ &= 24 + 2 = 26 \end{aligned}$$

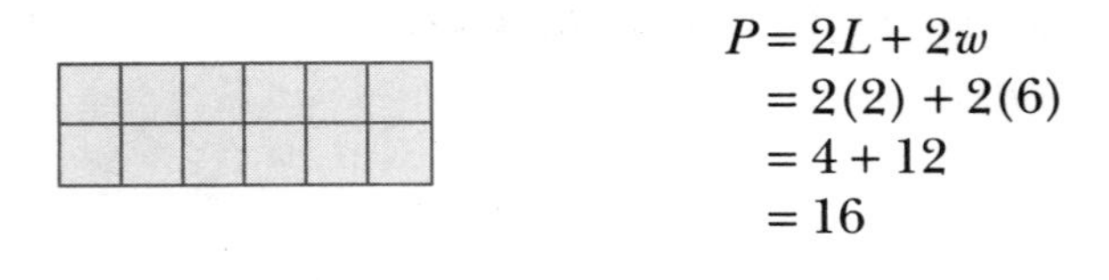

$$\begin{aligned} P &= 2L + 2w \\ &= 2(2) + 2(6) \\ &= 4 + 12 \\ &= 16 \end{aligned}$$

$$\begin{aligned} P &= 2L + 2w \\ &= 2(4) + 2(3) \\ &= 8 + 6 \\ &= 14 \end{aligned}$$

The rectangle with the least perimeter is 3 cm by 4 cm.

Look back.

The rectangle with the least perimeter should have dimensions that are close together. The lengths 3 cm and 4 cm are close, so the answer seems reasonable. A square would have less perimeter, but $\sqrt{12}$ is about 3.5. Nathan couldn't use the tiles to make a square with 3.5 cm sides.

Look ahead.

Think about other questions you could answer using the information in the problem. For example: If the length and width of Nathan's rectangle were doubled, what would be the new area and perimeter?

$$\begin{aligned} \textit{Original Area} &= 3 \times 4 \\ &= 12 \end{aligned}$$

$$\begin{aligned} \textit{New Area} &= 2 \times 3 \times 2 \times 4 \\ &= 6 \times 8 \\ &= 48 \end{aligned}$$

$$\begin{aligned} \textit{Original Perimeter} &= 2(4) + 2(3) \\ &= 8 + 6 \\ &= 14 \end{aligned}$$

$$\begin{aligned} \textit{New Perimeter} &= 2(6) + 2(8) \\ &= 12 + 16 \\ &= 28 \end{aligned}$$

The new area would be 48 cm^2, or 2^2 times the area of the original. The new perimeter would be 28 cm^2, or double the area of the original.

Exercises

1. Complete each sentence in two different ways.
   (a) The perimeter of a polygon is _________________.

   (b) The area of a polygon is _________________.

2. Describe how the perimeter and area of a rectangle will change if the width remains constant, but the length is:
   (a) doubled (b) tripled (c) halved

3. A rectangle has an area of 40 ft^2.
   (a) What are the possible whole-number dimensions for the rectangle?

   (b) Which rectangle from part (a) has the greatest perimeter? the least?

   (c) If you doubled the area of the rectangle, what could the new side lengths be? the new perimeter?

4. A rectangle has an area of 30 $in.^2$.
   (a) What are the possible whole-number dimensions for the rectangle?

   (b) Which of the rectangles you found for part (a) has the greatest perimeter? the least?

5. Marianna wants to plant a 28 yd^2 garden so she can build the shortest possible fence around it. Using integers, what shape should she make the garden? Draw a sketch and label the dimensions.

6. How long is the belt around these pulleys if the pulley centers are 6 ft apart?

7. Without measuring, how could you find the greatest possible area that could be enclosed by a length of rope?

8. Find the greatest possible area that could be enclosed by 16.25 yd of rope. Draw the area and label the dimensions.

9. A rectangle has a perimeter of 200 in. What whole-number dimensions would result in the greatest area?

10. A farmer has 40 yd of fencing material in 1 yd units. He intends to build a chicken pen so it has the greatest possible area. Draw the pen and label the dimensions.

11. Draw a chicken pen with a 40 yd perimeter so it covers the least possible area. Label the dimensions.

12. Marianna wants to rope off a rectangular area of grass near her garden. She will use the walls of her house and garage to block off two sides. She has 10 yd of rope to use for the other two sides. What is the greatest area that Marianna could rope off?

13. Calculate the area of the figure to one decimal place.

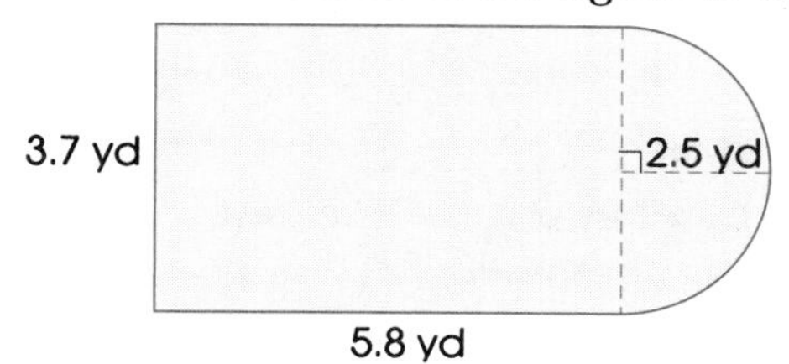

Save your work from Problems 14 to 16 to exchange with a classmate.

14. A rectangular yard is watered by two sprinklers that spray water in a circle. Create a word problem involving area and/or perimeter.

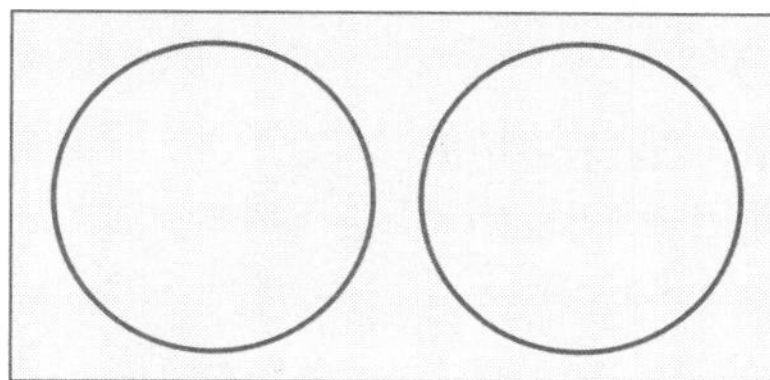

15. Create a word problem that involves these designs as part of the solution.

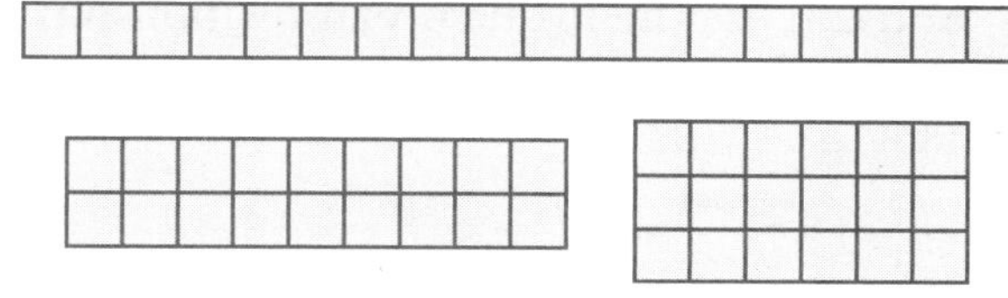

16. Use at least two sports field diagrams to create a word problem about area and/or perimeter.

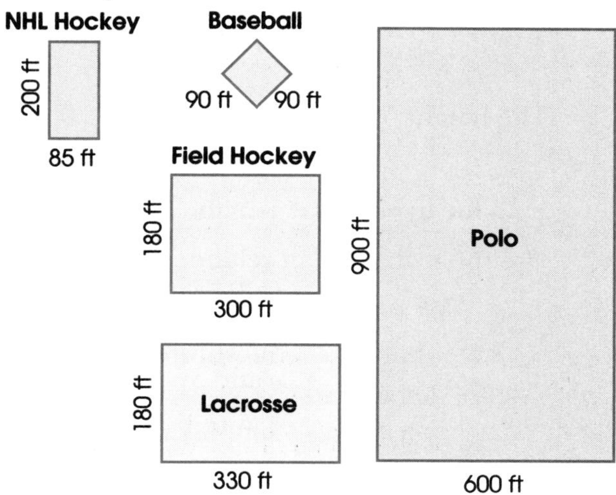

10.3 Area of a Triangle

You have learned how to calculate the **area** of any triangle, and how to work backward from a triangle's area to find its **side length(s)**, **height**, or **perimeter**.
Recall:
- Every triangle represents half of a parallelogram.
- To find the area of a parallelogram, multiply *base* × *height.*
- To find the area of a triangle, multiply $\frac{1}{2}$ × *base* × *height.*

Example 1

Find the area of this triangle.

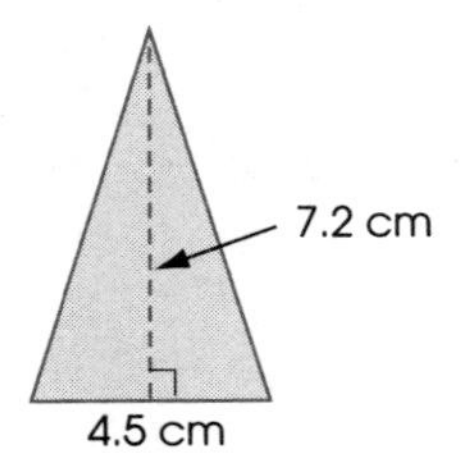

Solution

You know the base length and height of the triangle, so you can use the area formula.

$$\begin{aligned} Area &= \frac{1}{2} \times base \times height \\ &= \frac{1}{2} \times 4.5 \times 7.2 \\ &= 0.5 \times 4.5 \times 7.2 \\ &= 16.2 \end{aligned}$$

To multiply $\frac{1}{2}$ × 4.5 with a calculator, substitute 0.5 for $\frac{1}{2}$ (or divide 4.5 by 2).

The triangle has an area of 16.2 cm^2.

Example 2

The area of this triangle is 54.6 m^2. Find the length of the altitude to one decimal place.

Solution

The base length of the triangle is 9.3 ft and the area is 54.6 ft^2. Substitute these values into the area formula and solve the equation for height.

$$\begin{aligned} Area &= \frac{1}{2} \times base \times height \\ 54.6 &= \frac{1}{2} \times 9.3 \times height \\ 54.6 &= 0.5 \times 9.3 \times height \\ 54.6 &= 4.65 \times height && \text{Use 0.5 instead of } \tfrac{1}{2}. \\ 54.6 \div 4.65 &= height && \text{Divide both sides by 4.65.} \\ 11.7 &\doteq height \end{aligned}$$

The altitude measures about 11.7 ft.

Example 3

The area of this triangle is 61.5 m^2. Find the length of the hypotenuse to one decimal place.

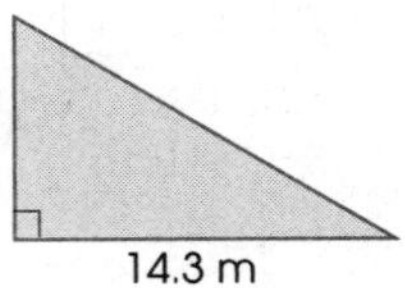

Solution

You can use the Pythagorean relationship to find the hypotenuse length if you know two side lengths.

Step 1

Find the height.

$$\begin{aligned} Area &= \frac{1}{2} \times base \times height \\ 61.5 &= \frac{1}{2} \times 14.3 \times height \\ 61.5 &= 0.5 \times 14.3 \times height \\ 61.5 &= 7.15 \times height \\ 61.5 \div 7.15 &= height && \text{Divide both sides by 7.15.} \\ 8.6 &\doteq height \end{aligned}$$

The height of the triangle is about 8.6 m.

Step 2

Find the hypotenuse length.

Use the Pythagorean relationship.

$$\begin{aligned} a^2 + b^2 &= c^2 \\ 8.6^2 + 14.3^2 &= c^2 && \text{Substitute known values for } a \text{ and } b. \\ 73.96 + 204.49 &= c^2 \\ 278.45 &= c^2 && \text{Simplify the left side.} \\ \sqrt{278.45} &= \sqrt{c^2} \\ 16.7 &\doteq c \end{aligned}$$

The hypotenuse measures about 16.7 m.

Exercises

1. Define each term:

   (a) altitude

   (b) height

   (c) base

   (d) right angle

   (e) vertex

2. Explain why two of the bases of a right triangle can also be altitudes.

3. Calculate the area of each triangle.

   (a)

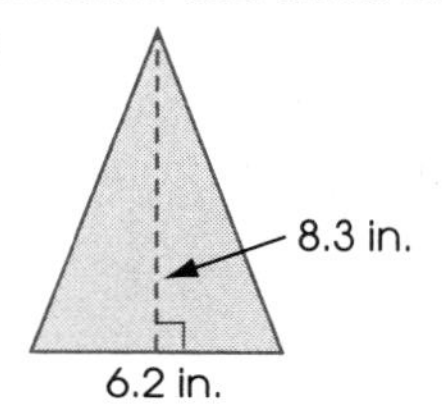

   (b)

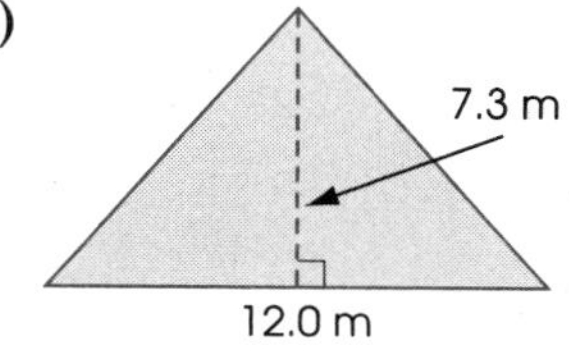

   (c)

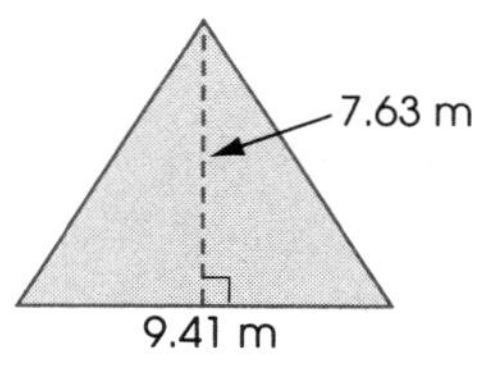

   (d)

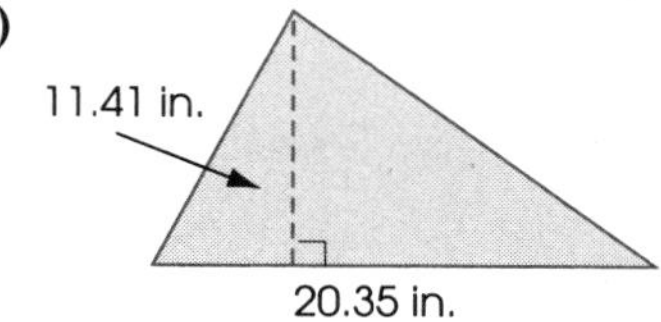

   (e)

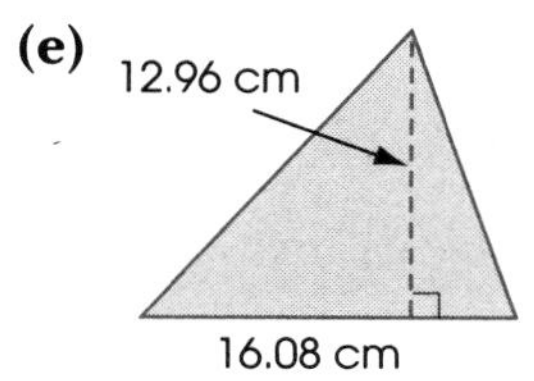

   (f)

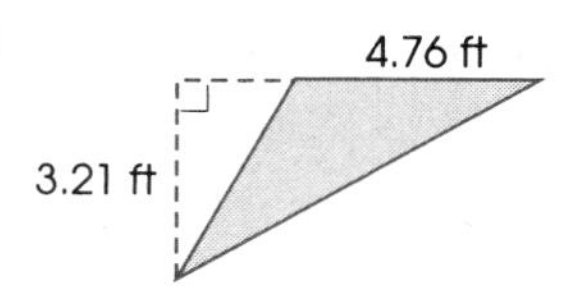

4. Find the length of each altitude.

   (a) *Area* = 17.8 m^2

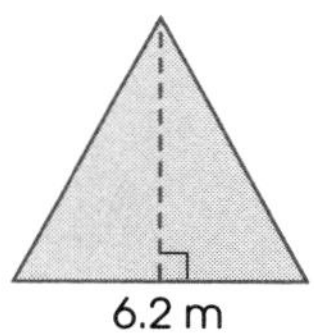

   (b) *Area* = 62.37 $in.^2$

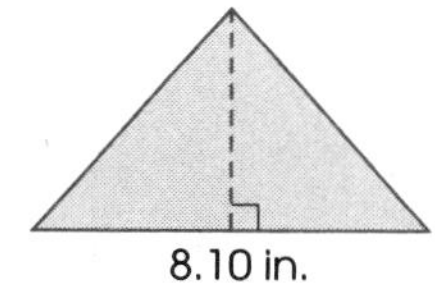

(c) $Area = 88.4\ cm^2$

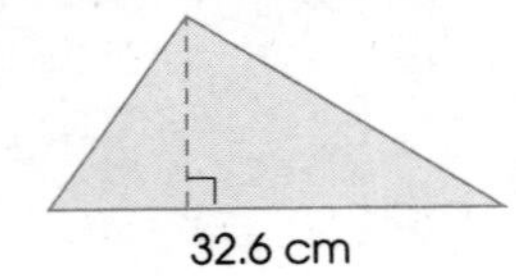

(d) $Area = 27.72\ in.^2$

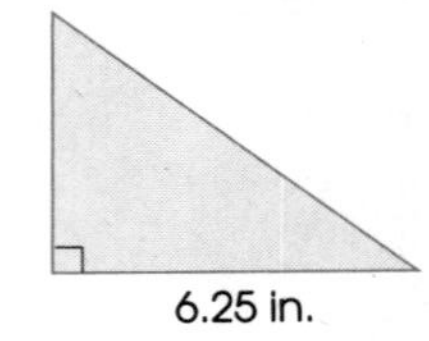

5. Find the base length of each triangle.

(a) $Area = 34.37\ cm^2$

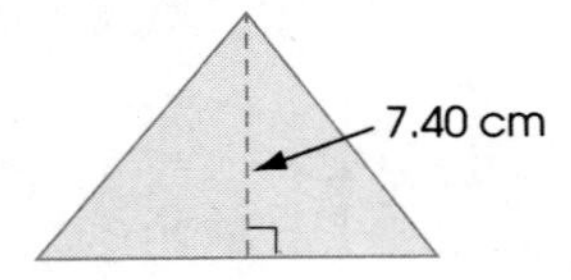

(b) $Area = 11.28\ in.^2$

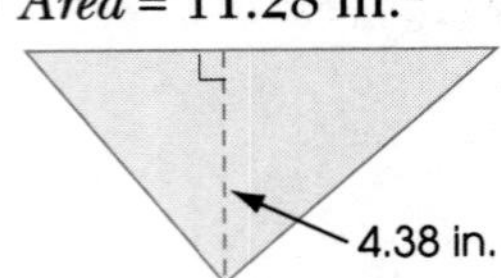

(c) $Area = 14.03\ ft^2$

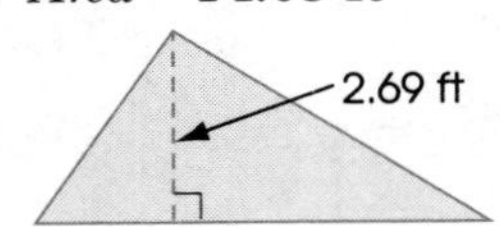

(d) $Area = 160.62\ m^2$

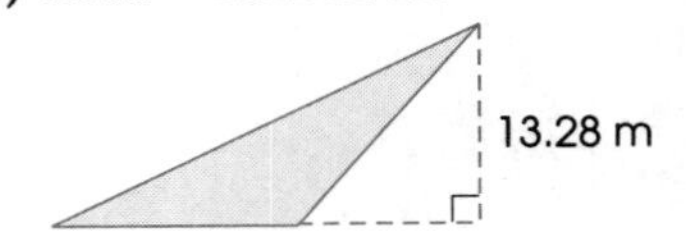

6. Find the hypotenuse length for each triangle.

(a) $Area = 3.5\ in.^2$

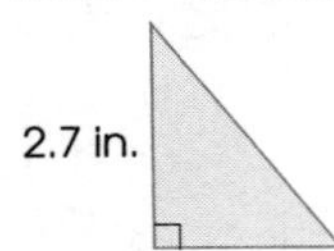

(b) $Area = 49.68\ m^2$

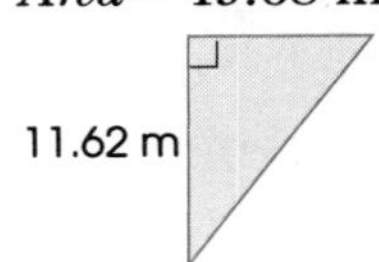

(c) $Area = 18.93\ ft^2$

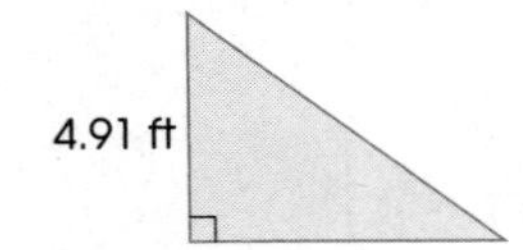

(d) $Area = 91.02\ cm^2$

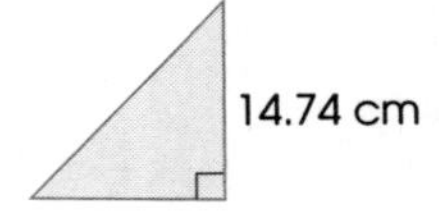

7. Draw two different triangles that have a perimeter of 18 cm. Label the side lengths. Calculate the area of each triangle. (Hint: Use a piece of string 18 cm long.)

8. A triangular pool has a base length of 43.5 ft. If the area of the pool is 1057.05 ft^2, what is the height?

9. Which figure has the greatest area?

(a) a square with sides 10.2 m long

(b) a rectangle 15 m long and 6.9 m high

(c) a parallelogram with base 14.6 m and height 7.2 m

(d) a triangle with base 16.5 m and height 12.4 m

(e) a trapezoid with bases 18.1 m and 10.4 m, and height 7.1 m

10. Find an object with one triangular face. Draw a diagram of the triangle. Label the side lengths and the height. Calculate the area of the triangle.

10.4 Composite Areas

In this lesson, you learned to find the areas of composite figures by looking for hidden rectangles, triangles, parallelograms, trapezoids, and circles. Recall these **area formulas**:

- $A_{rectangle} = length \times width$
- $A_{triangle} = \frac{1}{2} \times base \times height$
- $A_{parallelogram} = base \times height$
- $A_{trapezoid} = \frac{1}{2} \times height \times (sum\ of\ two\ parallel\ sides)$
- $A_{circle} = \pi \times radius^2$

Example

Estimate the area and then calculate.

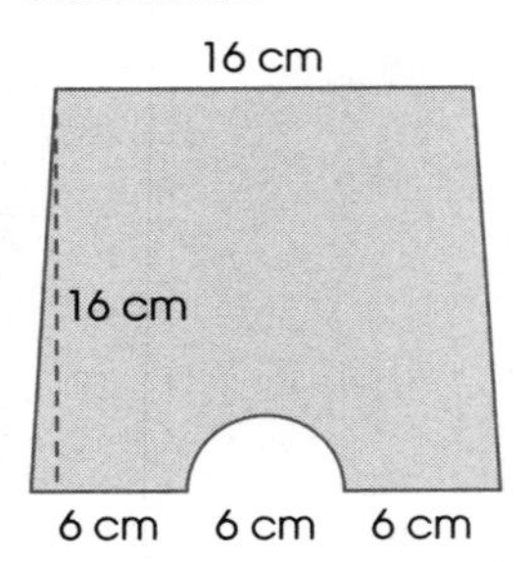

Solution

Step 1

Estimate the area.

A quick way to estimate the area of any composite figure is to find a rectangle that is about the same size. This figure contains a 16 cm by 16 cm square.

$$A_{square} = 16 \text{ cm} \times 16 \text{ cm} = 256 \text{ cm}^2$$

The actual area of the figure is:
$A_{square} + A_{2\ triangles} - A_{semicircle}$

Since the area added by the two triangles is partly offset by the removal of the semicircle, the area of the figure should be a bit more than 256 cm^2.

Step 2

Calculate the area.

This figure is a trapezoid with a semicircle removed from the base. The area of the figure must be equal to the area of the trapezoid minus the area of the semicircle.

Area of the Trapezoid

$$\begin{aligned} A_{trapezoid} &= \frac{1}{2} \times height \times (sum\ of\ parallel\ bases) \\ &= \frac{1}{2} \times 16 \times [16 + (6 + 6 + 6)] \\ &= \frac{1}{2} \times 16 \times (16 + 18) \\ &= \frac{1}{2} \times 16 \times 34 \\ &= 272 \text{ cm}^2 \end{aligned}$$

The area of the trapezoid is 272 cm^2.

Area of the Semicircle

If the diameter of the semicircle is 6 cm, then the radius must be half as long or 3 cm. Therefore, the semicircle has half the area of a circle with a 3 cm radius.

$$\begin{aligned} A_{semicircle} &= \frac{1}{2} \times \pi \times radius^2 \\ &\doteq \frac{1}{2} \times 3.14 \times (3)^2 \\ &= 14.13 \text{ cm}^2 \end{aligned}$$

Area of Composite Figure

$$\begin{aligned} A_{composite\ figure} &= A_{trapezoid} - A_{semicircle} \\ &= 272 \text{ cm}^2 - 14.13 \text{ cm}^2 \\ &= 257.87 \end{aligned}$$

The area of the entire figure is 257.87 cm^2.

Step 3

Check.

The calculated area, 257.87 cm^2, is very close to the estimated area, 256 cm^2, so the calculated area seems reasonable.

Exercises

1. Explain how to find the area of a composite figure.

2. Why is there often more than one approach you can use?

3. Complete the table to show which simple figures you could use to find the area of each composite figure.

(a)

2.2 in.
3 in.
2 in.
3 in.

(b)

3 in.
2 in.
1 in.

(c)

2 cm
2 cm

(d)

5 cm
3 cm
3 cm

	Composite Figure			
Hidden Shape	**(a)**	**(b)**	**(c)**	**(d)**
circle				
square				
triangle				
non-square rectangle				

4. Compare your answers for Problem 3 with a partner's. Can you find more than one way to determine the area for some figures?

5. Estimate the area of each figure in Problem 3.

6. Use the dotted lines to help you find the area of figure (a) from Problem 3. Write a complete solution and round the answer to two decimal places.

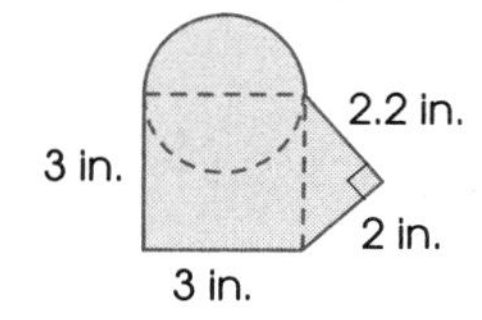

7. Write step-by-step plans you can use to find the areas of figures (b), (c), and (d) from Problem 3. Make your plans as efficient as you can.

8. Use the plans you wrote in Problem 7 to find the area of each figure. Compare your solutions with the estimates you made in Problem 5.

9. A 5 cm by 4 cm rectangle has a semicircular piece missing from one side. The radius of the semicircle is 1 cm. Is the area of the composite figure more or less than 15 cm^2? Explain how you know.

10. Estimate the area of this trapezoid. Then write three different plans you could use to calculate the area more accurately.

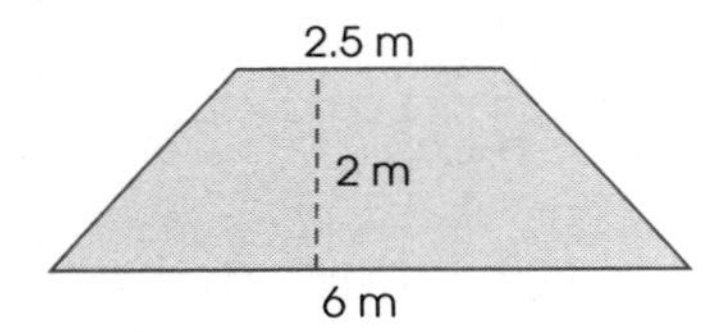

11. Choose the most efficient plan you wrote for Problem 10. Use this plan to find the area of the trapezoid. How does your result compare with your estimate?

12. A goat is tied by a 12 ft rope to a post located at the outside corner of a very large barn.

(a) Draw a diagram to show the area in which the goat could graze. (Assume the goat can graze on only two sides of the barn.)

(b) What is the area of the grazing space?

(c) How would this area change if the length of the rope were doubled?

(d) How would the grazing area increase from part (b) if the length of the rope were increased to 36 ft?

13. A restaurant has an L-shaped patio formed by two connected rectangles, each 19.5 ft by 42 ft. The manager wants to place flower pots at various locations on the patio. If each pot has a diameter of 2.5 ft, and the manager wants to keep 1575 ft^2 available for the tables and chairs, how many flower pots will fit on the patio?

14. Design a figure composed of at least three simple figures. Include dimensions in your design. Draw your figure on grid paper to make sure the dimensions you chose are possible.

(a) Estimate the area of your figure.

(b) Write at least two different plans you could use to find the area of your figure. Make sure you have labeled all the necessary dimensions.

(c) Choose the most efficient plan from part (b). Use this procedure to find the area of your figure and compare the results with your estimate.

10.5 Surface Area

In this lesson, you learned how to estimate and calculate surface areas of **right prisms and cylinders**:

- *Surface Area of a Cube* = 6 (*Area of one face*)
  = $6s^2$, where s is the side length of the square
- *Surface Area of a Rectangular Prism* = 2(*Area of front*) + 2(*Area of side*) + 2(*Area of top*)
  = $2(L \times W) + 2(H \times L) + 2(W \times H)$
- *Surface Area of a Triangular Prism*
  = 2(*Area of triangular base*)
  + (*Area of large rectangle or sum of areas for the 3 small rectangles*)
  $= 2\left(\frac{1}{2} \times base \times H\right) + (Perimeter\ of\ triangle \times L)$ or
  $= 2\left(\frac{1}{2} \times base \times H\right) + (base \times L) + (side\ a\ of\ triangle \times L) + (side\ b\ of\ triangle \times L)$
- *Surface Area of a Cylinder* = 2 × (*Area of circle*) + (*Area of curved rectangular surface*)
  $= (2 \times \pi r^2) + (2 \times \pi \times r \times h)$

Example 1

Calculate the surface area of this box.

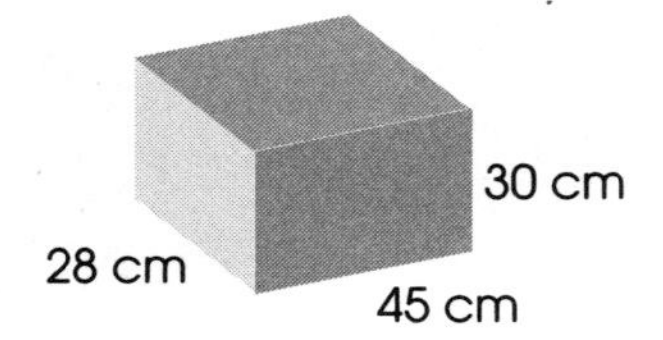

Solution

$$S.A. = 2(L \times W) + 2(H \times L) + 2(W \times H)$$
$$= 2(45 \times 28) + 2(30 \times 45) + 2(28 \times 30)$$
$$= 2520 + 2700 + 1680$$
$$= 6900\ cm^2$$

The surface area of the box is 6900 cm^2.

Example 2

A brand of chocolate is packaged in a box shaped like a triangular prism. Calculate the surface area of the box.

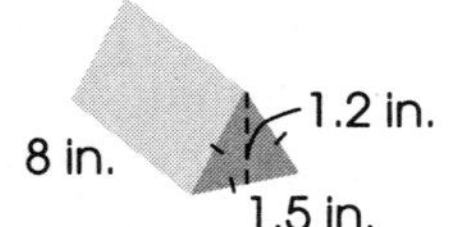

Solution

The prism has five faces. Two are congruent triangles and three are congruent rectangles.

$$SA = 2(triangle\ base) + 3(rectangle\ area)$$
$$= 2\left(\frac{b \times h}{2}\right) + 3(L \times W)$$
$$= 2\left(\frac{1.5 \times 1.2}{2}\right) + 3(8 \times 1.5)$$
$$= 1.8 + 36$$
$$= 37.8\ in.^2$$

The surface area of the box is 37.8 $in.^2$.

Example 3

Calculate the amount of aluminum needed to make this can. Round the answer to the nearest square inch.

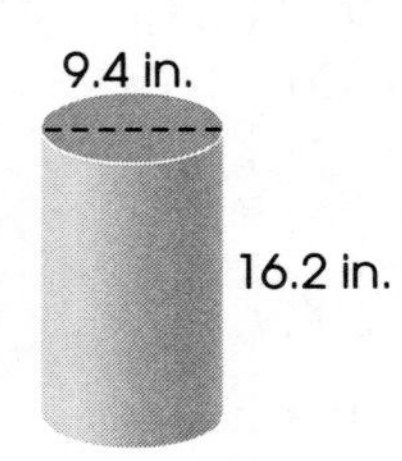

Solution

Draw a net of the cylinder.

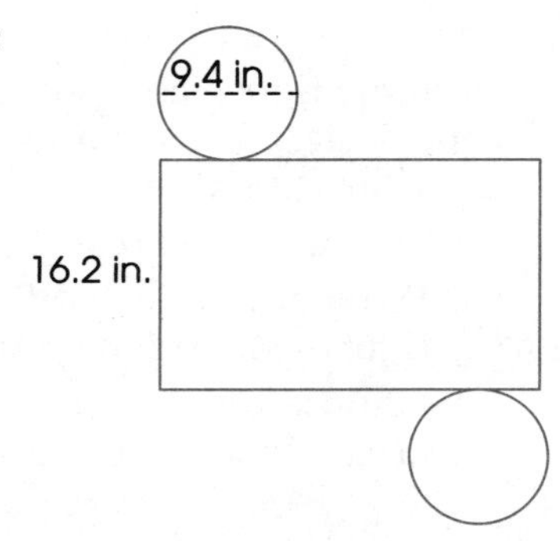

If the diameter of the circle is 9.4 in., then the radius is half as long, or 4.7 in.

$$SA = 2\ (area\ of\ circle) + (area\ of\ rectangle)$$
$$= (2 \times \pi r^2) + (circumference\ of\ circle \times rectangle\ height)$$
$$= (2 \times \pi r^2) + (2 \times \pi \times r \times h)$$
$$= (2 \times \pi \times 4.7^2) + (2 \times \pi \times 4.7 \times 16.2)$$
$$= (\pi \times 44.18) + (\pi \times 152.28)$$
= 617.197 if π was used
or 616.8844 if 3.14 was used

About 617 $in.^2$ of aluminum is needed to make the can.

Exercises

1. Explain how to find the surface area of a solid.

2. Calculate the surface area of this box.

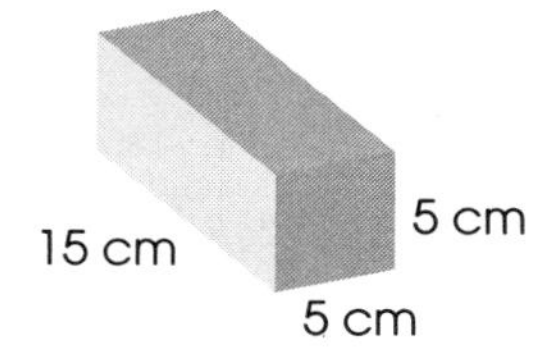

3. Calculate the surface area of a cube-shaped box with sides 12 in. long.

4. Estimate the surface area of the tent. Then calculate to the nearest tenth of a meter.

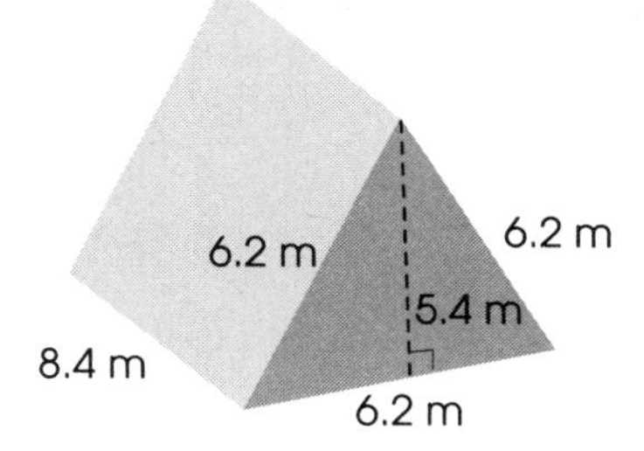

5. Estimate the surface area of the cylinder. Then calculate to the nearest tenth of a meter.

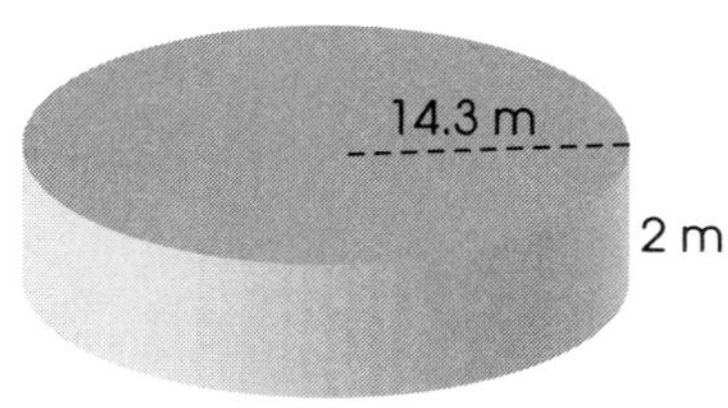

6. Estimate the surface area of the triangular prism. Then calculate.

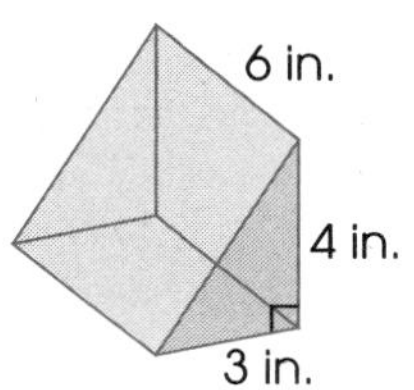

7. A cheese box is in the shape of a triangular prism. Calculate its surface area.

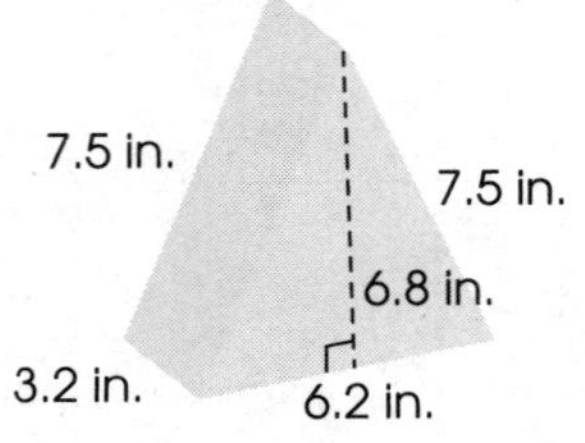

8. A can of tennis balls is 7.8 in. high with a base radius of 1.3 in. To the nearest square inch, how much metal is needed to make the can?

9. A cylindrical tin is 12.5 in. high and has a diameter of 8 in. How much paper would be needed to make a label to go around the tin? Round your answer to one decimal place.

10. Anna's aquarium has a rectangular base measuring 30 in. by 20 in. The height of the aquarium is 15 in. If the top of the aquarium is open, what is the total surface area of the glass?

11. A cereal box is 8.5 in. long, 1.2 in. wide, and 10.5 in. high.

(a) If cardboard costs 0.08¢ per square inch, what is the cost of the cardboard needed for the top, bottom, and sides of one box?

(b) Besides the top, bottom, and sides, what extra cardboard will the company need to make the box?

12. The outside of a drum is to be painted with waterproof paint. The diameter of the drum is 0.6 ft and the height is 1.2 ft.

(a) Calculate the surface area of the drum to two decimal places.

(b) How many pints of paint will be needed to cover 8 drums if 1 pt covers 6.5 ft^2? Calculate to the nearest pint.

13. Cans of tuna are 1.5 in. high with a diameter of 3.5 in. The cans are to be packed in cases of 12.

(a) Find the dimensions of the case you would need if the cans are to be packed in two rows, each stacked two cans high.

(b) Calculate the surface area of the case.

14. Create a word problem about the surface area of a cylinder or right prism.

Write a step-by-step solution and analyze each step to determine where errors might occur.

On another sheet of paper, rewrite the solution so that it contains an error. Ask a classmate to identify the error.

10.6 Volume of Composite Solids

In this lesson, you learned to find the volume of 3D composite objects – objects created by combining or separating two or more simple solids. Recall:

- To find the volume of a composite object, find and add the volumes of the **simple solids** that form it.
- If a solid is a **prism**, look for hidden shapes that can help you find the base area. Then multiply the base area by the height.

Example 1

Find the volume of this wedge.

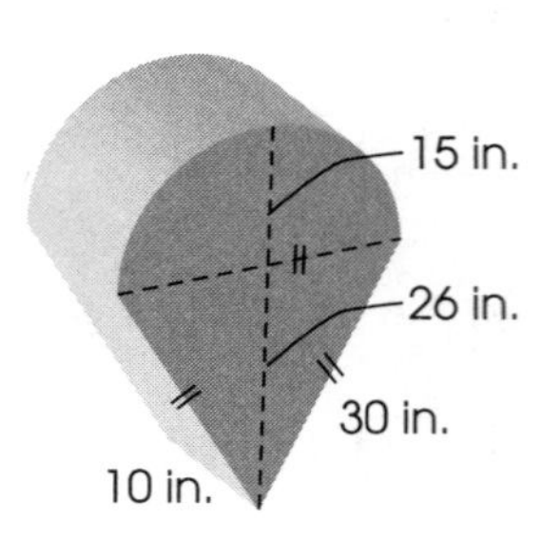

Solution

This solid is made of a triangular prism and half of a cylinder.

The volume of each part is equal to *base area* × *height.*

Step 1

Find the volume of the triangular prism.

$$\begin{aligned} V_{triangular prism} &= \textit{area of triangular base} \\ &\quad \times \textit{ prism height} \\ &= \frac{\textit{base} \times \textit{altitude}}{2} \times \textit{prism height} \\ &= \frac{30 \text{ in.} \times 26 \text{ in.}}{2} \times 10 \text{ in.} \\ &= 390 \text{ in.}^2 \times 10 \text{ in.} \\ &= 3900 \text{ in.}^3 \end{aligned}$$

Step 2

Find the volume of the half cylinder.

$$\begin{aligned} V_{half cylinder} &= \textit{area of semicircular base} \\ &\quad \times \textit{prism height} \\ &= \frac{\pi \times r^2}{2} \times \textit{prism height} \\ &= \frac{\pi \times (15 \text{ in.})^2}{2} \times 10 \text{ in.} \\ &= 353.25 \text{ in.}^2 \times 10 \text{ in.} \\ &= 3532.5 \text{ in.}^3 \end{aligned}$$

Step 3

Combine.

$$\begin{aligned} V_{composite\ solid} &= V_{triangular\ prism} + V_{half\ cylinder} \\ &= 3900 \text{ in.}^3 + 3532.5 \text{ in.}^3 \\ &= 7432.5 \text{ in.}^3 \end{aligned}$$

The volume of this solid is 7432.5 in.^{3}

Example 2

Find the volume of this object.

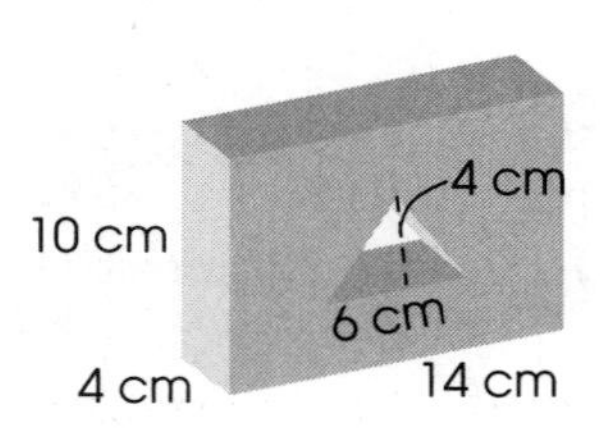

Solution

Step 1

Find the base area.

If you use a solid rectangle as the base, then the result will not include the area of the missing triangular prism. Instead, use one of the faces with a triangular hole as the base.

$$\begin{aligned} A_{base} &= A_{rectangle} - A_{triangle} \\ &= (14 \text{ cm} \times 10 \text{ cm}) - \left(\frac{6 \text{ cm} \times 4 \text{ cm}}{2}\right) \\ &= 140 \text{ cm}^2 - 12 \text{ cm}^2 \\ &= 128 \text{ cm}^2 \end{aligned}$$

Step 2

Multiply by the height.

$$\begin{aligned} V_{solid} &= \textit{base area} \times \textit{height} \\ &= 128 \text{ cm}^2 \times 4 \text{ cm} \\ &= 512 \text{ cm}^3 \end{aligned}$$

The volume of this solid is 512 cm^3.

Exercises

1. Complete each sentence.

(a) The volume of a prism is equal to the _______ of the base times the _______ of the prism.

(b) The units used to express the volume of an object are shaped like ___________.

(c) When a small object is removed from a larger one, the resulting volume will be _______ than the volume of the larger object.

2. Find the volume to the nearest cubic centimeter.

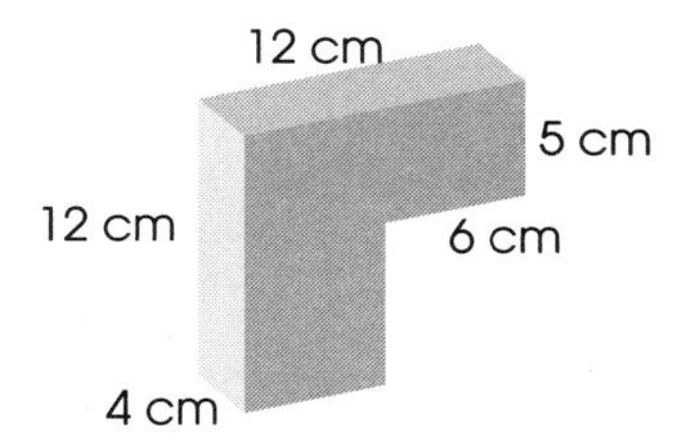

3. If 1 mL of milk occupies 1 cm^3, how many liters of milk will this carton hold? Give your answer to the nearest tenth of a liter.

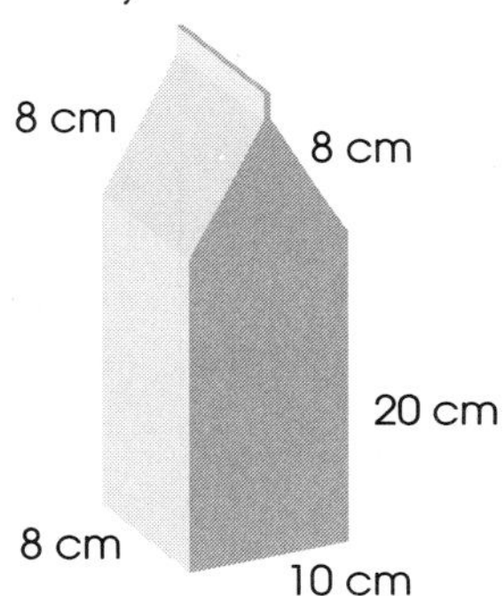

4. Linda calculated the volume of this figure. Find and correct any errors in her work.

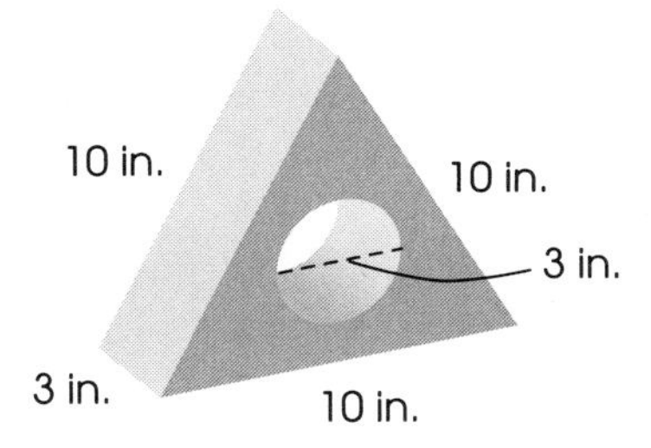

$V_{object} = V_{triangular\ prism} - V_{cylinder}$

$V_{triangularprism} = \dfrac{(10 \text{ in.})(10 \text{ in.})}{2}(3 \text{ in.}) = 150 \text{ in.}^3$

$V_{cylinder} = \pi(3 \text{ in.})^2(3 \text{ in.}) = 84.78 \text{ in.}^3$

$V_{object} = 150 \text{ in.}^3 - 84.78 \text{ in.}^3$
$= 65.22 \text{ in.}^3$

5. Find the volume to the nearest cubic centimeter.

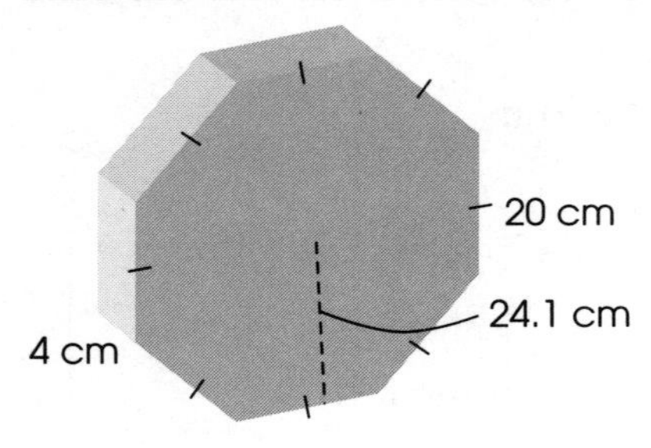

6. The volume of this fishbowl is 325.92 in.^{3}. Find the side length of the hexagon to the nearest tenth of an inch.

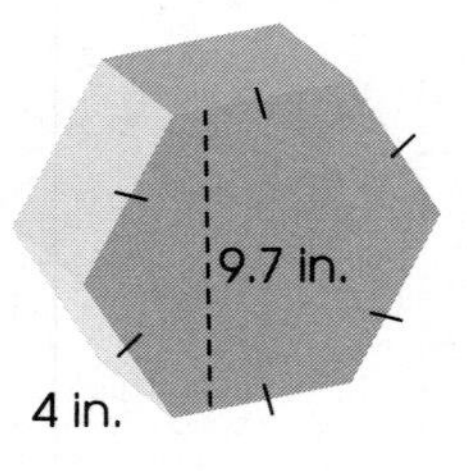

7. A square nut has been made of a metal alloy.

(a) Find the volume of metal, to the nearest cubic inch, that was used to make the nut.

(b) If 1 in.^{3} of metal weighs 6.88 oz, how much metal was used to make the nut?

(c) If the metal costs $5.45 for 1 lb, how much would it cost to make 1200 nuts?

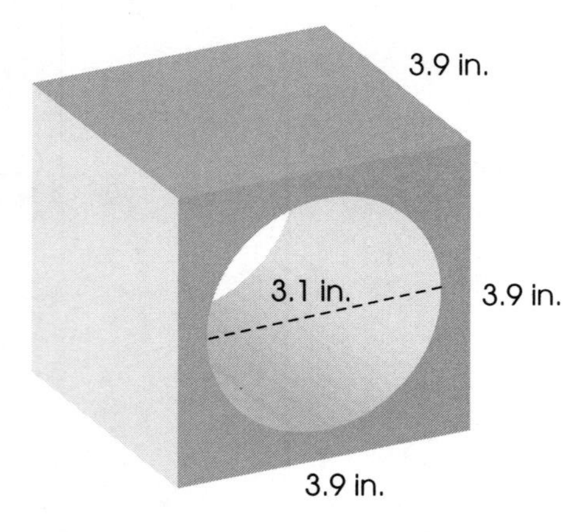

8. A cake has been made in the shape of the Superman logo.

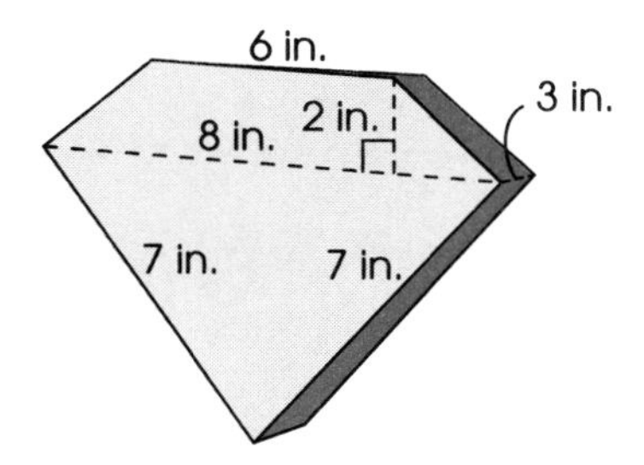

(a) Find the volume of the cake.

(b) If each person eats 8 in.^{3} of cake, how many people will the cake feed?

(c) If 2 eggs are required to make every 60 in.^{3} of cake, how many eggs are needed for three Superman cakes?

9. Create a problem based on a real-life situation that would require someone to find the volume of a composite 3D object. Write the solution.

10.7 Volume

You have used formulas for calculating the **volume**, V, of **prisms, pyramids, cones**, and **cylinders**. You have also found relationships among some of these formulas.

- To calculate the volumes of cylinders and prisms, use:
  $V = (\textit{area of base}) \times \textit{height}$
- To calculate the volumes of cones and pyramids, use:
  $V = \frac{1}{3} \times (\textit{area of base}) \times \textit{height}$
- When a cylinder and a cone have the same height and base, then:
  $\textit{Volume of cylinder} = 3 \times \textit{volume of cone}$
  $\textit{Volume of cone} = \frac{1}{3} \times \textit{volume of cylinder}$
- When a prism and a pyramid have the same height and base, then:
  $\textit{Volume of prism} = 3 \times \textit{volume of pyramid}$
  $\textit{Volume of pyramid} = \frac{1}{3} \times \textit{volume of prism}$

Example 1

Calculate the volume of the triangular prism.

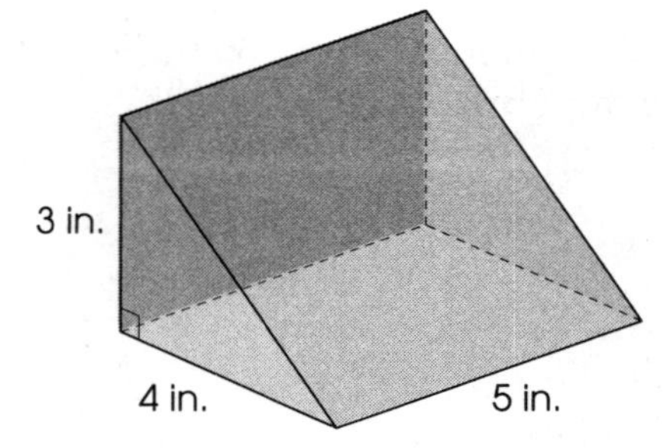

Solution

$$
\begin{aligned}
V &= (\textit{area of base}) \times \textit{height of prism} \\
&= \left(\frac{1}{2} \times 4 \times 3\right) \times 5 \\
&= 6 \times 5 \\
&= 30 \text{ in.}^3
\end{aligned}
$$

Example 2

Calculate the volume of the pyramid.

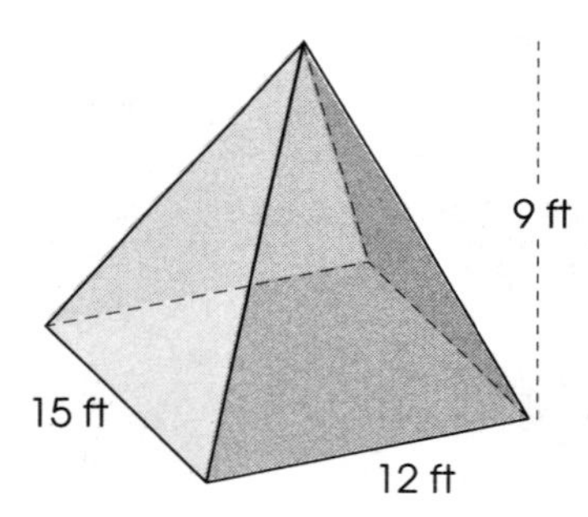

Solution

$$
\begin{aligned}
V &= \frac{1}{3} \times (\textit{area of base}) \times h \\
&= \frac{1}{3} \times (l \times w) \times h \\
&= \frac{1}{3} \times (15 \times 12) \times 9 \\
&= \frac{1}{3} \times 1620 \\
&= 540 \text{ ft}^3
\end{aligned}
$$

Example 3

A cone-shaped paper cup has a radius of 6 cm and a height of 10 cm. The cone is filled twice with water and emptied into a cylindrical glass with a radius of 6 cm and a height of 10 cm. What percent of the glass is filled?

Solution

Since the cone and the cylinder have the same radius and height, the cone's volume is equal to $\frac{1}{3}$ of the cylinder's volume. Two cones of water should fill $\frac{2}{3}$ or 67% of the cylinder. To check this conclusion, determine the volumes of the cone and the cylinder.

Cone Volume

$$V = \frac{1}{3} \times (\textit{area of base}) \times h$$
$$= \frac{1}{3} \times (\pi r^2) \times h$$
$$= \frac{1}{3} \times (3.14) \times (6^2) \times (10)$$
$$= \frac{1}{3} \times 1130.4$$
$$= 376.8 \text{ cm}^3$$

The volume of the cone is 376.8 cm^2.

Two cones would have a volume of 2×376.8 cm^2 or 753.6 cm^2.

Cylinder Volume

$$V = (\textit{area of base}) \times h$$
$$= (\pi r^2) \times h$$
$$= 3.14 \times 6^2 \times 10$$
$$= 3.14 \times 36 \times 10$$
$$= 1130.4 \text{ cm}^3$$

The volume of the cylinder is 1130.4 cm^3.

Two cones would fill $\frac{753.6}{1130.7}$ of the cylinder. You can use a calculator to divide 753.6 by 1130.7 to see that this is 0.667 or 66.7 hundredths or 66.7% of the cylinder.

Exercises

1. Complete.
   - **(a)** The general formula for calculating the volume of a cylinder or prism is ____________.
   - **(b)** If a cone and a cylinder have the same base and height, the volume of the cylinder is ____________ times the volume of the cone.
   - **(c)** If a pyramid and a prism have the same base and height, the volume of the prism is ____________ times the volume of the pyramid.

2. Calculate the volume.

(a)

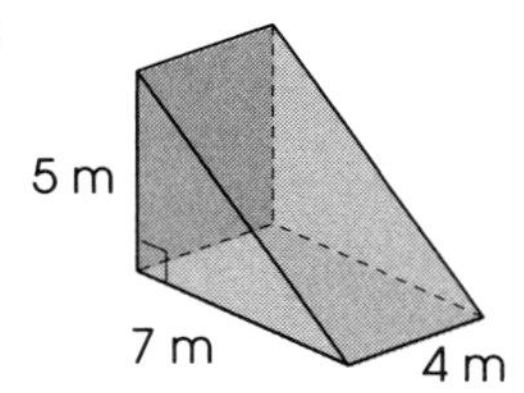

(b)

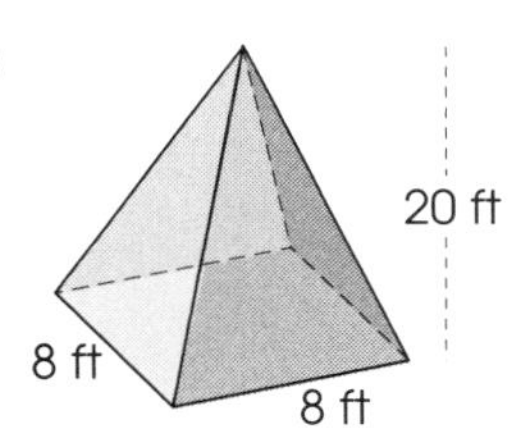

(c)

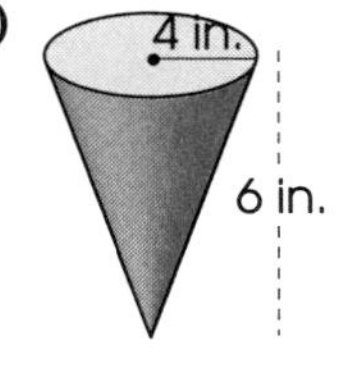

(d)

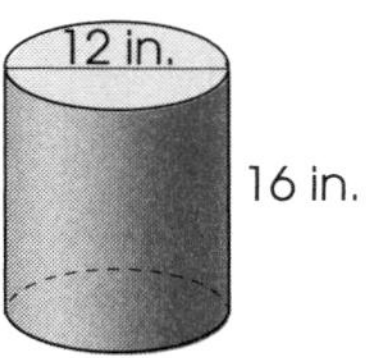

3. Correct Scott's solution.

$$V = \frac{1}{3} \times (\textit{area of base}) \times h$$
$$= \frac{1}{3} \times (2a)^2 \times (2a)$$
$$= \frac{1}{3} \times (2a^2) \times (2a)$$
$$= \frac{1}{3} \times 4a^3$$
$$= \frac{4}{3} a^3$$

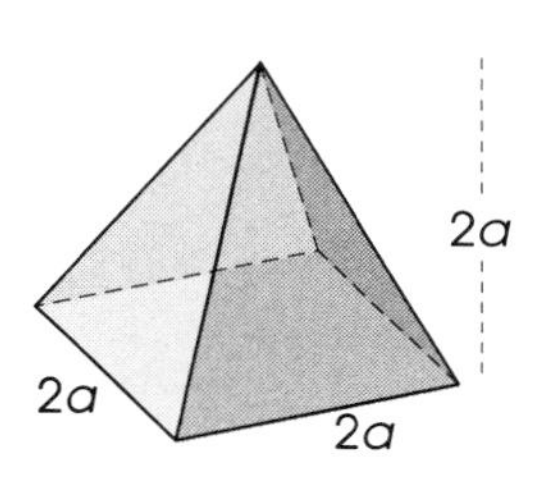

4. Calculate the volume of the cone to the nearest tenth if the cone is generated by rotating ΔPQR one full rotation about:

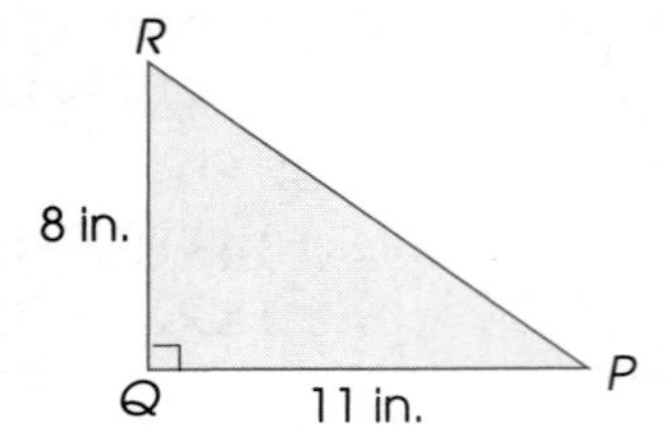

(a) side PQ

(b) side QR

5. A farmer plans to build a storage silo 5.2 ft high to hold 72 ft^3 of corn.

(a) If the silo is cylindrical, what should the diameter of the base be?

(b) If the silo is a square-based prism, how wide should the base be?

6. A cubic tank with 4.6 ft edges is filled with liquid. How much liquid will be left in the tank if some is drained off to fill a cylindrical tank with a radius of 2.2 ft and a height of 4.6 ft?

7. Find the volume.

(a)

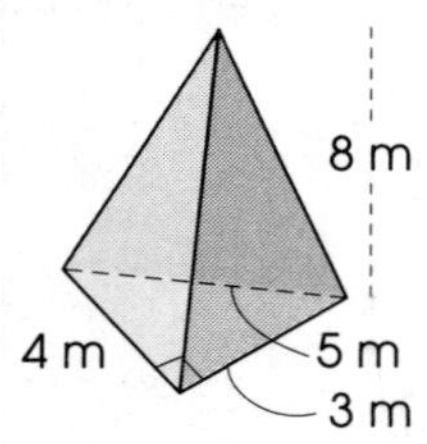

(b)

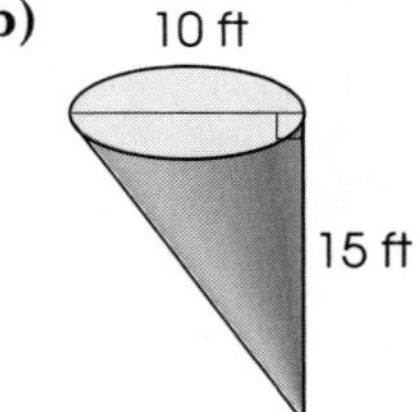

8. A square pyramid has a base with an area of 40 ft^2 and a volume of 100 ft^3. What is the height of the pyramid?

9. A cube with 36 in. edges is cut into 6 congruent pyramids. Each pyramid has one of the faces of the cube as its base.

(a) Find the volume of each pyramid.

(b) Find the volume of the cube.

10. What volume of concrete is required to make this arch?

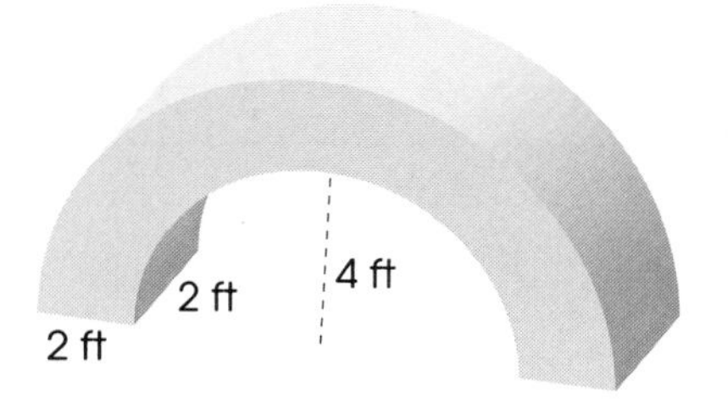

11. What percent of this rain gauge is filled?

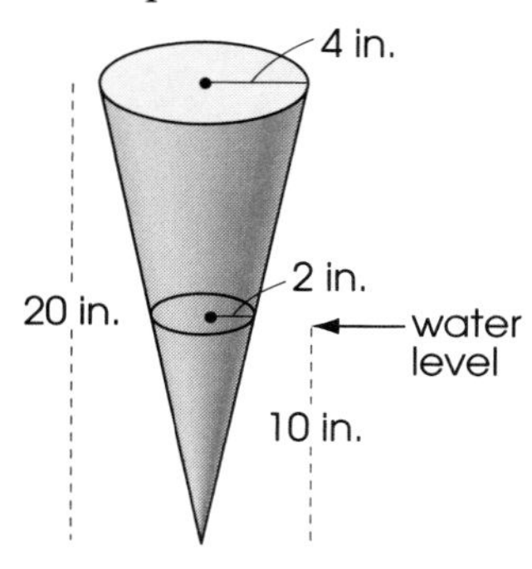

12. Create a volume problem based on this rectangular prism. Solve your problem, then exchange with a classmate.

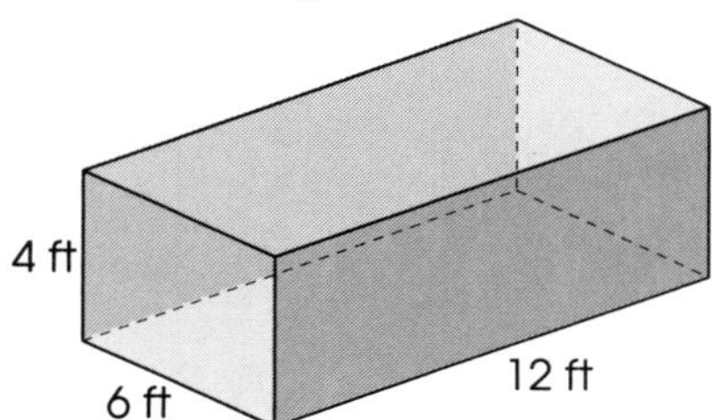

13. Draw a solid composed of connected cones, cylinders, pyramids, and/or prisms. Label the dimensions and find the volume of your object. Exchange with a classmate to check.

10.8 Surface Area and Volume

You have learned how to find the **surface area**, *SA*, of a cylinder and of various prisms and pyramids. Recall:

- For a **cylinder**, *SA* = *area of top and bottom* + *area of rectangular surface* = $2\pi r^2 + 2\pi rh$.
- For a **prism**, *SA* = *area of 2 bases* + *side area* × *number of sides*.
- For a **rectangular prism**, $SA = (2Lw + 2Lh + 2wh)$.
- For a **pyramid**, *SA* = *area of base* + *side area* × *number of sides*.

You have also seen that an object's side lengths, surface area, and volume are interdependent, and that equalizing the side lengths results in a shape where the least amount of surface area contains the greatest volume.

Example 1

A box is 8 in. long, 6 in. wide, and 3 in. high.
How much paper is needed to cover the sides?

Solution

The amount of paper needed is equal to the surface area of the box.

$$\begin{aligned} SA &= (2Lw + 2Lh + 2wh) \\ &= 2 \times (8 \times 6) + 2 \times (8 \times 3) + 2 \times (6 \times 3) \\ &= (2 \times 48) + (2 \times 24) + (2 \times 18) \\ &= 96 + 48 + 36 \\ &= 180 \end{aligned}$$

The side lengths of this box are measured in inches, so the area is 180 in.^2.

Example 2

Design a new box with the same volume as the box in Example 1, but with a smaller surface area. Make each dimension a whole number of inches.

Solution

Find the volume of the box in Example 1.

$$\begin{aligned} V &= L \times w \times h \\ &= 8 \times 6 \times 3 \\ &= 144 \text{ in.}^3 \end{aligned}$$

Recall that a box becomes more efficient as you equalize the dimensions.

Look at the original box's dimensions. Notice that doubling the 3 and halving the 8 brings the dimensions closer together. This should not change the volume, since $(2 \times 4) \times 6 \times 3 = 4 \times 6 \times (3 \times 2)$.

	Length	Width	Height
Original Box	8 in.	6 in.	3 in.
New Box	4 in.	6 in.	6 in.

Check the volume of the new box.

$$\begin{aligned} V &= L \times w \times h \\ &= 4 \times 6 \times 6 \\ &= 144 \text{ in.}^3 \end{aligned}$$

Calculate the new surface area.

$$\begin{aligned} SA &= (2 \times 4 \times 6) + (2 \times 4 \times 6) + (2 \times 6 \times 6) \\ &= (2 \times 24) + (2 \times 24) + (2 \times 36) \\ &= 48 + 48 + 72 \\ &= 168 \text{ in.}^2 \end{aligned}$$

The original surface area was 180 in.^{2}. The new surface area is 168 in.^{2}. The new box design has the same volume as the original box, but has a surface area that is 180 – 168 = 12 in.^{2} less.

Example 3

Design a box that has twice the volume of the box in Example 1, with the least possible surface area. Change only one dimension of the original box.

Solution

You can create a box with twice the volume by doubling one dimension. Choose the dimension so the side lengths of the new box will be as close as possible to those of the original box.Check the new volume.

$$\begin{aligned} V &= 8 \times 6 \times 6 \\ &= 288 \text{ in.}^3 \end{aligned}$$

	Length	Width	Height
Original Box	8 in.	6 in.	3 in.
New Box	8 in.	6 in.	6 in.

$288 \div 144 = 2$, so the volume of the new box is twice the volume of the original.

Calculate the new surface area.

$$\begin{aligned} SA &= 2 \times (8 \times 6) + 2 \times (8 \times 6) + 2 \times (6 \times 6) \\ &= 2 \times 48 + 2 \times 48 + 2 \times 36 \\ &= 96 + 96 + 72 \\ &= 264 \text{ in.}^2 \end{aligned}$$

Since a cube with a volume of 288 in.^{3} is not possible with dimensions in whole inches, this must be the least possible surface area for the box.

Exercises

1. Choose the term(s) that correctly complete(s) each statement.

(a) If the dimensions of an object are increased, then its volume increases **faster/slower** than its surface area does.

(b) If one dimension of a rectangular prism is doubled, the prism's **volume/surface area** also doubles.

(c) If all three dimensions of a cube are doubled, then its volume increases **two/four/six/eight/ten/twelve** times, and its surface area increases **two/four/six/eight/ten/twelve** times.

2. Calculate the surface area and volume of a cube-shaped box with sides 20 in. long.

3. How much soup would fit into a can that has a height of 14.5 in. and a diameter of 9 in.? How much paper would you need to make the label?

4. A birthday gift is 55 in. long, 40 in. wide, and 5 in. high. The sheet of paper you want to use to wrap it measures 75 in. by 100 in. Is the paper large enough to wrap the gift? Explain.

5. Soup cans 11.3 in. high with a diameter of 9 in. are to be shipped in boxes of 12.

   (a) Find the dimensions of the box needed to ship the cans in one layer with three rows.

   (b) Find the dimensions of the box needed to ship the cans in two layers of two rows.

   (c) Which box uses less cardboard?

6. A cylinder just fits inside a 10 in. × 10 in. × 10 in. cubical box. Find the surface area and volume of the cylinder.

7. What happens to the surface area of a rectangular prism if all three of its dimensions are doubled? tripled? multiplied by x?

8. The volume of a rectangular prism is 24 in.^{3}.

   (a) Find the dimensions of the six possible prisms that have whole-number dimensions.

   (b) What are the dimensions of the prism with the least surface area?

9. (a) Complete this statement: The rectangular prism that has the least possible surface area for a given volume is always a ________________.

   (b) Describe a situation in which you could apply the rule from part (a).

10. Three identical tennis balls with an 8 in. diameter are stacked in a cylindrical container. For this container, calculate its:

(a) volume

(b) surface area

11. A box of pins measures 1 in. × 2 in. × 3 in. Six pin boxes are to be packed together and wrapped in transparent plastic.

(a) Show three possible arrangements of the boxes.

(b) Calculate the total surface area of each arrangement.

(c) Which arrangement is best for packing the boxes? Explain.

12. A cylindrical water tank is 3 m high and 1 m in diameter.

(a) Find the tank's capacity in liters if 1000 cm^3 holds 1 L.

(b) Find the tank's surface area.

(c) Design a tank that could contain about the same amount of water, but that could be made from less metal.

13. A ramp shaped like a triangular prism is 2.5 yd wide, and reaches a loading dock 1.2 yd high. The ramp starts 3 yd from the dock. All sides of the ramp are to be covered with pressure-treated plywood. Calculate the amount of plywood sheathing required to cover the ramp.

14. Use a standard-sized sheet of paper (8.5 in. by 11 in.) to build the box or cylinder with the greatest possible volume.

15. Create a word problem about volume and surface area. Write a step-by-step solution and analyze each step to determine where errors might occur. On a clean sheet of paper, rewrite the solution so that it contains an error. Exchange with a classmate and identify the errors.

11 Polygons and Circles

11.1 Parts of a Circle

In this lesson, you learned how to calculate the dimensions of a circle from given lengths. Recall:

- The **diameter** of a circle is equal to 2 times the length of the **radius**.
- The **circumference** of a circle is equal to π (or about 3.14) times the length of the **diameter**.

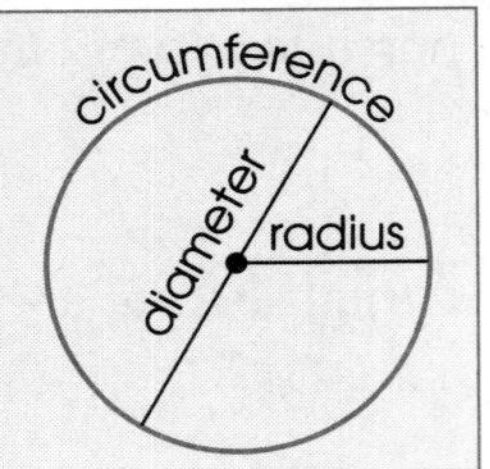

Example 1

A circle has a radius of 4 in. What is its circumference?

Solution

The circumference of a circle is about 3.14 times the length of the diameter. Before you can calculate the circumference, you must find the length of the diameter.

Step 1: **Draw a diagram.**

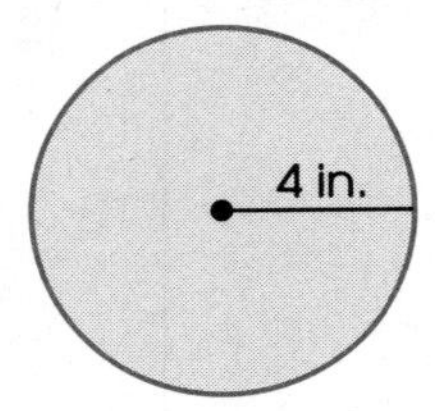

Step 2: **Find the length of the diameter.**
The diameter is the maximum distance across the circle, or twice the radius.

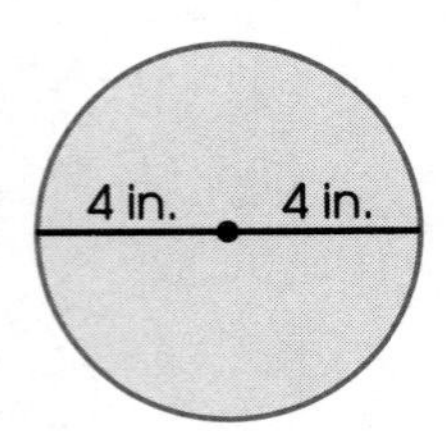

The diameter of the circle is 2×4 in. or 8 in.

Step 3: **Find the circumference.**
The circumference of the circle is about 3.14 times the length of the diameter.

3.14×8 in. = 25.12 in.

The circumference of the circle is about 25.12 in.

Example 2

Erin ties a rope to a stick. While her brother holds the stick, she extends the rope to its full length and marks out a circle with the stick as its centre.

If the circle has a circumference of 17.27 m, how long is the rope?

Solution

Step 1: **Draw a diagram.**

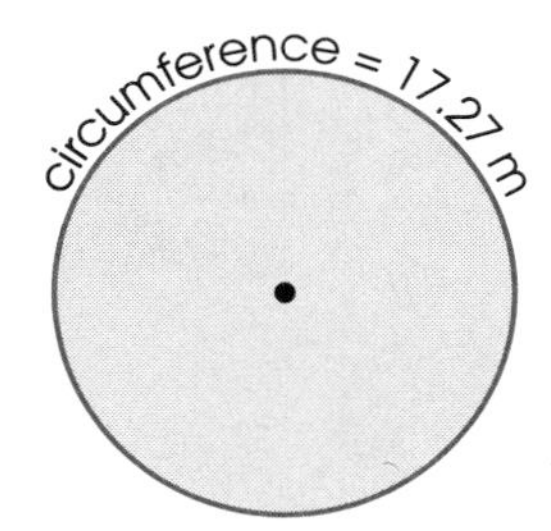

The length of the rope must be equal to the radius of the circle.

Step 2: **Find the length of the diameter.**

If $C = 3.14d$, then the diameter must be equal to $\frac{C}{3.14}$.

$\frac{17.27}{3.14} = 5.5$

The diameter of the circle is 5.5 m.

Step 3: **Find the length of the radius.**

The radius is half as long as the diameter.

5.5 m ÷ 2 = 2.75 m

The rope must be 2.75 m long.

Step 4: **Check.**

The circumference of a circle is about 3 times the length of the diameter, or 6 times the length of the radius.

6×2.75 is a bit less than 6×3 m or 18 m.

The actual circumference of the circle is 17.27 m, so the answer is likely correct.

Exercises

1. Write a mathematical expression to describe each relationship.

(a) diameter and radius

(b) diameter and circumference

(c) radius and circumference

2. Measure the circumference and diameter of six circular objects. Record the measurements in the table.

Object	Circumference	Diameter

3. Explain the relationship between diameter and circumference as shown by the table in Problem 2.

For Problems 4 to 9, round answers to two decimal places as necessary.

4. Find the circumference for a circle with each diameter.

(a) $d = 3.0$ in.

(b) $d = 2.94$ m

(c) $d = 1.809$ km

(d) $d = 0.51$ ft

5. Find the circumference for a circle with each radius.

(a) $r = 6.4$ ft

(b) $r = 7$ mm

(c) $r = 5.2$ in.

(d) $r = 1.9$ km

6. Find the diameter for a circle with each radius.

(a) 2.5 cm

(b) 80.6 in.

(c) 5 mm

(d) 1 mi

7. Find the radius for a circle with each diameter.

(a) 6.2 mm

(b) 8.5 ft

(c) 14 mi

(d) 9 cm

8. Find the diameter for a circle with each circumference.

(a) 2.55 ft

(b) 80.3 in.

(c) 40 mm

(d) 1 km

9. Find the radius for a circle with each circumference.

(a) 7.9 m

(b) 5 in.

(c) 100 mm

(d) 0.221 mi

10. A landscaper is making a circular garden with 25 ft of edging. The edging is 2 in. high. There is a straight pathway from the edge of the garden to the center. Draw a diagram of the garden to find the length of the path.

11. Explain how you could use Circle A to find the circumference of Circle B.

A

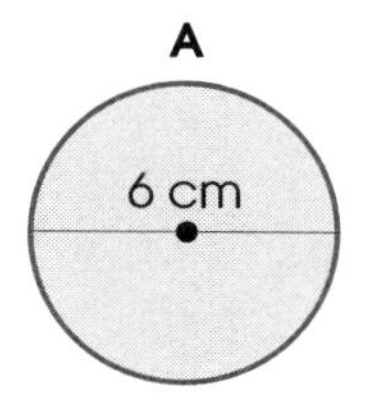

B

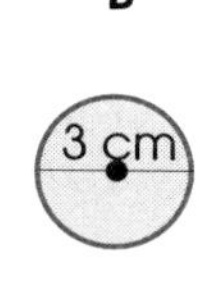

12. What radius is needed to construct a circle with each circumference?

(a) 15.7 cm

(b) 28.26 in.

(c) 12.56 in.

(d) 13.188 cm

13. Construct each circle from Problem 12.

14. If the large circle has a diameter of 14 cm, what is the circumference of the small circle?

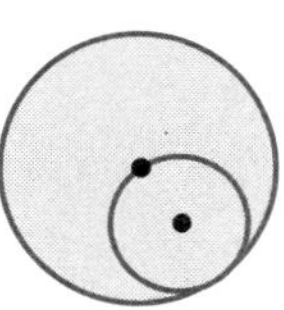

15. Use a pair of compasses to construct this design with congruent circles.

(a) Find the diameter and circumference of each circle.

(b) What is the height of the design?

(c) What is the height of the flower shape in the center?

16. Construct your own circle design. Find the radius, diameter, and circumference of each different circle you used.

11.2 Circle Problems

In this lesson, you solved problems about **circle measurements**. Recall the **four steps**:

- Think about the problem.
- Make a plan.
- Solve the problem.
- Look back.

Example 1

The dome of the Albert Einstein Planetarium in Washington has a radius of 35 ft. If a cable is to be placed around the base of the dome, what is the minimum length of cable that would be required?

Solution

Step 1: Think about the problem.

Given Information
The bottom of the dome is shaped like a circle.
The radius of the circle is 35 ft.

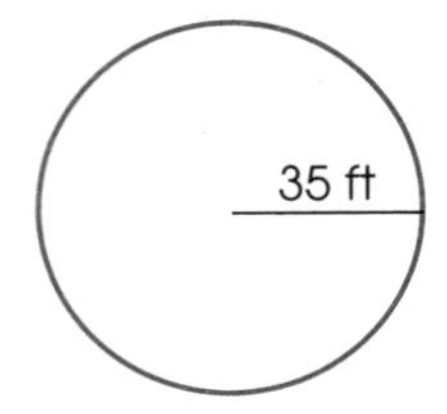

Needed Information
What is the minimum length of cable needed to go around the circumference of the circle?

Step 2: Make a plan.

If you know the diameter of a circle, you can use the formula $C = \pi \times d$ to find the circumference.

The diameter is the distance across the circle. This distance is twice as long as the radius.

$$\begin{aligned} \textit{diameter} &= 2 \times \textit{radius} \\ &= 2 \times 35 \text{ ft} \\ &= 70 \text{ ft} \end{aligned}$$

Now you can use the formula $C = \pi \times d$ to solve the problem.

Step 3: Solve the problem.

$$\begin{aligned} C &= \pi \times d \\ &\doteq 3.14 \times 70 \text{ ft} \\ &\doteq 219.8 \text{ ft} \end{aligned}$$

The minimum amount of cable needed to go around the dome is about 219.8 ft.

Step 4: Look back.

The problem asked for the minimum length of cable needed to go around the dome. A length of 219.8 ft seems reasonable because the distance around the dome should be a bit more than 3 times the diameter, or a bit more than 210 ft.

Exercises

1. To find the circumference of a circle you can't measure directly, you need to know either the radius or the diameter. Why?

2. Complete each statement.

   (a) There are ☐ radii in the diameter of a circle.

   (b) There are about ☐ diameters in the circumference of a circle.

3. **(a)** What is π?

   (b) How does π help you find the circumference of a circle?

4. What is the relationship between the side length of a square and the diameter of the largest possible circle that fits inside it?

5. A circular window has a diameter of 32 in.

   (a) What is the radius of the window?

   (b) What is the circumference?

6. The diameter of a bicycle wheel is 26 in. About how many rotations will the wheel make over a distance of 172 ft?

7. A forester uses a special measuring tape called a diameter meter. When the forester stretches the tape around a tree, the tape shows the circumference of the tree as well as the diameter.

   If the diameter of a tree is 12.5 in., what is the circumference?

8. A rectangular piece of pipe insulation is used to wrap a pipe 3.5 in. in diameter. What is the width of the insulation?

9. In archery, the target face can be made up of a bull's-eye surrounded by four colored rings. A sporting goods store sells a target with rings 4 in., 8 in., 12 in., and 16 in. in diameter.

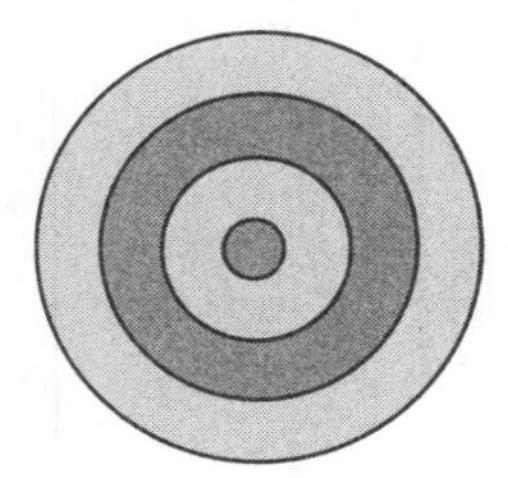

(a) Estimate the circumference of each circle.

(b) Make a table to show the changes in diameter and the related changes in circumference.

(c) How does changing the diameter affect the circumference of the circle?

10. The diameter of a gear, from point to point, is 6.5 in. How far will a point on the gear travel in:

(a) one rotation of the gear?

(b) ten rotations?

11. A furniture store sells a circular oak dining room table with a diameter of 5 ft. To accommodate dinner guests, the table can be enlarged another 1 ft around the edge.

(a) What is the circumference of the table without the extension?

(b) What is the circumference of the table with the extension in place?

(c) If guests sit at intervals of 32 in. around the table, how many people can fit around the table, with and without the extension?

(d) A table requires at least 4 ft of clearance space to allow people to walk behind seated guests. What is the smallest possible square dining room that would accommodate the extended table?

12. List as many ways as you can to find the circumference of a circle. Test your methods on some circles you have made.

13. Some problems are easier to solve if you use a table. Create a table problem about circle measurements. Write the solution on a separate piece of paper and then exchange problems with a classmate and solve.

14. Create a problem about circle measurements and write the solution. Then rewrite the solution on another sheet of paper so it contains an error. Exchange problems with a classmate and find the errors.

11.3 Polygons and Circles

In this lesson, you learned to identify and classify polygons and circles according to their **properties**. Recall:

- A **polygon** is a closed figure with straight sides. A **regular polygon** has congruent sides and angles.
- A **circle** is a curved figure formed by a set of points equidistant from a center point.
- A **quadrilateral** is a polygon with four sides. Types of quadrilaterals include parallelograms, rhombuses, rectangles, squares, trapezoids, and trapeziums.

Example 1

Which figure is a polygon? Explain how you know.

Name the polygon and describe its characteristics.

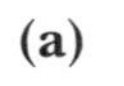

(a)

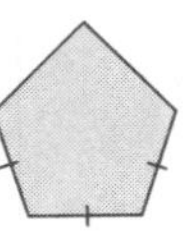

(b)

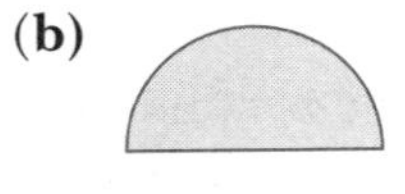

Solution

Figure A is a polygon because it is a closed figure with straight sides. Figure B has a curved side, so it is not a polygon.

The polygon has five sides, so it is a pentagon.

Only three sides are congruent, so it is an irregular pentagon.

Characteristics

1. There are five sides.
2. Three sides are congruent.
3. There are no parallel sides.
4. There are no right angles.
5. You can draw five diagonals inside the pentagon.
6. The diagonals form a five-pointed star.

Example 2

Select the statements that describe rectangles.

1. There are two pairs of parallel sides.
2. There is only one pair of parallel sides.
3. There are no pairs of parallel sides.
4. There are four congruent sides.
5. There are two pairs of congruent sides.
6. There is at least one right angle.
7. The diagonals bisect each other.
8. The diagonals intersect at a right angle.
9. The diagonals are congruent.

Solution

Draw a rectangle.

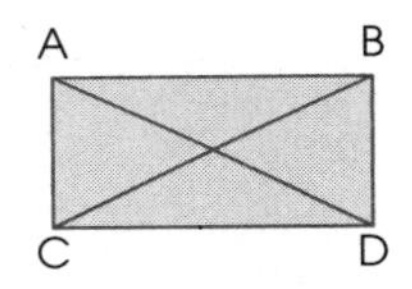

These statements describe all rectangles:

1. There are two pairs of parallel sides.
5. There are two pairs of congruent sides.
6. There is at least one right angle.
7. The diagonals bisect each other.
9. The diagonals are congruent.

These statements describe some rectangles, but only if the rectangle is also a square:

4. There are four congruent sides.
8. The diagonals intersect at a right angle.

Exercises

1. What is the difference between a regular polygon and an irregular polygon?

2. Explain why a circle is not a polygon.

3. What unique characteristics does a circle have?

4. Identify the polygons. Explain how you eliminated each nonpolygon.

(a) 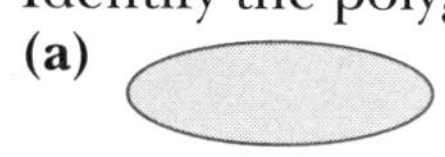
(b) 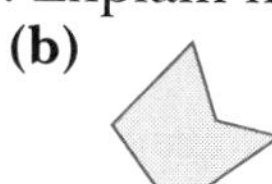
(c) 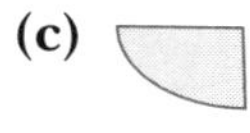
(d) 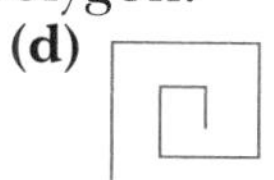

(e)
(f) 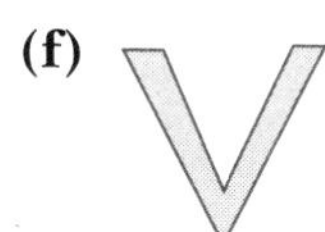
(g) 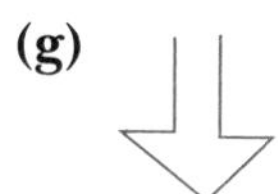
(h) 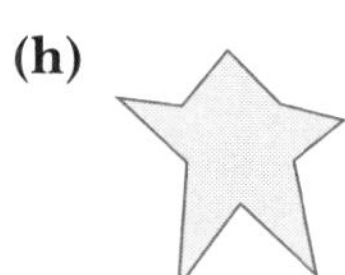

5. Name each quadrilateral.

(a) 
(b) 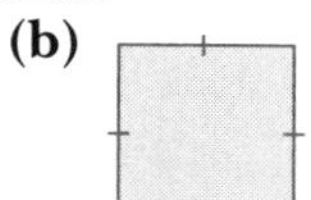
(c) 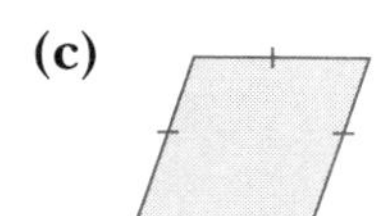
(d) 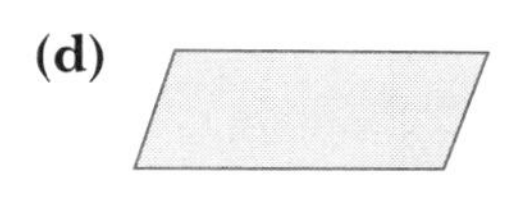

(e)
(f) 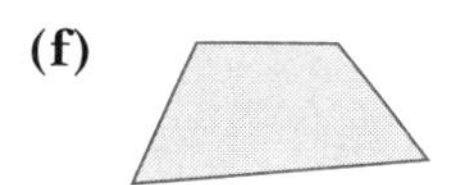
(g) 
(h) 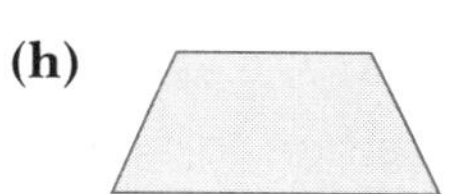

6. Complete the table to show the properties of each quadrilateral from Problem 5.

Properties	(a)	(b)	(c)	(d)	(e)	(f)	(g)	(h)
no congruent sides								
only one pair of congruent sides								
two pairs of congruent sides								
no pairs of parallel sides								
only one pair of parallel sides								
two pairs of parallel sides								
contains at least one right angle								

7. Construct a similar table to show properties of the diagonals of each quadrilateral. Use these headings:
   (a) diagonals are equal
   (b) diagonals bisect each other
   (c) diagonals intersect at a right angle
   (d) at least one diagonal is outside the figure

8. Is each statement true or false? Give reasons for your choice.
   (a) Every parallelogram is a quadrilateral.

   (b) Every rectangle is a square.

   (c) Every trapezoid is a parallelogram.

   (d) Every trapezium is a kite.

   (e) Every square is a rhombus.

   (f) Every trapezoid is a rectangle.

9. Arrange these rectangles in order from the one most like a square to the one least like a square.

1 2 3 4 5

   (a) Measure the length and height of each rectangle in millimeters. Find the ratio of length to height in decimal form to the nearest hundredth.

   (b) Put the ratios in order from least to greatest. Compare with the way you ordered the rectangles. What do you notice? Can you explain why this happens?

(c) What is the length-to-height ratio for a square?

10. This nested diagram shows how three types of quadrilaterals are related. Explain what the diagram means.

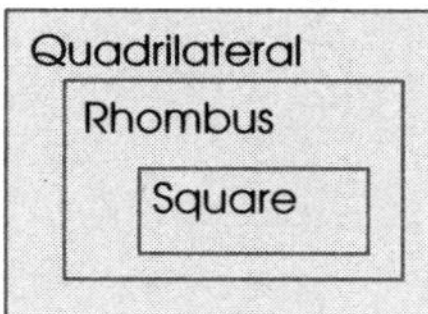

11. Can you draw a nested diagram to show the relationships for quadrilaterals, squares, rectangles, and parallelograms?

12. Work in a small group. Create a poster that illustrates each shape described in this lesson. Name each shape and describe its properties. Include definitions of these terms:

(a) polygon
(b) quadrilateral
(c) regular polygon
(d) irregular polygon
(e) circle

13. Create a true-or-false quiz about the properties of polygons and quadrilaterals. Include at least ten questions. Exchange quizzes with a classmate.

14. Choose a photograph from a personal collection, a postcard, or a newspaper or magazine. List the shapes you see in your picture and describe the properties of each one. Then exchange pictures with a classmate.

11.4 Similar Triangles

Two geometric figures are **similar** if they have the exact same shape, but not necessarily the same size. In this lesson, you determined that two triangles are similar if:

- at least two pairs of corresponding angles are congruent
- three pairs of corresponding sides are proportional.

A knowledge of similar triangles can often be useful in determining a length or angle that cannot be measured directly.

Example 1

Show why the two triangles are similar. Name the corresponding angles and the corresponding sides.

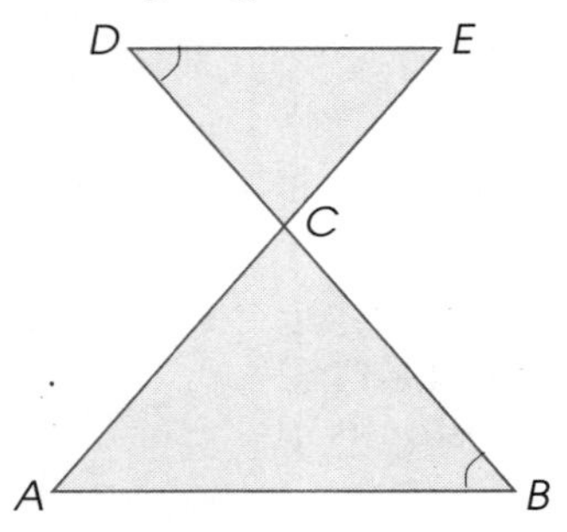

Solution

Two triangles are similar if two pairs of corresponding angles are congruent.

The diagram shows that $\angle EDC \cong \angle ABC$.

$\angle DCE \cong \angle BCA$, since these angles are vertically opposite.

Since two pairs of corresponding angles are congruent, the third pair must also be congruent if the angles in each triangle are to sum to 180°. Therefore, the two triangles are similar.

To find corresponding angles and sides, it helps to rotate $\triangle DCE$ so its position matches that of $\triangle BCA$.

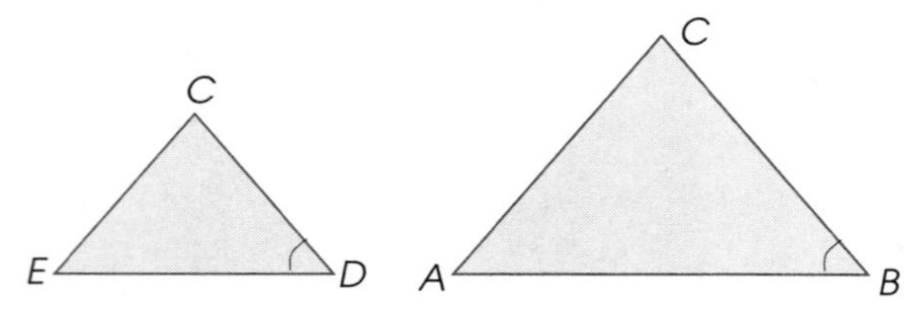

Corresponding angles that are congruent:

$\angle CED \cong \angle CAB$, $\angle EDC \cong \angle ABC$, $\angle ECD \cong \angle ACB$

Corresponding sides that are proportional:

$$\frac{EC}{AC} = \frac{ED}{AB} = \frac{CD}{CB}$$

Example 2

To determine the width of a canyon, AB, surveyors made the measurements shown below. Find the width of the canyon.

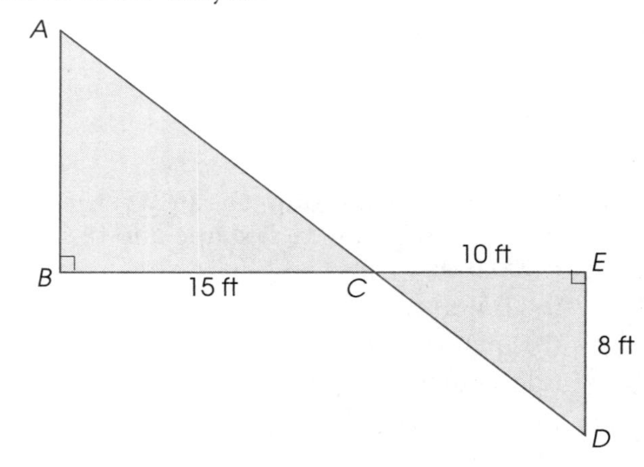

Solution

If $\triangle ABC$ is similar to $\triangle DEC$, you can use proportional side lengths to find the length of side AB. The two triangles must be similar, since the two pairs of corresponding angles are congruent.

$\angle ABC = \angle DEC = 90°$

$\angle BCA \cong \angle ECD$ because the angles are vertically opposite.

Solve the proportion to find length AB:

$\frac{AB}{DE} = \frac{BC}{EC}$ Set up the proportion using two known values and one unknown.

$\frac{AB}{8} = \frac{15}{10}$ Substitute the known values.

$8 \times \frac{AB}{8} = 8 \times \frac{15}{10}$ Multiply both sides by 8 to isolate the unknown.

$AB = \frac{120}{10}$ Divide.

$AB = 12$

The width of the canyon is 12 ft.

Exercises

1. Construct two similar triangles. Write six true statements about your triangles.

2. When you name two similar angles, sides, or triangles, it helps to order the letters so corresponding vertices are in the same position. For example, $\triangle KLM \cong \triangle NOP$. Complete the following statements for this pair of similar triangles.

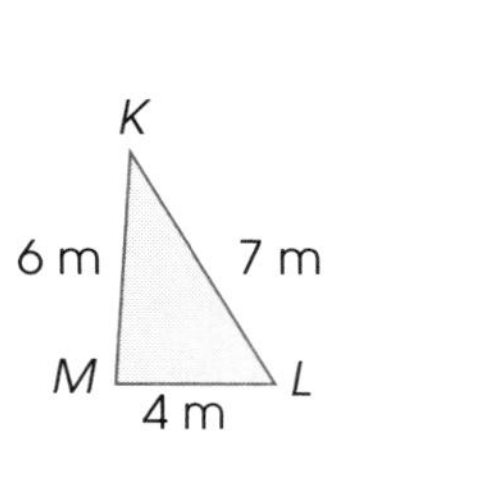

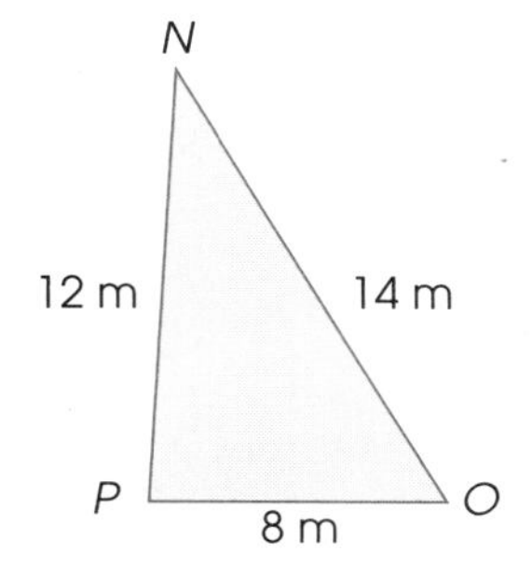

(a) $\angle K \cong \angle$ ☐ **(b)** $\angle L \cong \angle$ ☐

(c) $\angle M \cong \angle$ ☐ **(d)** side $KL \cong$ side ☐

(e) side $LM \cong$ side ☐ **(f)** side $KM \cong$ side ☐

3. Each pair of triangles is similar. Find the unknown side lengths and angle measures.

(a)

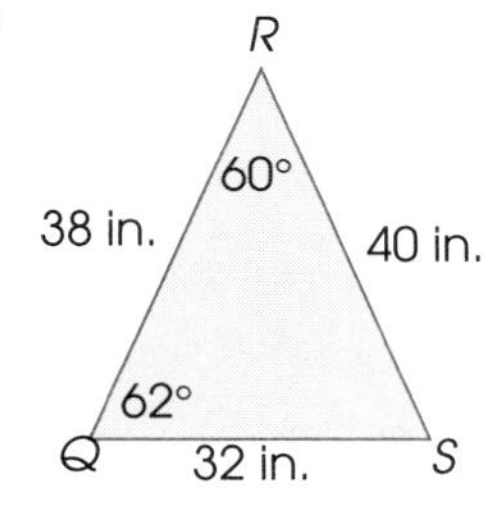

A
20 in.
B
C

(b)

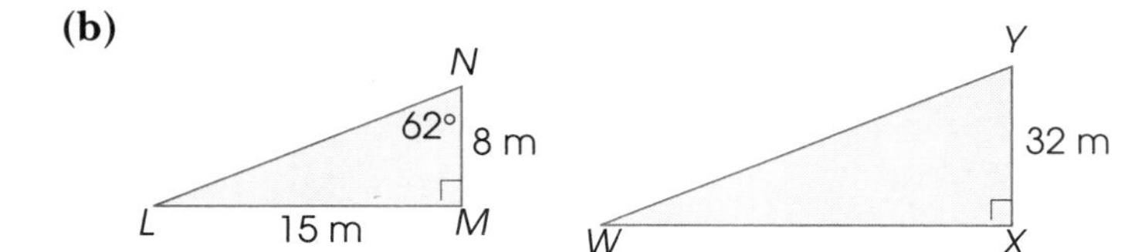

4. Explain how you know that each pair of triangles is or is not similar.

(a)

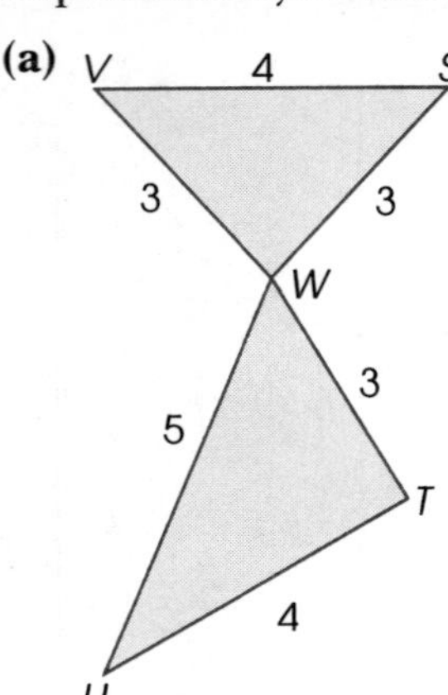

(b)

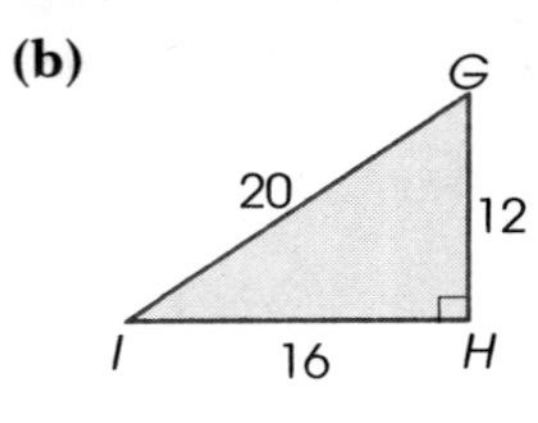

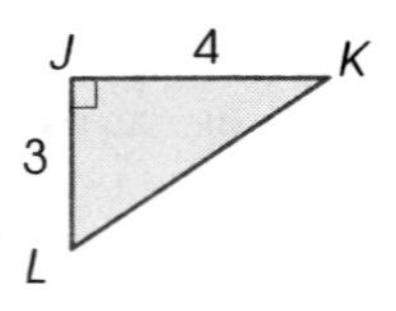

5. $\triangle ABC$ is similar to $\triangle EDC$.

(a) How long are sides d and e?

(b) What size is $\angle x$?

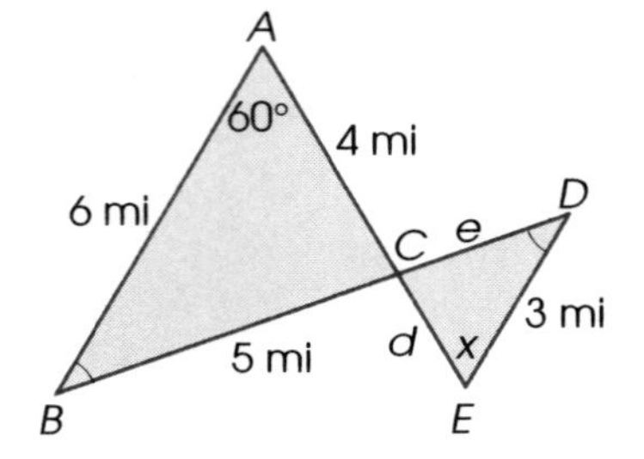

6. How wide is distance AB across the river?

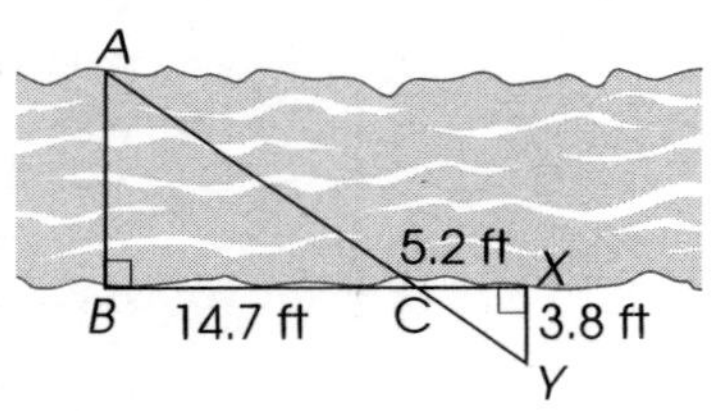

7. What is the depth of the ditch?

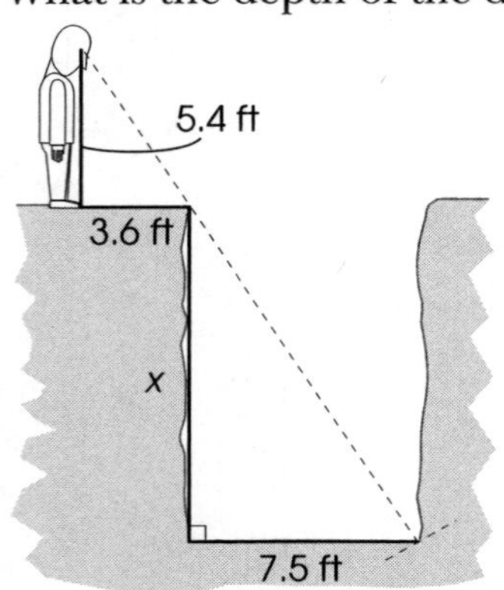

8. Construct a triangle that is similar to $\triangle ABC$. Label the side and angle measures. Show your calculations.

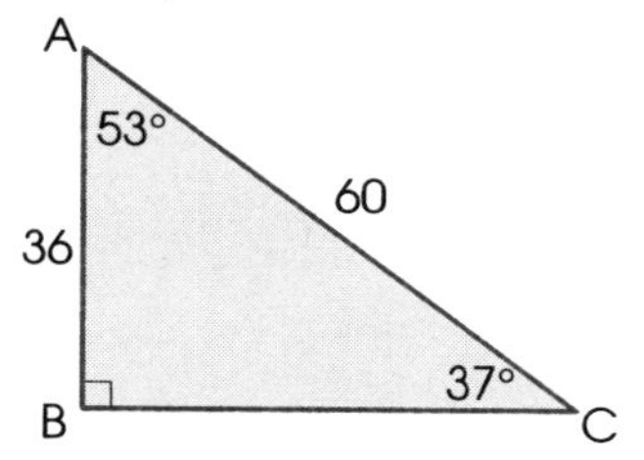

9. Two ladders are leaning against a building as shown. Create a problem about the two similar triangles they form. Test your problem by solving it, then exchange problems with a classmate.

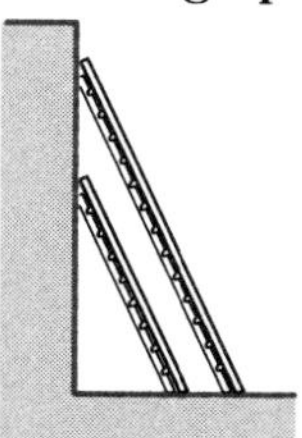

10. Create a ten-question true or false quiz about the sides and angles in these two similar triangles. Exchange with a classmate.

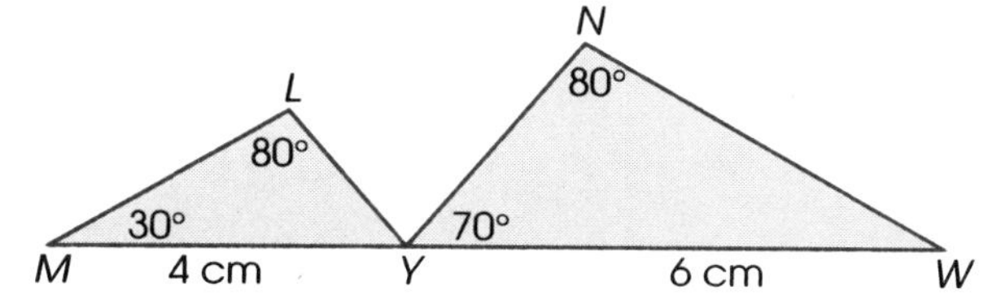

11.5 Congruence and Similarity

This lesson introduced **congruent** triangles. These special similar triangles are not only the same shape, but also the same size.

- All triangles contain six measurable parts — three sides and three angles.
- In congruent triangles, all six pairs of corresponding parts are congruent.
- The SSS (Side, Side, Side) Congruence Relation states that if all three pairs of corresponding sides of two triangles are congruent, then the two triangles are congruent.

Example 1

State whether the triangles are congruent. List the corresponding sides and angles.

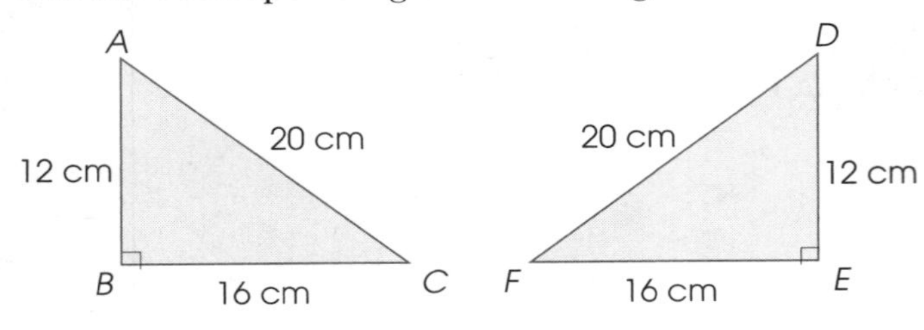

Solution

$AB = DE = 12$ cm; $BC = EF = 16$ cm; $AC = DF = 20$ cm

Because the triangles have three pairs of corresponding congruent sides, they are congruent. Since the triangles are congruent, they also have three pairs of corresponding congruent angles.

$\angle ABC = \angle DEF$; $\angle BAC = \angle EDF$; $\angle ACB = \angle DFE$

Example 2

Construct a triangle congruent to $\triangle ABD$ without tracing or measuring any unlabeled dimensions on the original. Use a ruler, but not a protractor.

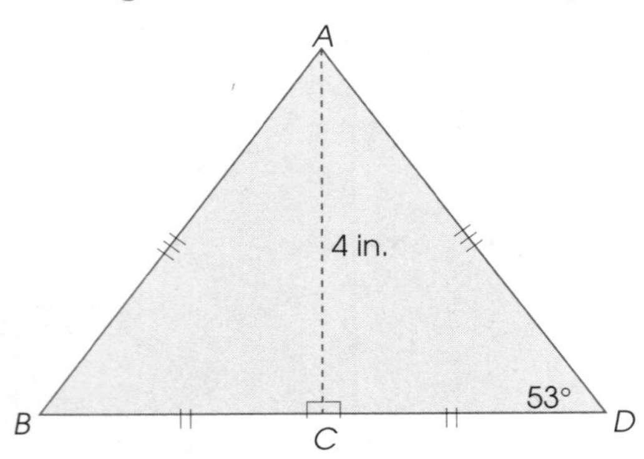

Solution

Since you know one side length and one angle measure of $\triangle ACD$, you can use trigonometry to find length CD. Since $CD \cong BC$, this will help you draw the base of the triangle.

Find CD.

$$\frac{AC}{CD} = \tan 53^\circ$$
$$\frac{4}{CD} = 1.327$$
$$4 = 1.327 \times CD$$
$$3 = CD$$

The length of CD is 3 in.

Find BC.

Since $\triangle ACB$ and $\triangle ACD$ share two pairs of congruent sides and have a third side in common, they must be congruent. Therefore, BC must also be 3 in. long.

Construct the triangle.

Use 4 in. for AC and 3 in. each for BC and CD. Use the corner of the ruler to construct the right angle.

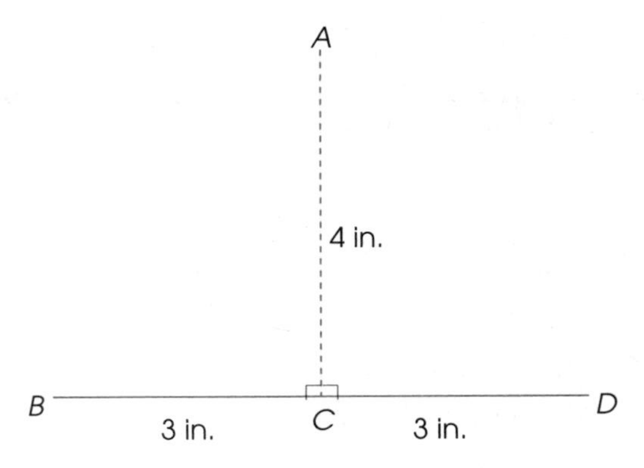

Finish the triangle by joining AB and AD.

Check.

Test for congruence by fitting your triangle over the original. If the triangles are congruent, the sides and angles will match exactly.

Exercises

1. Complete. Remember to order the letter names so congruent vertices are in the same position.

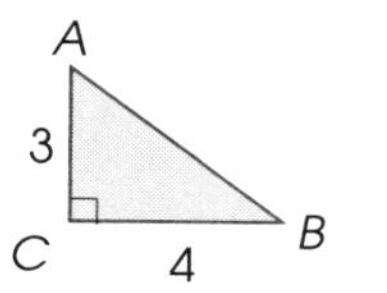

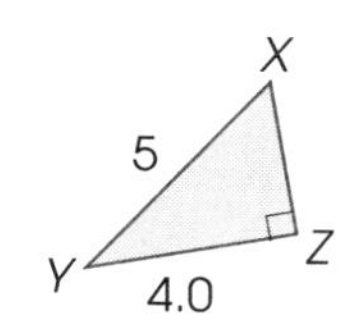

   (a) $\triangle ABC \cong \triangle$_____

   (b) side $AB \cong$ side _____

   (c) The length of XZ is _______.

   (d) The size of $\angle XZY$ is _______.

2. Are the triangles congruent, similar but not congruent, or neither? Explain.

   (a)

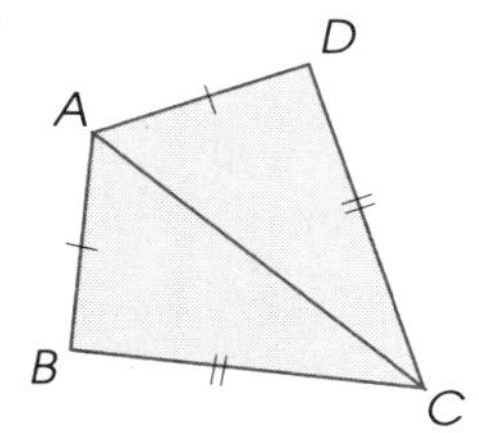

   (b)

3. The triangles are congruent. Find all the missing side and angle measures.

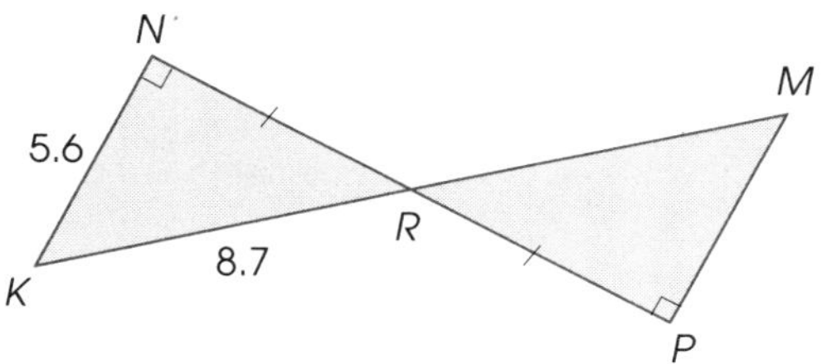

4. Find the missing side lengths.

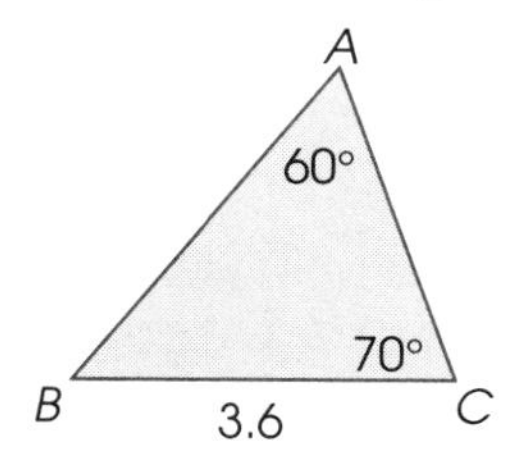

5. In $\triangle ABC$, $AB = 15$, $BC = 15$, and $AC = 30$. In $\triangle DEF$, $DF = 60$, $DE = 30$, and $EF = 40$. Are the triangles congruent, similar but not congruent, or neither? Explain.

6. Find and correct any errors.

$XY = RT = 8$
$YZ = TS = 10$
$XZ = ST = 5$

Since all three pairs of corresponding sides are congruent (SSS), then $\triangle XYZ \cong \triangle RTS$.

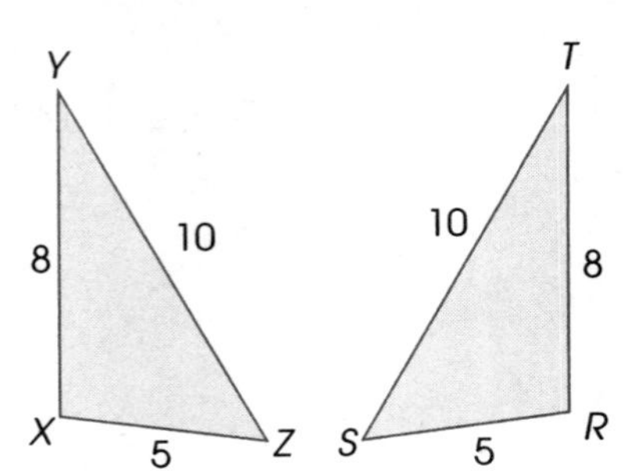

7. Complete the congruence statement. List the congruent angles and sides.

$\triangle EFG \cong$ __________

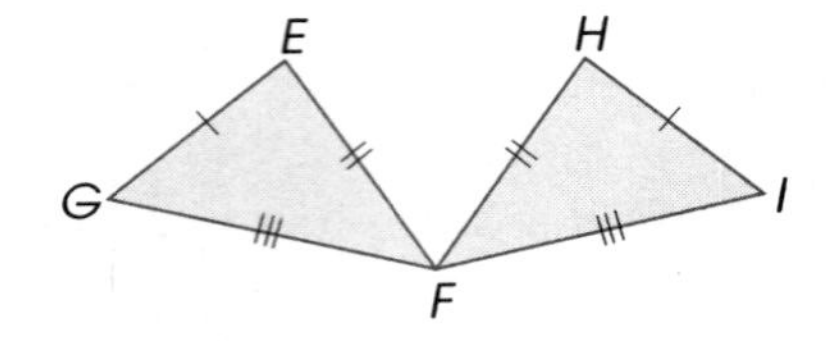

8. Construct a triangle:
(a) congruent to $\triangle JKL$.
(b) similar to $\triangle JKL$, but not congruent.

Label all side lengths and angle measures.

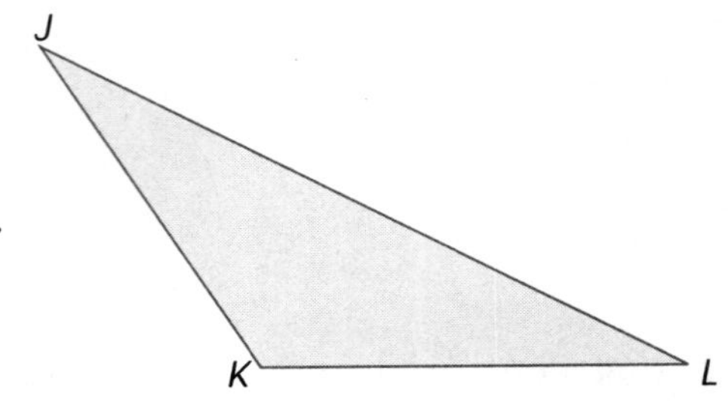

9. Give examples to show why each statement is true or false.
(a) All similar triangles are congruent.

(b) All congruent triangles are similar.

10. Compare the triangles.

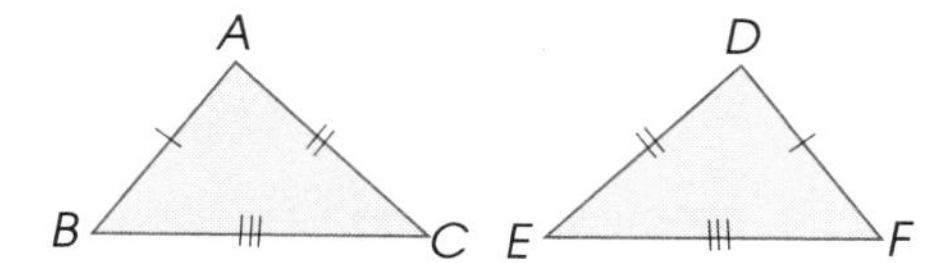

(a) Explain how you know the triangles are congruent.

(b) Write six true statements about the triangles. Compare your statements with a partner's.

11. Construct the triangles. Label all the side lengths and angles.

$\triangle KLM \cong \triangle XYZ$
$ML = YZ$
$\angle XYZ = 18^\circ$
$KM = 34$ cm
$\angle ZXY = 69^\circ$

12. Show two different ways to divide the baseball diamond into congruent triangles. Explain how you know the triangles are congruent in each case. Name the triangles and list the corresponding sides and angles.

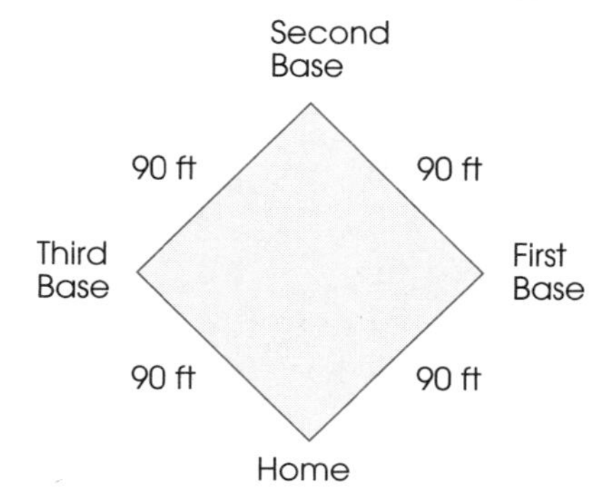

Selected Answers

1 WHOLE NUMBERS

1.1 Whole Numbers

1. (a) addition (b) factors (c) quotient (d) sum (e) difference (f) natural (g) whole
3. (a) 2736 (b) 777 (c) 27,598 (d) 7,452,860
5. (a) 69,600 (b) 69,600 (c) 70,000 (d) 100,000
7. (a) 3 (b) 202 (c) 118 (d) 792 (e) 3579 (f) 303
9. (a) 7623 (b) 12,760 (c) 12,987 (d) 554,005 (e) 79,104 (f) 6,085,206
11. (a) 13 (b) 11 (c) 33 (d) 31 R28 (e) 39 R43 (f) 99 R75
13. 406 mi
15. 480,000 pixels

1.2 Powers and Exponents

1. The base tells you what number is multiplied repeatedly; the exponent tells how often the number occurs; *example*: 5^4 means that 5 occurs 4 times. The repeated multiplication is $5 \times 5 \times 5 \times 5$. The standard form is 625.
3. There are 8 small cubes in the larger cube, so $2^3 = 2 \times 2 \times 2 = 8$. There are 9 small squares in the larger square, so $3^2 = 3 \times 3 = 9$.
5. (a) 4 (b) 3
7. (a) $4^2 = 16$ (b) $3^3 = 27$ (c) $2^5 = 32$ (d) $5^3 = 125$
9. (a) 2^3 (b) 3^3 (c) 8^2 (d) 10^2 (e) 10^3 (f) 12^2
11. (a) $<$ (b) $>$ (c) $<$ (d) $=$ (e) $>$
13. (b) 24×24, 24^2, 576

1.3 Divisibility Rules

1. 0
3. (a) $3 + 4 + 7 + 9 = 23$; Since 23 is not a multiple of 3, 3479 is not divisible by 3.
   (b) $3479 \div 3 = 1159.\overline{6}$; The decimal part of the quotient means that there is a remainder, so this supports the conclusion in part (a).
5. (a) no (b) yes (c) yes (d) yes
7. Paige is right, since these numbers represent groups of 3. If the remaining numbers also form groups of 3, then the number is divisible by 3. For example, for 366,954:
   You know that $3 + 6 + 6 + 9$ can be put into groups of 3 because all the numbers are multiples of 3. You only need to add $5 + 4$ to see that the rest of the number, 9, can also be put into groups of 3.
9. (a) no (b) yes (c) no (d) yes (e) no (f) yes (g) no
11. (a) 21 (b) yes (c) no (d) yes
13. (a) True; if you can divide a number into groups of 10, then you can split each group in half to make groups of 5.
    (b) False; if a number is made up of an odd number of groups of 5, you will have one group left over if you try to make groups of 10. Only even multiples of 5 are also multiples of 10.
15. (a) 1, 4, 7 (b) 0, 2, 4, 6, 8 (c) 1, 4, 7
17. Yes; a number is divisible by 12 if half the number is divisible by 6. Half of 324 is 162. Since 162 can be formed with groups of 6, then double as much, 324, can be formed with groups of 12. Therefore, 324 people can sit in rows of 12 with no empty seats.

1.4 Prime Factors and Exponents

1. *example:* A number is prime if it has only two factors — one and itself. The only even prime number is 2. If a number is odd, you can use divisibility rules to check for other factors.
3. *example:* 28
   Method 1: Step Factorization
   Method 2: Factor Tree
7. (a) $24 = 2^3 \times 3$ (b) $18 = 2 \times 3^2$ (c) $2^2 \times 3^2 \times 7$
9. (a) $2 \times 2 \times 2 \times 13$ (b) 6 (c) 75 (d) 24 (e) 168
13. (a) \$13 (b) 6 (c) 5

2 FRACTIONS AND DECIMALS

2.1 Fraction Conversions

1. (a) A proper fraction is less than 1 whole.
   (b) A mixed number is part whole number and part fraction.

(c) In an improper fraction, the numerator is greater than the denominator because the value is greater than 1 whole.

3. **(a)** $3, \frac{5}{4}$ **(b)** $4, \frac{6}{5}$ **(c)** $2, \frac{8}{5}$ **(d)** $6, \frac{5}{3}$

5. **(a)** $1\frac{2}{3}$ **(b)** $1\frac{6}{7}$ **(c)** $1\frac{1}{2}$ **(d)** $1\frac{1}{2}$
(e) $4\frac{4}{5}$ **(f)** $10\frac{1}{2}$ **(g)** $1\frac{2}{5}$ **(h)** $1\frac{1}{2}$
(i) $2\frac{3}{4}$ **(j)** $3\frac{7}{8}$

7. *example:* $3\frac{5}{8} = 3\times 8 \text{ eighths} + 5 \text{ eighths}$
$= 24 \text{ eighths} + 5 \text{ eighths}$
$= 29 \text{ eighths}$
$= \frac{29}{8}$

9. 12

2.2 Decimal Conversions

1. *example:* Divide the numerator by the denominator. If you get a remainder of 0, the decimal terminates. If you get a repeating pattern of remainders, the decimal repeats.

3. *example:* No; equivalent numbers represent the same amount, but they are written in different ways.

5. **(a)** 0.375 **(b)** $0.\overline{45}$ **(c)** 0.1875
(d) $0.1\overline{3}$ **(e)** $0.\overline{6}$ **(f)** $1.\overline{1}$

7. **(a)** 2.25 **(b)** $4.\overline{3}$ **(c)** 4.5
(d) 1.0625 **(e)** $2.58\overline{3}$ **(f)** $3.8\overline{3}$

9. **(a)** 0.8 **(b)** $0.\overline{3}$ **(c)** 0.2857
(d) $0.\overline{3}$ **(e)** 0.8 **(f)** 0.625
(g) 0.625 **(h)** 0.2857
Matching decimals: (a) and (e), (b) and (d), (c) and (h), (f) and (g)

11. **(a)** $\frac{1}{2} = 0.5, \frac{1}{4} = 0.25, \frac{1}{8} = 0.125$

(b) $\frac{1}{16}$ and $\frac{1}{32}$ will also terminate, since each fraction represents half of a terminating fraction.

(c) $\frac{1}{3} = 0.\overline{3}, \frac{1}{6} = 0.1\overline{6}, \frac{1}{9} = 0.\overline{1}$; $\frac{1}{12}$ and $\frac{1}{15}$ will also repeat. Each decimal is produced by dividing a repeating decimal.

13. **(a)** Entertainment: $\frac{25}{200} = \frac{1}{8} = 0.125$

Clothes: $\frac{50}{200} = \frac{1}{4} = 0.25$

Snacks: $\frac{20}{200} = \frac{1}{10} = 0.1$

(b) $\frac{105}{200} = \frac{21}{40} = 0.525$

2.3 Ordering Numbers

1. When the numbers are fractions, you compare numerators or denominators. When they are decimals, you compare decimal places. If the numbers are in the same form, you can compare the same parts.

3. *examples:*

(a) $\frac{3}{4}$ is greater because it is more than halfway between 0 and 1.

(b) $\frac{2}{5}$ is less than halfway between 0 and 1.

(c) $\frac{7}{9}$ is greater, because 7 ninths is more than 5 ninths.

(d) $\frac{2}{5}$ is greater, because fifths are bigger than fifteenths.

(e) $\frac{1\times 25}{4\times 25} = \frac{25}{100} = 0.25$

(f) $1\frac{7}{8} = 1.875$, so 1.901 is more.

(g) $1\frac{3}{7} = \frac{7}{7} + \frac{3}{7} = \frac{10}{7}$

(h) 2.6 is more than 2.5, so 2.658 is more than 2.568.

7. **(a)** $\frac{3}{8} < 0.4$ **(b)** $1.6 < \frac{12}{3}$
(c) $2\frac{1}{3} < \frac{5}{2}$ **(d)** $3.222 < \frac{17}{5}$
(e) $\frac{5}{4} > 1.2$ **(f)** $1\frac{1}{8} < 1.13$

9. *examples:*
(a) $\frac{5}{8}$, 0.6 **(b)** $\frac{13}{18}$, 0.75 **(c)** $\frac{11}{24}$, 0.495

11. **(a)** $<$ **(b)** $>$ **(c)** $>$ **(d)** $<$

13. No; if both people eat more than half, then they would need more than one sandwich.

2.4 Terminating and Repeating Decimals

3. *example:* You can't use place value to convert a repeating decimal because there is no final digit.

5. **(a)** $2.32 = 2\frac{8}{25}$ **(b)** $2.05 = 2\frac{1}{20}$
(c) $2.8 = 2\frac{4}{5}$ **(d)** $0.44 = \frac{11}{25}$
(e) $3.35 = 3\frac{7}{20}$ **(f)** $0.09 = \frac{9}{100}$

7. **(a)** $\frac{7}{100}$ **(b)** $\frac{9}{20}$ **(c)** $\frac{3}{25}$ **(d)** $1\frac{17}{100}$

(e) $2\frac{3}{4}$ (f) $5\frac{17}{20}$ (g) $3\frac{11}{50}$ (h) $4\frac{1}{50}$

9. *example*: The result is always a repeating decimal. If the fraction is less than 1 whole, the repeating digit is the same as the number of ninths. If the fraction is greater than 1 whole, separate the wholes first. Then the repeating digit will be the same as the number of ninths that are left.
11. (a) 0.95 (b) 0.625 (c) $3.\overline{2}$
(d) 2.75 (e) $5.\overline{2}$ (f) $4.\overline{3}$

2.5 Multiplication with Decimals

1. *example*: Multiplication is repeated addition. For example, 4×2.5 means 4 groups of 2.5, or $2.5 + 2.5 + 2.5 + 2.5$.
3. (a) 7.82 (b) 5.59
5. $3.2 \times 4.5 = 14.4$
7. (a) 400 (b) 730 (c) 60 (d) 0.6
9. (a) 5400 (b) 1.689 (c) 846
(d) 1.3 (e) 1.869 (f) 8004
11. greater
13. $2.15
15. (a) yes (b) $11.06

2.6 Division with Decimals

1. Dividing a number by a divisor is like repeatedly subtracting the divisor from the number to see how many times you can do it. For example, $2.8 \div 0.4$ means "How many times can you subtract 0.4 from 2.8?"
$0.4\overline{)2.8}$ becomes $4\overline{)28}$, so the quotient is 7.
This means you can subtract 0.4 from 2.8 exactly 7 times.
3. (a) $68\overline{)156.4}$ (b) $32\overline{)275.2}$
5. (a) 48; There are four 0.25s in 1, so there are 4×12 or 48 in 12.
(b) 156; There are two 0.5s in 1, so there are 78×2 or 156 in 78.
(c) 150; There are five 0.2s in 1 so there are 30×5 or 150 in 30.
(d) 1689.4; There are ten 0.1s in 1, so there must be 168.94×10 or 1689.4 in 168.94.
7. (a) 5.2 (b) 43.5
(c) 0.67 (d) 101.25
11. The error is in the way the long division was set up. When the divisor was multiplied by 10 to get 275, the dividend should also have been multiplied by 10 to get to 3355. Therefore, the quotient should be 12.2, not 1.22.
13. (a) Yes; 12 lb priced at $0.50 would cost exactly $6, so 9.75 lb for $0.49 must cost less.
(b) $4.78
15. $2.50

2.7 Order of Operations

1. Rules for ordering operations are important because they ensure that everyone who uses the same expression will get the same answer. For example, if you use the standard rules, the expression $4 + 6 \div 2$ has a value of $4 + 3$, or 7. If you work from left to right, the same expression has a value of $10 \div 2$, or 5.
3. (a) 31.05 (b) 139.84 (c) 3.48
(d) 20.48 (e) 26.86 (f) 4.275
5. (a) $\times, \div$ (b) $-, \times$ or $-, \div$
(c) $\div, +, \div$ or $\times, \div, +$ (d) $-, +, \times$ or $-, +, \div$
7. $4 + 3 \times 6 - 4 \div 2 = 20$; $(4 + 3) \times 6 - 4 \div 2 = 40$; $(4 + 3 \times 6 - 4) \div 2 = 9$; $(4 + 3) \times (6 - 4) \div 2 = 7$; $4 + 3 \times (6 - 4 \div 2) = 16$
9. *examples*:
(a) $4 + (3 \times 6) - 8 = 14$
(b) $84 \div 2 \times 12 =$ M+ $833 -$ RM $= 329$
(c) $14.2 - 7.8 + (3.5 \times 6.1) = 27.75$

3 INTEGERS

3.1 An Introduction to Integers

1. –12 is less, since it is farther from 0. If you owe –$12, you owe more than if you owe –$8.
3. –6; *example*: If you are using integers to represent steps on a ladder, then steps up will be positive and steps down will be negative.
5. (a) +3 (b) –4 (c) +6
(d) 0 (e) –7
7. (a) –6 (b) +5 (c) +7
(d) –5 (e) +3 (f) –4
9. *examples:*
(a) 0, +1 (b) –8, –7 (c) 0, –1
11. (a) –6, –3, 0, +9, +10
(b) –29, –25, –15, 0, +2
(c) –15, –9, –6, –3, 0
(d) –4, –3, –2, +2, +5
13. (a) $-5 < -4$, $-5 < +4$
(b) $+3 > -4$, $-3 > -4$
(c) $0 > -7$
15. hamburgers

3.2 Adding Integers

1. (a) 2 black + 5 black; start at +2 and move right 5; 7

(b) 5 black + 3 red; start at 5 and move left 3; 2
(c) 6 red + 1 red; start at −6 and move left 1; −7
(d) 3 red + 5 black; start at −3 and move right 5; 2

3. (a) 9 (b) 4 (c) −5
(d) −9 (e) −4 (f) −5

5. (a) and (d) sum to +3; (b), (c), (e), and (f) sum to −3

7. (a) 2 (b) −2 (c) 3 (d) −10
(e) −9 (f) −14 (g) 2 (h) 1
(f) is the least sum and (c) is the greatest

9. 5°C

11.

−4	1	0
3	−1	−5
−2	−3	+2

3.3 Subtracting Integers

1. You can't take 8 red tiles from 5 black ones. Adding zeros increases the number of red tiles, making it possible to subtract 8.

5. +1 or −4; 0 or −5; (−1) or −6; (−2) or −7; next one is (−5) + (−3) = −8

7. (b), (c), and (d) all equal +5.

9. (a) +2 (b) −4 (c) +1 (d) −13
(e) −9 (f) +10 (g) +10
(h) −19 (i) +7 (j) −13

11. 1929°C

13. +4 yd

3.4 Multiplying and Dividing Integers

1. (a) a whole number that exactly divides another number; for instance, 4 is a factor of 24
(b) the result of multiplication; for instance, the product of 2 and 3 is $2 \times 3 = 6$
(c) the result of dividing one number by another; for instance, the quotient of $12 \div 3$ is 4
(d) a number by which another number is divided; for instance, in $12 \div 3 = 4$, the divisor is 3
(e) a number that is divided; for instance, in $12 \div 3 = 4$, the dividend is 12

3. (a) $(+4) \times (-3) = -12$ (b) $(+3) \times (-4) = -12$

5. (a) +4 (b) −3 (c) −2
(d) −5 (e) +15 (f) −14

7. (a) × (+3); +81, +243 (b) ÷ (+2); −20, −10
(c) ÷ (−3); −3, +1 (d) ÷ (−5); +5, −1

9. $(-2) \times (+5)$; $(+2) \times (-5)$

13. (a) $[(+4) + (-5)] \times [(-3) - (-3)]$
(b) $(-8) \div [(-2) + (+6)] - (-2)$

4 RATIONAL NUMBERS AND SQUARE ROOTS

4.1 Calculating Square Roots

1. *example:* The square root of a number represents the side length of a square whose area is equal to the number.

3. *example:* The side length of a square is the square root of the area. If you don't recognize the area as a perfect square, you can use a number line to estimate the side length or find the exact side length by using the [√] key on a calculator.

5. (a) 5 × 5 (b) 7 × 7 (c) 12 × 12 (d) 11 × 11

7. (a) 3 (b) 12 (c) 13 (d) 15
(e) 4 (f) 11 (g) 6 (h) 10
(i) 7 (j) 14 (k) 9 (l) 8

9. *examples:* (a) 6.3 (b) 5.5 (c) 8.9

11. (a) 4.1 (b) 5.3 (c) 10.8
(d) 18.7 (e) 14.8 (f) 20.0

13. (a) 4.9999956
(b) There are additional digits in the decimal that do not show in the display.

15. (a) −9 (b) 2 (c) 0

4.2 Adding and Subtracting Fractions

3. (a) $\frac{2}{5}$ (b) $\frac{3}{5}$ (c) $\frac{3}{8}$
(d) $\frac{5}{16}$ (e) $\frac{5}{6}$ (f) $\frac{3}{4}$

5. (a) $\frac{1}{8}+\frac{3}{8}=\frac{4}{8}=\frac{1}{2}$ (b) $\frac{3}{12}+\frac{7}{12}=\frac{10}{12}=\frac{5}{6}$

7. (a) $\frac{3}{4}$ (b) $\frac{11}{13}$ (c) $\frac{4}{17}$ (d) $1\frac{2}{11}$

9. (a) 1 (b) 8 (c) $4\frac{1}{2}$ (d) $5\frac{5}{12}$
(e) $2\frac{7}{10}$ (f) $\frac{3}{4}$ (g) $\frac{2}{3}$ (h) $1\frac{1}{3}$

11. (a) $1\frac{1}{2}$ hours more on Day 1
(b) Shannon worked $\frac{3}{4}$ of an hour more than Jessica and $1\frac{1}{2}$ hours more than Linda.

4.3 Adding Fractions (Unlike Denominators)

1. (a) the smallest number that is a multiple of several other denominators
(b) *example:*
Method 1: List all the multiples, in order, for each denominator. Circle the first

multiple that is common to both lists. Method 2: If one denominator is a multiple of the other, use the greater denominator.

3. **(a)** 8 **(b)** 40 **(c)** 30
(d) 18 **(e)** 21

5. *example:*
1. Use a list of multiples to find the LCD for the two denominators.
2. Find out what you have to multiply the denominator of the first fraction by to get the LCD. Then multiply the numerator by this amount. Write the equivalent fraction.
3. Find out what you have to multiply the denominator of the second fraction by to get the LCD. Then multiply the numerator by this amount. Write the equivalent fraction.
4. Add the two fractions.
5. Rewrite the sum in simplest terms.

7. **(a)** about $\frac{1}{2}$; $\frac{1}{2}$ **(b)** about $\frac{4}{5}$; $\frac{23}{30}$
(c) a bit less than $1\frac{3}{4}$; $1\frac{5}{8}$
(d) a bit more than $3\frac{1}{2}$; $3\frac{3}{4}$
(e) about 5; $5\frac{1}{8}$ **(f)** about $3\frac{1}{2}$; $3\frac{8}{15}$

9. **(a)** *example:* $\frac{5}{12}+\frac{1}{12}$ **(b)** *example:* $\frac{7}{8}+\frac{10}{16}$
(c) There are an infinite number of solutions, because there are an infinite number of denominators.

4.4 Subtracting Fractions (Unlike Denominators)

1. *examples:* Similarities: When you add or subtract, you express the fractions with a common denominator and then add or subtract the numerators. Differences: Addition always means putting two parts together. Subtraction can mean either taking something away or comparing. When you subtract mixed numbers, you sometimes have to regroup the first one in order to be able to subtract.

7. **(a)** $\frac{3}{4}$ **(b)** $\frac{11}{15}$ **(c)** $\frac{1}{2}$ **(d)** $\frac{7}{8}$

11. **(a)** $\frac{1}{4}$ **(b)** 5 **(c)** $2\frac{3}{10}$ **(d)** $\frac{7}{10}$

4.5 Multiplying Fractions

1. **(a)** *example:* When you add, subtract, or multiply, you need to write the result with a single denominator. When you are adding or subtracting, the parts have to be the same size to make this possible. When you are finding a part of a part, the answer will have a single denominator no matter how big the parts are.
(b) *example:* The answer will still be correct but the numbers will be larger, so it will be more difficult to do the multiplication and simplification.

3. *example:*
1. Express the whole number as a fraction with 1 in the denominator.
2. Find $\frac{\textit{product of the numerators}}{\textit{product of the denominators}}$.
3. Express in simplest terms.
For example: $5\times\frac{3}{4}=\frac{5}{1}\times\frac{3}{4}=\frac{15}{4}=3\frac{3}{4}$

5. **(a)** $\frac{1}{10}$ **(b)** $\frac{1}{6}$ **(c)** $\frac{3}{10}$ **(d)** $\frac{5}{8}$

7. **(a)** $\frac{1}{2}$ **(b)** $\frac{3}{20}$ **(c)** $\frac{3}{20}$ **(d)** $\frac{9}{16}$
(e) $\frac{7}{18}$ **(f)** $3\frac{1}{2}$ **(g)** 2 **(h)** $1\frac{1}{5}$

9. **(a)** $9\frac{1}{3}$ **(b)** $9\frac{3}{5}$ **(c)** $5\frac{1}{3}$ **(d)** $3\frac{15}{16}$
(e) $3\frac{3}{10}$

13. The product is not in simplest terms.
$\frac{420}{40}=10\frac{1}{2}$

4.6 Dividing Fractions

1. **(a)** a number being divided
(b) a number by which another number is divided
(c) the result of dividing one number by another
(d) the multiplier of a number that gives 1 as a result; for example, the reciprocal of 2 is $\frac{1}{2}$ and vice versa.

3. *examples:*
(a) $\frac{1}{9}$ is a bit more than $\frac{1}{10}$, so a bit less than 7
(b) $\frac{3}{4}$ is a bit less than 1, so a bit less than $\frac{1}{5}$, maybe $\frac{1}{7}$

(c) 3 sets of $\frac{2}{3}$ is 2, so you could make a bit more than 9 sets from 7, maybe $10\frac{1}{2}$

(d) 3 sets of $\frac{3}{8}$ is a bit more than 1, so you could make about 36 sets from $12\frac{1}{4}$

5. *example:*

1. Write the mixed numbers as improper fractions and then follow the steps outlined in the answer for Problem 4.
For example:
$1\frac{1}{2} \div 3\frac{3}{5} = \frac{3}{2} \div \frac{18}{5} = \frac{3}{2} \times \frac{5}{18} = \frac{15}{36} = \frac{5}{12}$

7. (a) 20 (b) 9 (c) $1\frac{1}{2}$

9. The reciprocal of $1\frac{2}{3}$ is not $1\frac{3}{2}$, since $1\frac{2}{3} \times 1\frac{3}{2} \neq 1$. It would be better to express the fractions as mixed numbers, then find the reciprocal.
$1\frac{2}{3} = \frac{5}{3}$, so the reciprocal is $\frac{3}{5}$.
$2\frac{7}{8} \times \frac{3}{5} = \frac{23}{8} \times \frac{3}{5} = \frac{69}{40} = 1\frac{29}{40}$

11. $\frac{1}{2}$ a page

13. 0; *example:* $0 \div \frac{1}{2}$ means "How many halves can you make from 0?" The answer will always be 0, because you can't make any halves, or any other fraction, from 0.

4.7 Problem Solving with Fractions

1. *example:* Similarities: Operations have the same meaning and are performed the same way. Differences: You need to convert mixed numbers to improper fractions before you can multiply or divide, and you can do this for addition and subtraction too, if you wish. If you are adding or subtracting mixed numbers directly, you have to work with two different parts — the wholes and the fractions. If you are subtracting, you may need to regroup.

3. $\frac{5}{12}$

5. $\frac{3}{4}$

7. 7

9. When a fraction is multiplied by the denominator, the result is the numerator. This happens because you are multiplying and dividing by the same number. For example:
$20 \times \frac{19}{20} = (20 \times 19) \div 20 = 19$

11. (a) $12\frac{7}{9}$ minutes (b) $7\frac{2}{9}$ minutes

4.8 Adding and Subtracting Rational Numbers

1. *example:* combining profits and losses written in dollars and cents

3. (a) right, right (b) right, left
(c) left, right (d) left, left
(e) right, left (f) right, right
(g) left, left (h) left, right

5. (a) 0.5 (b) 0.2 (c) 0.333
(d) 0.1 (e) 0.167 (f) 0.25
(g) 0.125 (h) 0.667 (i) 0.8
(j) 0.375 (k) 0.75 (l) 0.833

7. *examples:*
(a) 0.4 + 0.9 = 1.3
(b) –7 + (–11) is –18, plus another –0.56 gives –18.56
(c) 1.3 and –1.3 cancel out, leaving 0.026
(d) 5 – 10 = –5, so answer is a bit farther from 0 than –5
(e) –3 and –5 make –8, plus another –0.87 makes –8.87
(f) –4 + 2 = –2

9. (a) 137.4 (b) correct
(c) 0 (d) correct

11. (a) All the digits are wrong. The correct difference is –2.522.
(b) *example:* If Kyra starts at 6.213 and goes left 6.213, the result is 0. To find out how much less than 0 the difference is, Kyra can subtract 6.213 from 8.735.
8.735 – 6.213 = 2.522, so the answer is 2.522 below 0, or –2.522.

4.9 Multiplying and Dividing Rational Numbers

1. *examples:* calculating losses on stock shares, finding an average temperature

3. *example:*
0.25 means the same as $\frac{1}{4}$. $\frac{1}{4}$ of –36 tenths is –9 tenths, so the answer is –0.9.

5. *examples:*
(a) Think $1\frac{1}{2} \times$ (–15 hundredths) →
–15 hundredths + –7.5 hundredths
= –22.5 hundredths or –0.225 → –22.5
(b) Think "How many sets of 1 tenth can you make from about 6 tenths?" Estimate 6.
(c) Think about 2 × 6 tenths, or about 1.2.
(d) Think "How many sets of 7 tenths can you make from about 49 tenths? Estimate 7 and then make the answer negative. Estimate –7.

(e) The answer will be positive, so disregard the negative signs. Think "How many sets of about 9 tenths can you make from about 81 tenths?" Estimate 9.
(f) The answer will be positive, so disregard the negative signs. Estimate 3×37 tenths $\rightarrow$ 111 tenths $\rightarrow$ 11.1.

7. **(a)** correct **(b)** –44.44, not 44.44
(c) 10, not 10.36 **(d)** 7, not 0.7
(e) 42.496, not –42.496 **(f)** –40.036, not 44

9. **(a)** $22.25 **(b)** $22.18
(c) $14.99 **(d)** $285.39

11. 12

13. 36 h

5 THE LANGUAGE OF ALGEBRA

5.1 Patterns and Relations

1. *examples*: repeating: 1, 2, 3, 1, 2, 3, 1, 2, 3, …; changing: 1, 3, 5, 7, 9, …

3. **(a)** start by adding 1, then add 2, 3, 4, etc.; 22, 28, 35
(b) start by subtracting 2, then subtract 4, 6, etc.; 70, 58, 44
(c) multiply 10×1, then $\times 2$, then $\times 3$, etc.; 1200, 7200, 50,400
(d) divide each term by 2 to get the next; 32, 16, 8
(e) terms are groups of two alphabet letters, alternately from the end and beginning of the alphabet; EF, TS, GH
(f) increasing powers of 3; 3^6, 3^7, 3^8
(g) increasing odd powers of 4; 4^1, 4^5, 4^9
(h) the number added to n decreases by 1 each time; $n + 5$, $n + 3$, $n + 1$
(i) the number of 2s in this number increases by 1 each time; 12221, 122221, 1222221
(j) letters are in alphabetical order with one more letter each time; numbers reflect the number of letters in the preceding term; GHIJ, 4, KLMNO

5. *example*: $5 + 3s$

7. **(a)** $150h$ **(b)** $100w$ **(c)** $150h - 100w$

9. No; the 1st, 8th, 15th, 22nd, and 29th must be Tuesdays, so the 28th is a Monday.

11. **(a)** $4n$ **(b)** $n + 12$ **(c)** $12 - x + y$
(d) $\frac{n}{16}$ **(e)** $(40 + 2w)$ in.

13. 4 ways

5.2 Graphing Relations

1. **(a)** Advantages: easy to make; shows data in order
Disadvantages: hard to interpolate and extrapolate
(b) Advantages: gives you a picture of the data; easy to see if the relationship is a straight line, broken line, or curve; easy to interpolate and extrapolate Disadvantages: harder to make; can only extrapolate as far as the edges of your graph

3. **(a)** 1.5 gal **(b)** 5.5 min

5. The trip lasted a bit more than 2 min. The car started out at 30 mph and traveled at that speed for 40 s. Then it increased speed to 50 mph over the next 40 s. Then it slowed down steadily and came to a stop.

7. **(b)** Midland to Abilene; This section of the graph is steepest because the bus covers more distance in one minute.
(c) *example*: whether the times included any stops

5.3 Writing and Evaluating Expressions

1. with numbers or shapes, with a table of values, with an algebraic expression
For example: with numbers: 1, 3, 5, 7, 9, …
with a table of values:

Place in the Pattern	Number
1	1
2	3
3	5
4	7
5	9

with an expression: *number* = 2 × *place* – 1

3.

d	$4d + 2$
1	6
2	10
3	14
4	18
5	22
6	26

4.

s	$3s - 9$
–2	–15
–1	–12
0	–9
1	–6
2	–3
3	0
4	3

5. **(a)** 9 **(b)** –2 **(c)** 18
(d) –1 **(e)** 12 **(f)** –8
(g) 7 **(h)** 1

7. The multiplication should be done before the subtraction:

$$\begin{aligned} 4n - n &= 4 \times 3 - 3 \\ &= 12 - 3 \\ &= 9 \end{aligned}$$

9. **(a)**

Number of Squares	Number of Triangles
1	1
2	3
3	5
4	7

(b) *Number of triangles* = $2s - 1$

5.4 Solving Equations by Addition and Subtraction

1. *example:* The value on the left side is equal to the value on the right side. So you have to find a solution that makes both sides the same. If you do an operation on one side, you have to do the same operation on the other side so the values stay in balance.
3. **(a)** 2 **(b)** 10 **(c)** –11 **(d)** –10 **(e)** –13
5. *examples:*
   (a) $x + (+2) = (+6)$; $x + (-3) = (+1)$
   (b) $m + (+3) = (+1)$; $m + (-4) = -6$
7. **(a)** correct **(b)** number is –2
   (c) number is –7 **(d)** correct
9. 17
11. –3°
13. –16°F

5.5 Solving Equations by Division and Multiplication

1. **(a)** A quantity, represented by a symbol, that can take on any one of a set of values. In the equation $5a = 30$, a is the variable.
   (b) A statement that two mathematical expressions have the same value. For instance, $5a = 30$ is an equation.
   (c) The set of values that result in a true statement when replacing the unknowns in an equation. For instance, 6 is the only value of a that makes $5a = 30$ a true statement, so 6 is a solution to the equation.
3. Incorrect; Leslie should have divided each side by 4.
5. **(a)** 7 **(b)** 4 **(c)** 6
   (d) 7 **(e)** –8 **(f)** –4
7. **(a)** 7 **(b)** –7 **(c)** –64 **(d)** 4
   (e) 48 **(f)** 5 **(g)** –100 **(h)** 160
9. 42
11. 11 in.
13. ÷ 5

5.6 Substitution and Relations

1. **(a)** a collection of symbols representing numbers and operations, but not containing an equals sign; for example, $6y + 2$
   (b) a symbol used to represent a number
   (c) to replace one thing with another; in algebra, to replace part of an expression with another expression or value, or to replace a variable with a specific number
   (d) to determine the value of an expression or an unknown quantity
   (e) a pair of numbers in which the first number describes the horizontal displacement and the second number describes the vertical displacement
3. **(a)** 21 **(b)** –3
5. **(a)** $2w + 1t$
   (b) Cougars 19, Saints 26, Tigers 38, Robins 48
7. **(a)** \$2200 **(b)** \$3200 **(c)** \$4200
9. **(a)**

Number of Subscriptions (s)	Dollars Earned in One Week (d)
5	115
10	130
15	145
20	160

5.7 Translating Written Phrases

1. **(a)** a symbol used to represent a number
   (b) a collection of symbols representing numbers and operations, but not containing an equal sign; for example, $6y + 2$
   (c) a mathematical sentence that shows equality where the symbol = is used; for example, $4 + 4 = 8$
3. **(a)** $n + 16$ **(b)** $5 - n$
   (c) $9 + 2n$ **(d)** $6n - 12$
   (e) $n \div 10 - 5$ or $\frac{n}{10} - 5$ **(f)** $2n - 18$
   (g) $(n + 11) \div 3$ or $\frac{n+11}{3}$
   (h) $7n - 4$ **(i)** $3n + 13$
   (j) $(14 - 2n) \div 9$ or $\frac{14-2n}{9}$
5. **(a)** $n - 14 = 26$
   (b) $9n - 5 = 15$
   (c) $9 + 16n = 10$ **(d)** $n \div 6 = n + 12$
   (e) $3n + 17 = n + 20$ or $3(n + 17) = n + 20$
7. **(a)** $3m$ **(b)** $2y + 9$ **(c)** $a - 50$
   (d) $x + 7$ **(e)** $0.5r$ or $\frac{1}{2}r$ or $\frac{r}{2}$ or $r \div 2$
9. **(a)** $3s + 3$ **(b)** $s + 3s + 3$ or $4s + 3$

6 LINEAR EQUATIONS

6.1 Exploring Equations

1. *example:*
   (a) An equation is only true if the two sides balance. Therefore, if one side is changed, the other side must be changed in the same way.
   (b) There are no positive tiles to subtract, so the *Explorer* adds negative ones instead.
5. **(a)** $3x = 1$ **(b)** $-4 = 3x - 2$
9. **(a)** correct **(b)** $x = \frac{7}{3}$
   (c) $x = 6$ **(d)** correct

6.2 Linear Equations 1

1. **(a)** A variable is a letter that represents an unknown number. In $c + 8 = 14$, c is the variable.
   (b) You verify a solution by substituting the number you found into the original equation to see if the left side balances the right side. To verify that $c = 6$ in $c + 8 = 14$, you write:
   L.S. R.S.
   $6 + 8 = 14$ 14
   The solution $c = 6$ is correct because both sides equal 14.
   (c) To solve an equation, you isolate the variable on one side of the equal sign and the number calculations on the other. For example, to solve $c + 8 = 14$, you subtract 8 from both sides to get $c = 6$.
   (d) To check the solution in part (b), you substitute the value 6 for c in the equation.
5. **(a)** subtract 8
   (b) add –5 or subtract 5 **(c)** add 5
   (d) add 8 **(e)** add 4 **(f)** add –36
7. **(a)** 6 **(b)** 2 **(c)** 7
   (d) 11 **(e)** 6 **(f)** 8
9. **(a)** 3 **(b)** 2 **(c)** 7
   (d) 5 **(e)** 4 **(f)** 6
11. **(a)** $k = 12$ **(b)** $r = -3$ **(c)** $x = 0.7$
    (d) $n = 3$ **(e)** $m = 60$ **(f)** $x = -42$
13. $\frac{1}{3}x = 4.2$, $x = 12.6$
15. $1.4c = 217$ lb, $c = 155$ lb
17. **(a)** $x = 0$ **(b)** x can be any number
    (c) no solution

6.3 Linear Equations 2

1. **(a)** *example*: Similarities: You isolate the variable by performing matching operations on each side. You check the solution by substituting the value into the equation. Differences: You have to perform two operations to solve the equation. You start by isolating the variable term and then you isolate the variable.
3. **(a)** (v), $x = -54$ **(b)** (i), $x = 8.67$
   (c) (iii), $x = 6.25$ **(d)** (ii), $x = 6.25$
   (e) (iv), $x = 78$
5. **(a)** $b = 3$ **(b)** $m = 4$ **(c)** $k = 4$
   (d) $n = 4$ **(e)** $m = 6$ **(f)** $k = 2$
7. **(a)** $k = 343$ **(b)** $s = 512$ **(c)** $m = -12$
   (d) $c = 16$ **(e)** $n = -128$ **(f)** $b = -90$
9. **(a)** 17.8 mi **(b)** 6 h
   (c) Jeff: 16.5 h, Rick 11.5 h

6.4 The Coordinate Plane

5. **(a)** G **(b)** D
7. **(a)** (–4, 2) **(b)** (3, –5) **(c)** (–1, 0)
9. MALL
11. **(a)** *example:* (8, 4)
    (b) *example*: (8, –4), since this gives a reflection of the same rectangle

6.5 Graphing Linear Equations

1. **(a)** straight line **(b)** x-intercept
   (c) independent, dependent **(d)** horizontal
   (e) vertical
3. **(a)**

x	y	Ordered Pair (x, y)
–5	–4	(–5, –4)
0	–2	(0, –2,)
5	2	(5, 2)

   (b)

x	y	Ordered Pair (x, y)
–2	7	(–2, 7)
3	–3	(3, –3)
4	–5	(4, –5)

5. *examples*:
   (a)

x	y	Ordered Pair (x, y)
–2	0	(–2, 0)
0	–2	(0, –2,)
1	–3	(1, –3,)

   (b)

x	y	Ordered Pair (x, y)
–2	4	(–2, 4)
0	1	(0, 1)
2	–2	(2, –2)

   (c)

x	y	Ordered Pair (x, y)
0	0	(0, 0)
1	3	(1, 3)
2	6	(2, 6)

   (d)

x	y	Ordered Pair (x, y)
–4	0	(–4, 0)
–4	1	(–4, 1)
–4	2	(–4, 2)

   (e)

x	y	Ordered Pair (x, y)
0	–2	(0, –2)
1	–1	(1, –3)
2	0	(2, 0)

   (f)

x	y	Ordered Pair (x, y)
0	0	(0, 0)
1	1	(1, 1)
2	2	(2, 2)

7. The ships will both pass through point (3, –3).

7 POLYNOMIALS

7.1 Terms of Polynomials

1. **(a)** a monomial or a sum or difference of monomials; whereas a monomial is a number, variable, or the product of numbers and variables

(b) *examples*: monomial $3x^2$; binomial $3x^2 + 2x$; trinomial $3x^2 + 2x + 1$

3. Find the terms, coefficients, variables, and then constants.

5. **(a)** constants **(b)** coefficients **(c)** terms

7. **(a)** $-3y^2 + xy + 4x + 2$

(b) $3y^2 + y - x^2 - 3x + 2xy + 5$

9. binomial; the expression consists of 2 terms

11. *examples*:

(a) $3 + 4x + 5y^2$

(b) $2x^2 + 3y$

13. *example*: $\frac{3}{x} + 7$

7.2 Evaluating Polynomials

1. **(a)** variable: a letter or other symbol used to represent a number

constant: a value that does not change

(b) a variable can represent different numbers, but a constant can only represent one

3. Perform operations within parentheses first, then perform calculations of exponents, multiply or divide from left to right, and add or subtract from left to right.

5. **(a)** –87 **(b)** 7 **(c)** –43

(d) 3.5 **(e)** 15.625 **(f)** 11

(g) 11.5 **(h)** $\frac{5}{6}$ **(i)** –5

7. **(a)** 5050 **(b)** 50.5

9. **(a)** James charges $46.50 and Tacia charges $49.50 for 3 hours, so James is cheaper.

(b) James charges $18 for one hour and Tacia charges $12 for one hour, so Tacia is cheaper.

11. $s = 25$; $d = 22.5$

$s = 50$; $d = 70.0$

$s = 75$; $d = 142.5$

$s = 100$; $d = 240$

$s = 125$; $d = 362.5$

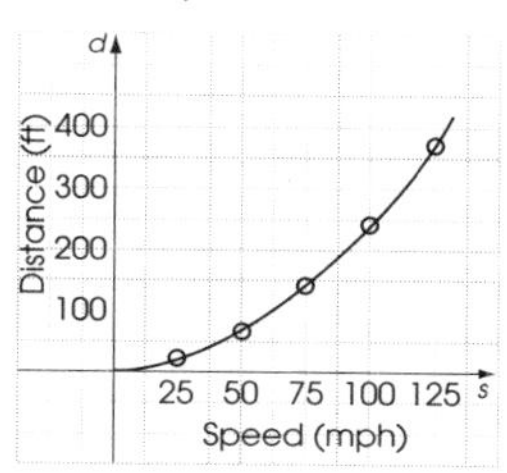

Pattern: as the speed increases, the distance needed to stop the car increases. To calculate the next increment of 25, take the difference of the distance of the two previous speeds and add 25.

$(d_b - d_a) + 25 = d_n$

7.3 Adding and Subtracting Polynomials with Tiles

1. Take the polynomial and change each term to its opposite by multiplying each term by –1.

5. **(a)** $9x + 6$ **(b)** $5 - 6x$

(c) $-2x^2 + 5x - 4$ **(d)** $2 - 3x + 3x^2$

7. $4x^2 - 4x - 8$

9. **(a)** $(-8x + 6)$ **(b)** $(8x - 11)$

11. $(4x^2 + 2x - 2) - (2x^2 - 3x + 1) = 2x^2 + 5x - 3$

13. *examples*:

$(-2x^2 + 4x - 1) + (3x^2 + 2x - 4) = x^2 + 6x - 5$

$(-2x^2 + 4x - 1) - (3x^2 + 2x - 4) = -5x^2 + 2x + 3$

$(x^2 + 6x - 5) - (-2x^2 + 4x - 1) = 3x^2 + 2x - 4$

$(3x^2 + 2x - 4) - (x^2 + 6x - 5) = 2x^2 - 4x + 1$

$(-5x^2 + 2x + 3) + (3x^2 + 2x - 4) = -2x^2 + 4x - 1$

$(-2x^2 + 4x - 1) - (-5x^2 + 2x + 3) = 3x^2 + 2x - 4$

7.4 Multiplying Polynomials

1. **(a)** a monomial or the sum or difference of two or more monomials

(b) the combination of multiplying over addition/subtraction, when simplifying an expression

(c) a method to multiply two binomials: multiply First terms, Outside terms, Inside terms, Last terms

3. The tiles should be positive because the length and width of the model need to represent the factors of the polynomial. So if you add three positive units, the factors are $(x - 3)(x - 1)$, which represents polynomial $(x^2 - 4x + 3)$.

7. **(a)** $6x^2 + 8x$ **(b)** $3a - 6a^2$

(c) $35 - 5x$ **(d)** $-6d^2 + 18d$

9. **(a)** $x^2 + 6x + 9$ **(b)** $m^2 - 12m + 36$

(c) $4x^2 + 28x + 49$ **(d)** $1 - 6y + 9y^2$

11. **(a)** $3x^2 + 12x$ **(b)** $3x^2 + 20x + 12$
(c) $8x + 12$

13. **(a)** $0.6x - 0.8x^2$ **(b)** $8h^3 + 28h^2$
(c) $3x^3y + 12xy^2 - 3xy$ **(d)** $2a^2 + 5ab - 12b^2$
(e) $x^2 - 6xy + 9y^2$ **(f)** $x^3 + 3x^2 + 3x + 1$

15. Form the factors on the quadrant and fill in the model to complete a rectangle.
(a) $2y^2 - y - 1$ **(b)** $6x^2 - 4x - 42$

7.5 Powers, Bases, and Exponents

1. the number to be multiplied
3. the numerical non-variable portion that is to be multiplied by the power
5. **(a)** a **(b)** 3 **(c)** 12
(d) $12 \times a \times a \times a$ **(e)** $12a^3$
7. **(a)** $3 \times x \times x \times x \times y \times y \times y \times y$
(b) $6 \times r \times r \times s \times s$
(c) $\frac{3}{7} \times \frac{3}{7} \times \frac{3}{7} \times \frac{3}{7} \times \frac{3}{7}$
9. **(a)** 43 **(b)** 121
11. \$21,589
13. **(a)** no **(b)** yes
(c) negative base: $(-7)^4$, $(-7)^3$;
negative coefficient: -7^4, -7^3
(d) If the exponent is odd, the answer will be negative.
15. **(a)** 48 **(b)** 95 **(c)** 6 **(d)** 86

7.6 Laws of Exponents: Product Laws

1. **(a)** To multiply powers with the same base, add the exponents: $x^m \times x^n = x^{m+n}$
(b) To simplify a power of a product, apply the exponent to each term in the product: $(x \times y)^m = x^m \times y^m$
(c) To simplify a power of a power, multiply the exponents: $(x^m)^n = x^{mn}$
3. **(a)** $4^5 \times 3^5$ **(b)** $2^0 \times 8^0$
(c) $2^8 \times 5^8$ **(d)** $(-2)^5 \times 5^5$
(e) $4.1^2 \times 2.3^2$ **(f)** $7^4 \times 7^4$
(g) $x^3 \times y^3$ **(h)** $c^5 \times d^5$
5. **(a)** 108^4 **(b)** 29.33^8
7. **(a)** 531,441 **(b)** 16,777,216 **(c)** 11,390,625
(d) 16,384 **(e)** 48,828,125 **(f)** 38,416
(g) 531,441 **(h)** 117,649 **(i)** 11,390,625
9. **(a)** $n = 2$ **(b)** $n = 6$ **(c)** $n = 3$
11. $(3^4)^8$ or $(3^8)^4$
13. *examples*:
(a) $8^2 \times 8^3 = 8^5$
(b) $(-3^3)^4 = (-3)^{12}$
(c) $1.5^2 \times 1.5 = 1.5^3$
15. $10^6 \times 10^6 = 10^{12}$
17. $(5^3)^4$ or 5^{12}

19. Step 1: Power of a Power Law
Step 2: Product Law
Step 3: Product Law
Step 4: Power of a Product Law
Step 5: Power of a Product Law
Step 6: Simplifying exponent
Step 7: Product Law

8 RATIO AND PERCENT

8.1 Percent

5. *example*: makes them easier to compare
7. **(a)** $\frac{4}{5} = 80\%$ **(b)** $\frac{1}{2} = 50\%$ **(c)** $\frac{1}{4} = 25\%$
(d) $\frac{1}{10} = 10\%$ **(e)** $\frac{5}{6} = 83\%$ **(f)** $\frac{5}{7} = 71\%$
(g) $\frac{7}{11} = 64\%$ **(h)** $\frac{8}{9} = 89\%$
9. \$4.50
11. 30%
13. They are the same, since both are equal to $\frac{27 \times 72}{100}$.
15. It is a 75% decrease.
17. $\frac{1}{16}$ or 6.25%

8.2 Rate and Ratio

3. **(a)** 2 : 3 **(b)** 1 candy bar/50¢
(c) 1 lunch/\$8 **(d)** \$20/allowance
(e) 60 mph **(f)** \$6/h **(g)** 7 : 6
5. **(a)** 6 **(b)** 3 **(c)** 15 **(d)** 9
7. 3
9. 5 people
11. 8¢
13. **(a)** 529 : 1946
(b) 295 : 973
(c) Because there are other animals that are pets, e.g., rabbits, hamsters, birds.

8.3 Proportions

1. Two pairs of numbers are in proportion if the ratio formed by the first pair equals the ratio formed by the second pair; *example*: $\frac{1}{2} = \frac{12}{24}$
3. **(a)** 10 **(b)** 6 **(c)** 6
(d) 12 **(e)** 6 **(f)** 3
5. **(a)** 3 : 5 **(b)** 9 nickels
(c) 2 : 4 or 1 : 2 **(d)** 12 dimes
7. \$13.33
9. \$21.00
11. about 27

8.4 Mental Mathematics

Note: Where answers are estimates, sample estimates have been provided.

1. *examples:* Think about the problem: What information do I need to solve the problem? What do I need to find?
   Make a plan: What calculations must I make to solve the problem? What is the most efficient way to calculate?
   Solve the problem: If I am estimating, how should I change the numbers to make estimation easier? Have I made the calculations correctly?
   Look back: Did I answer the right question? Is the answer reasonable?

3. **(a)** $\frac{11}{50}$ **(b)** $\frac{3}{50}$ **(c)** $\frac{1}{20}$
   (d) $\frac{99}{100}$ **(e)** $\frac{9}{50}$ **(f)** $\frac{3}{5}$

5. **(a)** = **(b)** > **(c)** > **(d)** >
   (e) < **(f)** =

7. *examples:*
   (a) 1.5 **(b)** 50 **(c)** \$1.00
   (d) \$15 **(e)** 40 **(f)** 12

9. *examples:*
   (a) \$50 **(b)** \$210

11. *example:* about $\frac{2}{3}$

13. *examples:* about \$50 + \$3.50 or \$53.50

15. about 17%

8.5 Applications of Percent

1. *example:* 100% is 1 whole. Since 24.6 is less than 100, 24.6% is less than 1 whole. Since 154.2 is greater than 100, 154.2% is more than 1 whole.

3.

Percent	Decimal	Fraction
$55\frac{1}{2}\%$	0.555	$\frac{111}{200}$
$72\frac{1}{2}\%$	0.725	$\frac{29}{40}$
$33\frac{1}{3}\%$	0.333	$\frac{1}{3}$
145%	1.45	$1\frac{9}{20}$

5. **(a)** $\frac{9}{25}$ **(b)** $\frac{7}{12}$ **(c)** $1\frac{31}{200}$
   (d) $\frac{291}{400}$ **(e)** $1\frac{63}{250}$ **(f)** $1\frac{29}{80}$

7. 8 blocks

9. **(a)** = **(b)** > **(c)** >
   (d) = **(e)** < **(f)** >

11. $316\frac{36}{79}$

9 MEASUREMENT

9.1 Units of Measurement

3. *examples:*
   (a) The American system is more commonly used in the United States, and I already know how some units are related. For example, I know that there are 12 inches in a foot and 3 feet in a yard. The system is more familiar to me, so it's easy to estimate because I know the sizes of the units.
   (b) The units in the metric system are less familiar to me, but it's easier to convert between units in the metric system because all the units are related by multiples of ten. The prefix on the name of the unit indicates how the size of the unit compares with the base unit. The metric system is also used in most countries other than the United States, and it is the measurement system used in most areas of science.

5. **(a)** 48 in. **(b)** 2 ft **(c)** 288 in.
   (d) 15 ft **(e)** 0.25 mi **(f)** 5 lb
   (g) 24,800 lb **(h)** 6 pt **(i)** 5 gal
   (j) 384 fl oz

7. **(a)** $2\frac{9}{16}$ in. or 6.51 cm
   (b) $\frac{15}{16}$ in. or 2.38 cm
   (c) $8\frac{1}{2}$ in. or 21.59 cm
   (d) *example:* $6\frac{3}{4}$ in. or 17.145 cm

9. $C = \frac{5(18)}{9}$
   $C = \frac{5(2)}{1}$
   $C = 10$ Therefore, 50°F = 10°C.

9.2 Classifying Angles

1. Complementary: $\angle EGF$ and $\angle FGA$, $\angle BGC$ and $\angle DGC$
   Supplementary: There are two pairs of supplementary angles on each side of every line in the diagrams. This means that there are four pairs for each line, or 12 pairs in all.
   For line EB, the pairs are $\angle EGA$ and $\angle AGB$, $\angle EGF$ and $\angle FGB$, $\angle EGD$ and $\angle DGB$, $\angle EGC$ and $\angle CGB$.
   For line *AD* the pairs are $\angle DGE$ and $\angle EGA$, $\angle DGF$ and $\angle FGA$, $\angle DGB$ and $\angle AGB$, $\angle DGC$

and $\angle AGC$.
For line FC, the pairs are $\angle FGA$ and $\angle AGC$, $\angle FGB$ and $\angle BGC$, $\angle FGE$ and $\angle EGC$, $\angle FGD$ and $\angle DGC$.

3. *angle* = 35°, *complement* = 55°, *supplement* = 145°

5. $x = 34$

7. **(a)** 146° **(b)** 66° **(c)** 90° **(d)** 101° **(e)** 3° **(f)** 51°

9. **(a)** complementary, 90°
(b) supplementary, 180°

11. $\angle X = 86°$, $\angle Y = 74°$

9.3 Angles and Parallel Lines 1

1. **(a)** two straight lines that are always the same distance apart
(b) to meet or cross over
(c) a line that intersects two or more other lines

3. **(a)** $\angle 4$ is vertically opposite $\angle 1$; $\angle 5$ corresponds to $\angle 1$; $\angle 8$ is vertically opposite $\angle 5$.
(b) $\angle 3$ is vertically opposite $\angle 2$; $\angle 6$ corresponds to $\angle 2$; $\angle 7$ is vertically opposite $\angle 6$.

5. **(a)** corresponding **(b)** vertically opposite
(c) supplementary **(d)** corresponding

7. *examples*:
(a) 35° because it is vertically opposite $\angle m$
(b) 130° because $\angle n$ corresponds to $\angle r$
(c) 41° because $\angle s$ and $\angle t$ are supplementary, so they have a sum of 180°
(d) 56° because $\angle m$ corresponds to $\angle q$
(e) $\angle s$ because $\angle o$ corresponds to $\angle s$; $\angle n$ because $\angle n$ is vertically opposite $\angle o$; $\angle r$ because it corresponds to $\angle n$
(f) $\angle t$ because $\angle t$ is vertically opposite $\angle q$; $\angle m$ because it corresponds to $\angle q$; $\angle p$ because it corresponds to $\angle t$

9. **(a)** 45° because $\angle a + 135° = 180°$
(b) 135° because $\angle b$ is vertically opposite the angle labeled 135°
(c) 135° because $\angle c$ corresponds to the angle labeled 135°
(d) 135° because $\angle d$ is vertically opposite $\angle c$, which has been proven to measure 135°
(e) 45° because $\angle e$ corresponds to $\angle a$, which has been proven to measure 45°

9.4 Angles and Parallel Lines 2

1. **(a)** angles lying between a pair of parallel lines
(b) angles lying outside a pair of parallel lines
(c) angles formed on opposite sides of a transversal where it crosses two parallel lines

3. *examples*:
1. $\angle 2$ and $\angle 7$ are alternate interior angles
2. $\angle 2$ is vertically opposite $\angle 5$, which corresponds to $\angle 7$
3. $\angle 3$ is supplementary to both $\angle 2$ and $\angle 7$, so $\angle 2$ must equal $\angle 7$

5. **(a)** 32° because $\angle b$ and $\angle g$ are alternate exterior angles
(b) 123° because $\angle c$ and $\angle e$ are same-side interior angles
(c) 136° because $\angle d$ and $\angle e$ are alternate interior angles
(d) 63° because $\angle a$ and $\angle g$ are same-side exterior angles

7. **(a)** parallel because the same-side exterior angles are supplementary
(b) parallel because the same-side interior angles sum to 180°
(c) not parallel because the same-side exterior angles do not sum to 180°
(d) parallel because the alternate interior angles are equal

9. *example*: George could draw a diagonal to join one pair of parallel sides and check to see if the alternate interior angles are 45° and then repeat with a diagonal that joins the other two parallel sides.

9.5 The Pythagorean Relationship

Answers for some problems may vary if rounded numbers are used in the calculations.

1. The square of the hypotenuse of a right triangle is equal to the sum of the squares of the other two sides: $a^2 + b^2 = c^2$.
You can cut the two smaller squares to fit exactly over the large one.

3. *example*: Square the two shorter sides and find the sum of the squares. If the result is equal to the square of the longest side, the triangle is a right triangle.

5. **(a)** $n \doteq 5.39$ **(b)** $p \doteq 5.25$ **(c)** $m \doteq 2.62$
(d) $b \doteq 12.22$ **(e)** $g \doteq 18.20$ **(f)** $r \doteq 14.44$

7. No, since the hypotenuse or longest side must always be opposite the right angle.

9. When you substitute 7.33 for x, the left side is equal to 72.2189 and the right side is equal to 72.25. This happens because 7.33 is a rounded value for x.

10 AREA, PERIMETER, AND VOLUME

10.1 Area and Perimeter 1

1. (a) a rectangle with four equal sides
   (b) a right-angled parallelogram
   (c) a parallelogram with equal sides
   (d) a quadrilateral with parallel opposite sides
   (e) a quadrilateral having one pair of opposite sides that are parallel and unequal
3. This shows that the parallelogram covers the same area as a rectangle with the same base and height. This is why the formula for finding the area of a parallelogram is $A = b \times h$.
5. (a) $P = 71.6$ in., $A = 320.41$ in.^{2}
   (b) $P = 25.8$ m, $A = 36.98$ m^2
   (c) $P = 33.8$ ft, $A = 52.43$ ft^2
   (d) $P = 48$ cm, $A = 129.6$ cm^2
   (e) $P = 22.6$ cm, $A = 16.2$ cm^2
   (f) $P = 63.5$ ft, $A = 198.5$ ft^2
7. 10,784.64 yd^2
9. 5
11. 46.5 in.

10.2 Area and Perimeter 2

1. (a) the sum of the side lengths, or the length of the borders
   (b) the space within the polygon calculated by an area formula specific to each type of polygon, or the amount of surface the polygon covers
3. (a) 10 ft × 4 ft, 1 ft × 40 ft, 2 ft × 20 ft, 5 ft × 8 ft
   (b) greatest perimeter is 1 ft × 40 ft; least perimeter is 5 ft × 8 ft
   (c) 1 ft × 80 ft with perimeter 162 ft, 2 ft × 40 ft with perimeter 84 ft, 4 ft × 20 ft with perimeter 48 ft, 5 ft × 16 ft with perimeter 42 ft, 8 ft × 10 ft with perimeter 36 ft
5. (a) circle with a radius of about 2.99 yd

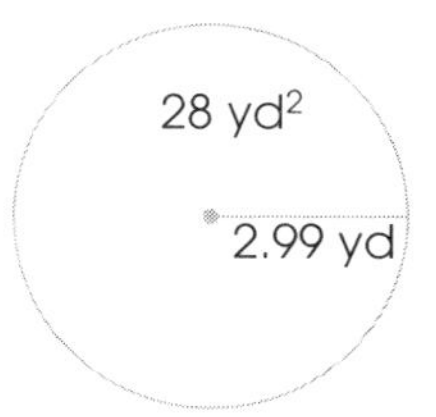

7. Form a circle.
9. 50 in. × 50 in. = 2500 in.^{2}
11. 1 yd
    19 yd
    $A = 19$ yd^2
13. 31.27 yd^2

10.3 Area of a Triangle

1. (a) for a triangle (or any polygon), the line drawn from one vertex perpendicular to the opposite side
   (b) the length of the altitude of a geometrical figure, such as a triangle
   (c) in geometry, the bottom line of a geometrical figure, such as a triangle
   (d) a 90° angle
   (e) the point of intersection of the sides of a polygon; a triangle has 3 vertices
3. (a) 25.73 in.^{2} (b) 43.8 m^2
   (c) 35.90 m^2 (d) 116.10 in.^{2}
   (e) 104.20 cm^2 (f) 7.64 ft^2
5. (a) 9.29 cm (b) 5.15 in.
   (c) 10.43 ft (d) 24.19 m
9. The parallelogram in (c) has the greatest area.
   (a) 104.04 m^2 (b) 103.5 m^2
   (c) 105.12 m^2 (d) 102.3 m^2
   (e) 101.175 m^2

10.4 Composite Areas

1. Separate the composite figure into simple figures like rectangles, triangles, and circles. Find the area of each simple figure. Then add or subtract to find the composite area.
3. **Composite Figure**

Hidden Shape	**(a)**	**(b)**	**(c)**	**(d)**
circle	√	√	√	√
square	√			√
triangle	√		√	√
non-square rectangle		√		√

Answers for these problems may vary depending on the rounding strategies used.

5. *examples:*
   (a) The area might be a bit less than double the area of the 3 in. by 3 in. square, or about 15 in.^{2}.
   (b) The area of the rectangle is 6 in.^{2} and the semicircle covers about $\frac{1}{6}$ of the rectangle, or about 1 in.^{2}. So the area is about 6 in.^{2} – 1 in.^{2} or 5 in.^{2}.
   (c) The area is a bit more than double the area of the 2 cm^2 triangle, or about 5 cm^2.
   (d) The triangles added to the top of the 3 cm by 3 cm square are offset by the semicircle removed from the base, so the area is close to the area of the square, or 9 cm^2.
7. *example:*
   (b) $A_{figure} = A_{rectangle} - A_{semicircle}$
   (c) $A_{figure} = A_{triangle} + A_{semicircle}$; Use the

Pythagorean relationship to find the radius of the circle.

(d) $A_{figure} = A_{3\ cm\ by\ 3\ cm\ square} - A_{semicircle} + 2 \times A_{triangle\ with\ base\ 1.5\ cm\ and\ height\ 2\ cm}$

9. more than 15 cm^2; The area of the rectangle is 20 cm^2, and the area of the semicircle is only half of 3.14 cm^2, so the figure has an area of about 18.5 cm^2.

11. $A_{trapezoid} = \frac{1}{2} \times height \times (sum\ of\ parallel\ bases)$

$= \frac{1}{2} \times 2\ m \times (2.5\ m + 6\ m)$

$= \frac{1}{2} \times 2\ m \times (8.5\ m)$

$= 8.5\ m^2$

13. 12 pots

10.5 Surface Area

Some answers may vary slightly depending on whether you use π or 3.14.

1. Find the area of each surface on the solid and then add the areas. If some areas are congruent, you can use multiplication to simplify the process.

3. 864 $in.^2$

5. 1464.5 m^2

7. 110 $in.^2$

9. 314.2 $in.^2$

11. **(a)** $0.18

(b) The company will need extra cardboard to make parts overlap so the box is more secure.

13. **(a)** 3 in. by 7 in. by 10.5 in. **(b)** 252 $in.^2$

10.6 Volume of Composite Solids

1. **(a)** area, height **(b)** cubes **(c)** less

3. 1.8 L

5. 7712 cm^3

7. **(a)** 29.88 $in.^3$

(b) 205.60 oz or 12.85 lb

(c) $84,037.15

10.7 Volume

1. **(a)** *area of base* × *height*

(b) 3 **(c)** 3

3. $V = \frac{1}{3} \times (2a)^2 \times 2a = \frac{1}{3} \times 2^2 a^2 \times 2a = \frac{1}{3} \times 8a^3 = \frac{8}{3}a^3$

5. **(a)** 2.1 ft **(b)** 6.95 ft

7. **(a)** 16 m^3 **(b)** 392.7 ft^3

9. **(a)** 7776 $in.^3$ **(b)** 46,656 $in.^3$

11. 12.5%

10.8 Surface Area and Volume

1. **(a)** faster **(b)** volume **(c)** eight, four

3. 922.45 $in.^3$; 409.98 $in.^2$

5. **(a)** 11.3 in. × 27 in. × 36 in.

(b) 22.6 in. × 27 in × 18 in.

(c) $SA_{box\ B} = 3006\ in.^2$;

$SA_{box\ A} = 3367.8\ in.^2$

Since $SA_{box\ A} > SA_{box\ B}$, box *B* uses less cardboard.

7. surface area increases 4 times, 9 times, x^2 times

9. **(a)** cube

11. *examples*:

(a) 9 in. long × 2 in. high × 2 in. wide

6 in. long × 3 in. high × 2 in. wide

3 in. long × 12 in. high × 1 in. wide

(b) $2(9 \times 2) + 2(2 \times 2) + 2(2 \times 2) = 52\ in.^2\ (SA)$

$2(6 \times 3) + 2(3 \times 2) + 2(6 \times 2) = 72\ in.^2\ (SA)$

$2(3 \times 12) + 2(12 \times 1) + 2(3 \times 1) = 102\ in.^2\ (SA)$

(c) The last arrangement uses the least amount of plastic and is therefore the best.

13. 22.178 yd^2

11 POLYGONS AND CIRCLES

11.1 Parts of a Circle

1. **(a)** $d = 2r$ **(b)** $C = \pi d$ **(c)** $C = 2\pi r$

3. Each number in the Circumference column is about 3 times as great as the related number in the Diameter column.

5. **(a)** 40.19 ft **(b)** 43.96 mm

(c) 32.66 in. **(d)** 11.93 km

7. **(a)** 3.1 mm **(b)** 4.25 ft

(c) 7 mi **(d)** 4.5 cm

9. **(a)** 1.26 m **(b)** 0.80 in.

(c) 15.92 mm **(d)** 0.04 mi

11. The circumference of Circle B is half as large as the circumference of Circle A. For Circle A, $C = 18.84$ cm, so for Circle B, $C = 9.42$ cm.

15. **(b)** *height of design* = 2 × *diameter of circle*

(c) *height of flower* = *diameter of circle*

11.2 Circle Problems

1. Without this information, you can't tell how big the circle is, so you can't find the circumference. The circumference is equal to π × *diameter*, or π × (2 × *radius*).

3. **(a)** π is the number of diameters it takes to fit around the circumference of a circle.

(b) Once you know the diameter, you just multiply it by π to find the circumference.

5. **(a)** 16 in. **(b)** about 100.48 in.

7. about 39.25 in.

9. **(a)** *examples*: 12 in., 24 in., 36 in., 48 in.

(b)

Diameter (in.)	Circumference (in.)
4	12
8	24
12	36
16	48

(c) Each time the diameter increases by 4 in., the circumference increases by a bit more than 12 in.

11. **(a)** about 15.7 ft **(b)** about 18.84 ft **(c)** 7 with, 5 without **(d)** 14 ft × 14 ft or 196 ft^2

11.3 Polygons and Circles

1. On a regular polygon, all sides are congruent and all angles are congruent. On an irregular polygon, two or more sides and/or angles are not congruent.

3. A circle is a curve formed by a set of points which are all equally distant from a fixed point known as the center.

5. **(a)** rectangle **(b)** square **(c)** rhombus **(d)** parallelogram **(e)** kite **(f)** trapezium **(g)** chevron **(h)** trapezoid

9. in order: 4, 5, 1, 3, 2

(a) 1.6, 7.5, 2.6, 1.08, 1.57

(b) *example:* The order of the ratios is the same as the order of the rectangles in part (a). As the shape of the rectangle approaches a square, the ratios get closer to 1.

(c) 1

11.4 Similar Triangles

3. **(a)** $\angle S = \angle C = 58°$, $\angle B = 62°$, $\angle A = 60°$, $AB = 19$ in., $BC = 16$ in.

(b) $\angle L = \angle W = 28°$, $\angle Y = \angle N = 62°$, $WX = 60$ m, $LN = 17$, $WY = 68$ m

5. **(a)** $d = 2$ mi, $e = 2.5$ mi **(b)** $\angle x = 60°$

7. 11.25 ft

11.5 Congruence and Similarity

1. **(a)** XYZ **(b)** XY **(c)** 3 **(d)** 90°

3. $\angle PRM = 40°$, $\angle PMR = 50°$, $\angle NKR = 50°$, $\angle KRN = 40°$, $RM = 8.7$, $PM = 5.6$, $RN = RP = 6.7$

5. similar, since each side length on $\triangle ABC$ is doubled on $\triangle DEF$

7. $\triangle HFI$; $\angle EFG \cong \angle HFI$, $\angle EGF \cong \angle HIF$, $\angle GEF \cong \angle IHF$, $EG \cong HI$, $EF \cong HF$, $GF \cong IF$,

Picture Credits

Whole Numbers: CanaPress: Globe and Mail, Barrie Davis; CORBIS-BETTMANN
Powers and Exponents: Philip D'Agonstino
Prime Factors and Exponents: UPI/CORBIS-BETTMANN
Decimal Conversions: NBA Photo; G. Locke/First Light
Ordering Numbers: Wide World Photos
Terminating and Repeating Decimals: Hospitality & Tourism Centre, George Brown College, Toronto, Canada; Robyn Craig
Division with Decimals: Eurelios
Order of Operations: Tom Nelson/First Light; CORBIS-BETTMANN
An Introduction to Integers: IBM Corporation, Research Division; Sir Sanford Fleming College
Adding Integers: Courtesy of Skysport Balloon Co.; Photo by John E. Sokolowski
Subtracting Integers: Courtesy of British Airways
Multiplying and Dividing Integers: Motion Picture and Television Archives; Woods Hole Oceanographic Institute
Calculating Square Roots: The Persian and Oriental Rug Centre
Adding and Subtracting Fractions: Judo Ontario
Subtracting Fractions (Unlike Denominators): Courtesy of the Vancouver Symphony, photo by Dave Roels
Multiplying Fractions: Dick Hemingway; Photo courtesy of the Nunavut Secretariat
Dividing Fractions: CORBIS-BETTMANN
Problem Solving with Fractions: First Light
Multiplying and Dividing Rational Numbers: Dick Hemingway
Patterns and Relations: Egyptian Tourist Bureau; Luxor, Las Vegas
Graphing Relations: The Palace Pier, Toronto
Solving Equations by Addition and Subtraction: Dick Hemingway; Courtesy of George Brown College, Toronto
Solving Equations by Division and Multiplication: NASA; Eileen Jung
Substitution and Relations: Courtesy of Tom Perera; Bob Templeton
Translating Written Phrases: CNIB/Library for the Blind
Exploring Equations: The Weather Network
Linear Equations 1: CORBIS-BETTMANN; Dick Hemingway
Linear Equations 2: Art Resource
The Coordinate Plane: Dan Couto/First Light
Graphing Linear Equations: CAA; A Five Star Limousine Services
Terms of Polynomials: First Folio; Tessa Macintosh/NWT Archives
Evaluating Polynomials: BETTMANN; Mobius Encryption Technologies
Adding and Subtracting Polynomials with Tiles: BETTMANN; P. Beck/First Light
Multiplying Polynomials: Brown Brothers, Sterling, PA 18463; Dick Hemingway
Powers, Bases, and Exponents: BETTMANN; Institut Pasteur/First Light
Laws of Exponents: Product Laws: NASA; Harry Turner/National Research Institute
Percent: Archive Photos
Rate and Ratio: The Toronto Sun; Debbie Smith
Proportions: CORBIS-BETTMANN; Famous Players/Burlington 8 Theatre
Mental Mathematics: Photo by Dick Haneda/Toronto Zoo
Applications of Percent: IMAX; Blair's Dinnebito Trading Post
Units of Measurement: Bureau International des Poids et Mesures, CORBIS-BETTMANN
Classifying Angles: Photo courtesy of the Honshu-Shikoku Bridge Authority; CORBIS-BETTMANN
Angles and Parallel Lines 1: The Granger Collection; Ivy Images
Angles and Parallel Lines 2: Dick Hemingway; Placer Dome Inc.
The Pythagorean Relationship: Museum for Textiles, Toronto, Ontario, Canada
Area and Perimeter 1: Bob Swartman
Area and Perimeter 2: MYSTIQUE Creative
Composite Areas: Steve Matheson/Paper Magic; Steve Matheson/Paper Magic
Surface Area: The Toronto Sun
Volume of Composite Solids: Gard W. Otis/Nature by Otis; Robyn Craig
Volume: Ralph Clevenger/First Light; KHEOPS Pyramids
Volume and Surface Area: BETTMANN
Parts of a Circle: Glenbow Archives, Calgary, Alberta; Simon Fraser University
Circle Problems: CORBIS-BETTMANN
Similar Triangles: Dick Hemingway; NASA
Congruence and Similarity: First Folio; Jody Dole/The Image Bank